Optical Wireless Communication Theory and Technology

Series Editor

Xizheng Ke, School of Automation and Information Engineering, Xi'an University of Technology, Xi'an, Shaanxi, China

The book series Optical Wireless Communication Theory and Technology aims to introduce the key technologies and applications adopted in optical wireless communication to researchers of communication engineering, optical engineering and other related majors. The individual book volumes in the series are thematic. The goal of each volume is to give readers a comprehensive overview of how the theory and technology in a certain optical wireless communication area can be known. As a collection, the series provides valuable resources to a wide audience in academia, the communication engineering research community and anyone else who are looking to expand their knowledge of optical communication.

Xizheng Ke

UV Light Self Organizing Network Theory

Science Press
Beijing

Springer

Xizheng Ke
School of Automation and Information
Engineering
Xi'an University of Technology
Xi'an, Shaanxi, China

ISSN 2731-5967 ISSN 2731-5975 (electronic)
Optical Wireless Communication Theory and Technology
ISBN 978-981-97-7853-9 ISBN 978-981-97-7854-6 (eBook)
https://doi.org/10.1007/978-981-97-7854-6

Jointly published with Science Press
The print edition is not for sale in China mainland. Customers from China mainland please order the print book from: Science Press.
ISBN of the Co-Publisher's edition: 978-7-03-031322-5

Translation from the Chinese Simplified language edition: "紫外光自组织网络理论" by Xizheng Ke,

This Springer imprint is published by the registered company Springer Nature Singapore Pte Ltd.
The registered company address is: 152 Beach Road, #21-01/04 Gateway East, Singapore 189721, Singapore

Preface

As the radio spectrum becomes increasingly congested, attempts have been made to develop new communication systems to meet the need for ubiquitous communication. Wireless laser communication combines the advantages of microwave and optical fiber communication; it has the advantages of confidentiality, high capacity, and rapid deployment, and it does not require spectrum licenses. The straight-line propagation of lasers limits the application of wireless laser communication; solar-blind ultraviolet (UV) light can be used for non-line-of-sight communication, but atmospheric effects greatly limit its communication distance. This book applies the theory of self-organizing networks to UV communication, which can not only provide non-line-of-sight communication, but also extend the communication range. The theoretical basis of self-organizing networks is this book's contribution to the field.

This book summarizes work performed under the National High-Tech Development Program (2008AAJ159) and the National Natural Science Foundation of China (60977054). It consists of nine chapters. The first chapter introduces the basic concept of UV optical communication, and the second chapter describes the characteristics of UV optical channels. Subsequent chapters analyze media access protocols and routing protocols for UV self-organizing networks in detail. On the basis of an analysis of classical communication protocols, a basic framework for UV self-organizing networks is proposed.

This book is the result of collective research at the Optoelectronic Engineering Technology Research Center of Xi'an University of Technology by the doctoral student He Hua and postgraduate students Feng Yanling, Deng Lijun, Hou Zhaomin, Wu Changli, Liang Wei, Chen Xiang, Yang Peilin, and others. During post-doctoral mobile station work, He also participated in the work and contributed to the research on this topic. He, Feng, Deng, Hou, Wu, and Liang proofread the manuscript many times and created illustrations for the book. Their patient and meticulous work has removed many errors and infelicities from the book, and I sincerely thank them. My mentor, Professor Wu Zhensen, has always been very concerned with my research, and he has shared many valuable opinions, which quickly relieved my confusion. I am deeply grateful to him.

This book describes research funded by the National High-tech Development Plan (2008AAJ159) at the Optoelectronic Engineering Technology Research Center of Xi'an University of Technology, National Natural Science Foundation of China (60977054), National Defense Key Laboratory Fund (9140C3601010701), Science and Technology Project of the Shaanxi Provincial Department of Education (04JK247, 07JK332), Shaanxi Provincial Department of Education Industrialization Fund, Shaanxi Provincial Intellectual Property Office Fund, Guangdong Provincial Department of Communications Research Program (2007–26), Guangdong Provincial Ministry of Education Industry-University-Research Fund, and China Academy of Space Technology (CAST) Innovation Fund (CAST200828). I would like to express my gratitude to Xi'an Innovation and Technology Program (CXY1012) and the other projects and funds.

This book is a preliminary summary of research on UV self-organizing networks. Because of my limited knowledge, errors are inevitable. Readers are welcome to correct me.

I dedicate this book to my kind, hard-working mother. If she hears about it in Jiuquan, she will be gratified by my work.

Autumn, 2010

Xizheng Ke
Xi'an University of Technology
Xi'an, China

Contents

Chapter 1
Theoretical Basis of Ultraviolet Self-organizing Network

1.1 Wireless Laser and UV Communication

1.1.1 Wireless Laser Communication

With the development of information technology, people expect to obtain required information conveniently at any time, anywhere, and in any way. Because conventional wired or wireless communication can no longer meet increasing communication needs, new communication methods are being explored. Free space optical (FSO) communication is a technology that combines optical and wireless communication; it uses light as a carrier to transmit information without using a wired channel as a transmission medium. Compared with other types of wireless communication, it has the advantages of not requiring a frequency license, high frequency bandwidth, low cost, good confidentiality, low bit error rate, fast installation, resistance to electromagnetic interference, flexibility, convenient networking, and so on. The maximum transmission rate of FSO products is currently 2.5 Gbps; thus, it is very useful for short-distance high-speed networks and can also be used in exhibitions, short-term rental buildings, and temporary workplaces in the field, or at the scene of emergencies such as earthquakes or other occasions for temporary communication. Figure 1.1 shows a schematic diagram of a wireless laser communication system.

Wireless laser communication is a directional point-to-point communication method, as shown in Fig. 1.2. Table 1.1 lists the functions of each module in the system, which has three main parts: the laser transmission system, optical antenna, and laser receiving system. The laser is modulated by a signal. Next, it is collected by the transmitting antenna into a beam with a small divergence angle. The beam is transmitted through free space or the atmosphere, received by the receiving antenna, and concentrated on the detector for the detection of the communication signal. Laser communication technology has attracted considerable attention in various countries because of its good monochromaticity, strong directionality, concentrated optical power, resistance to eavesdropping, low cost, and fast installation. Wireless laser

X. Ke, *UV Light Self Organizing Network Theory*, Optical Wireless Communication Theory and Technology, https://doi.org/10.1007/978-981-97-7854-6_1

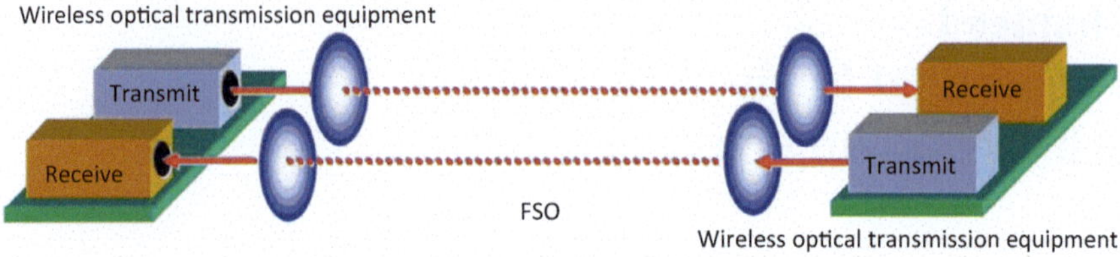

Fig. 1.1 Schematic diagram of atmospheric laser communication

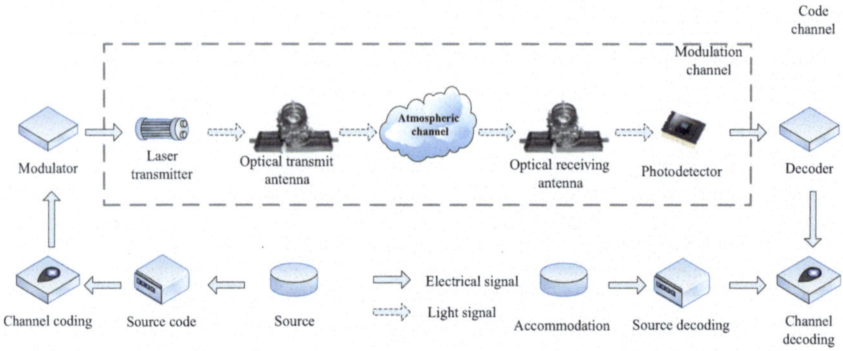

Fig. 1.2 Conceptual diagram of wireless laser communication transmitter/receiver

communication uses laser beams as a carrier to transmit information. Rain, fog, clouds, haze, and sandstorms in the atmosphere strongly affect the communication quality and code rate, which affect the maximum transmission rate of information and reduce the communication distance.

FSO communication has become a hot topic in information technology. Its role and status can be compared with those of optical fiber and microwave communication. It is an indispensable technology for building a future worldwide communication network [1, 2]. FSO systems can realize high-speed two-way transmission of information such as voice, data, and images. The main components of FSO technology include light source and high-bit-rate modulation technology; high-sensitivity and anti-jamming optical signal receiving technology; channel coding technology; antenna transceiver technology; acquisition, tracking, and pointing (ATP) technology; and atmospheric channel processing technology.

Communication transmission media include microwaves and optical fibers. Compared with wired communication, microwave communication can considerably reduce the use of non-ferrous metals, and it easily crosses complex terrain. It can flexibly form a communication network consisting of points and lines so that users on islands and in mountainous or rural areas can more easily use trunk lines for information exchange. However, compared with optical fiber communication systems, it has a narrow frequency band, low channel capacity, low code rate, and many other deficiencies. Optical fiber communication systems have a high line capacity and

Table 1.1 Functional modules in FSO communication system

Send/receive module	Function
Source/sink	Source: Generates the initial signal (digital or analog) Sink: The message arrives at the destination in the same format as the original signal
Source encoding/decoding	Source coding: Handles A/D conversion, data compression, and modulation coding of matching modulation channel for source information Source decoding: The original signal sent by the source is restored from the output sequence of the channel decoding
Channel coding/decoding	Channel coding: By adding redundant information, the receiver can overcome the effects of noise and interference when the signal is transmitted in the atmospheric channel Channel decoding: The pre-encoding sequence is recovered according to the encoding method adopted for channel encoding and the redundant information contained in the received data
Modem	Modulation: After the encoded signal is loaded into the modulator, the excitation current varies with changes in the signal and modulates the intensity of the laser through the driving circuit Demodulation: The preprocessing of the received signal includes signal amplification, filtering, and pulse width processing to restore the pre-modulation format
Semiconductor laser/photodetector	Semiconductor laser: The laser is driven by the excitation current to emit laser light Photodetector: The photodetector converts the received light pulse signal into an electrical signal
Optical transmit/receive antennas	Transmit antenna: The laser beam is transmitted to space through the optical antenna to reduce the divergence angle Receiving antenna: This antenna receives light beams from space and passes them to a photodetector

are not easily disrupted by external sources, but there must be a public channel for installing the optical cable. In harsh terrain, construction is difficult, slow, and expensive. Optical wireless communication combines the advantages of optical fiber and microwave communication; it not only has a high communication capacity, but also does not require the use of optical fibers. It uses lasers as the information carrier and does not require any wired transmission communication method; thus, it can be used for communication between space and ground. Its transmission is characterized by the fact that the beam travels in a straight line.

The propagation media of free space can be divided into three types according to their light penetration characteristics: the near-ground atmosphere, deep space far from the ground, and water. In addition, according to the transmission channel characteristics, FSO communication can be divided into atmospheric, interstellar, and underwater laser communication, as shown in Fig. 1.3.

FSO networks can have various topologies, including point-to-point, point-to-multipoint (star), ring, and grid-type structures. A combination of these structures

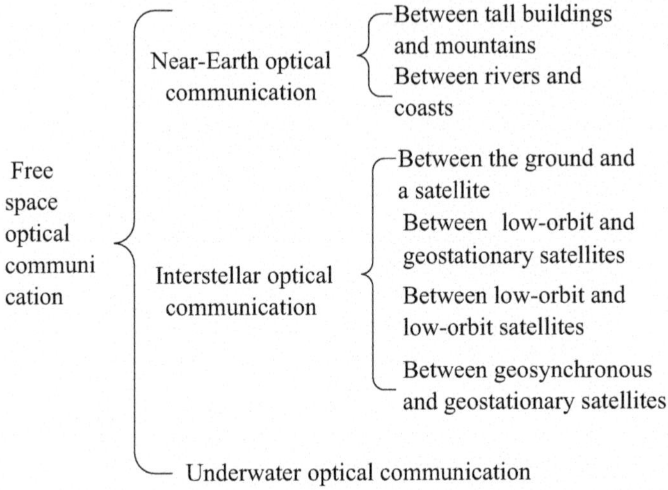

Fig. 1.3 Classification of free space laser communication according to channel characteristics

can be used, which can be selected according to user needs and practical applications. The most basic network topology structure for FSO networks is point-to-point structure (Fig. 1.4), which is used mainly for connections between buildings within an enterprise or dedicated line connections for broadband access. The advantage of this structure is that each link is independent, the structure is simple, and it is suitable for network access. The disadvantage is that links have no redundant protection measures, and there is a single point of failure.

Fig. 1.4 Schematic diagram of point-to-point topology

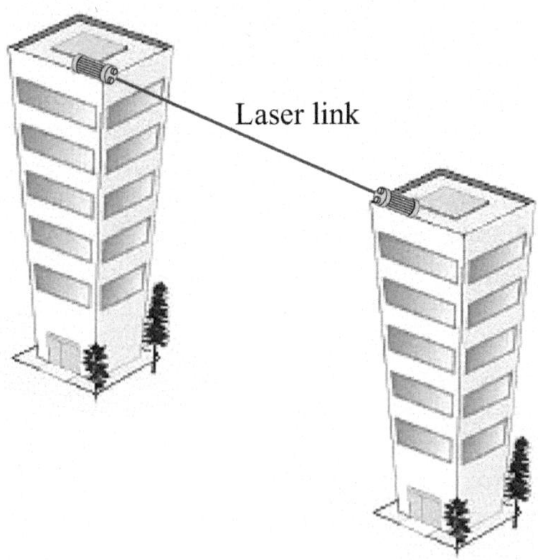

The advantage of a point-to-multipoint (star) structure (Fig. 1.5) is that services can be concentrated at one point (a hub or central node) and then connected to the core network. This network structure is more efficient and economical. The disadvantage is that less bandwidth can be provided; each link still has no redundancy protection, and the reliability is poor. In addition, to connect as many buildings as possible within the line of sight, the hub location is critical, and the hub is also very expensive. However, there is a type of point-to-multipoint structure that performs point-to-point transmission but uses a number of terminals for different directions to the central node; thus, it provides dedicated bandwidth, which can be expanded and can provide single-user service.

In a ring structure (Fig. 1.6), the optical links between all nodes are connected end to end and form a closed loop. The advantage of this structure is that after a failure occurs, the network can automatically restore services quickly without manual intervention.

The main advantage of the grid-type structure (Fig. 1.7) is that it can provide real-time detour routes through multiple network nodes to protect the service; that is, it

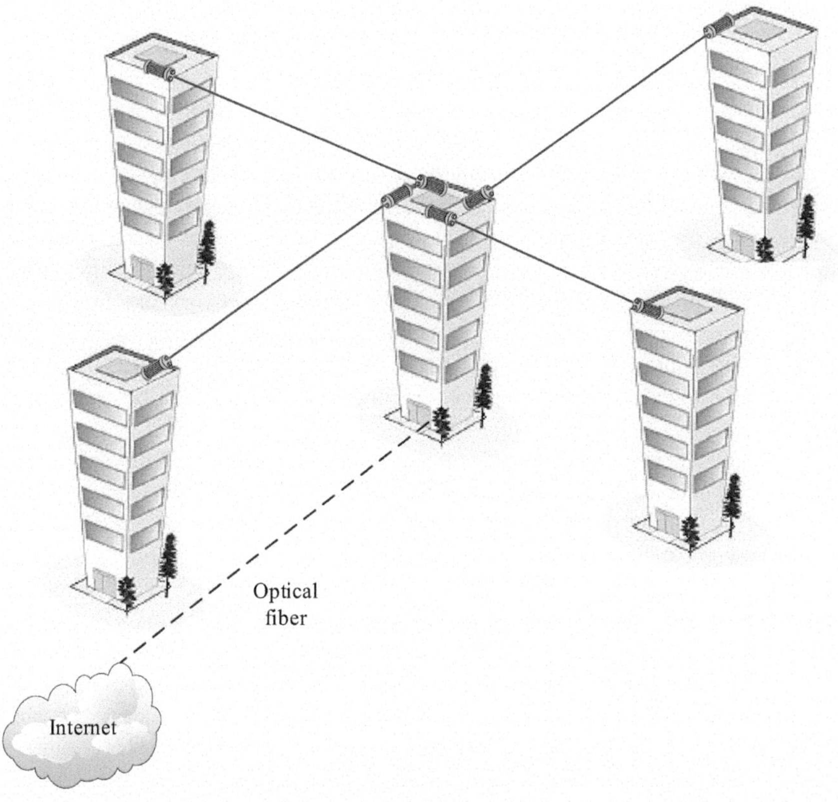

Fig. 1.5 Schematic diagram of point-to-multipoint topology

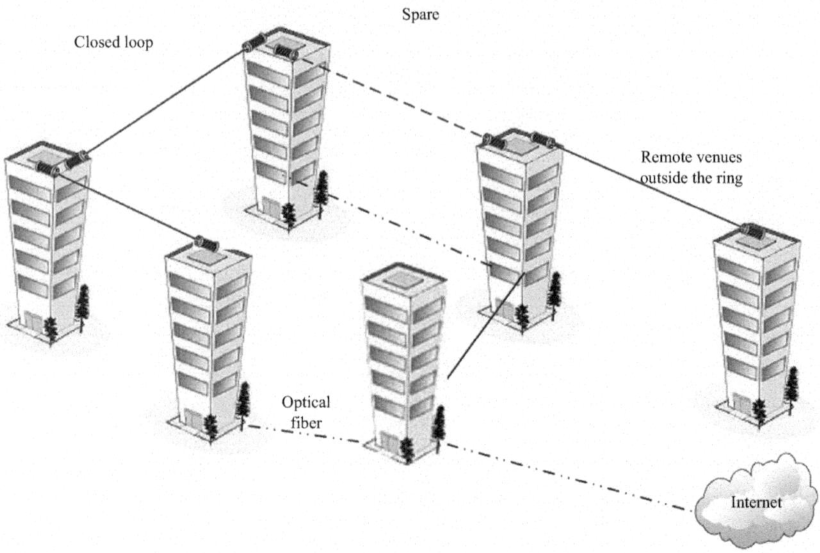

Fig. 1.6 Schematic diagram of ring topology

enables service recovery and redundancy. Under this structure, FSO communication can be used as a supplementary backup for wired optical fiber communication. It can also be used independently to provide partial access and to aggregate services to selected access points, and then provide centralized access to the optical fiber network; thus, it is suitable for carrier-class usage. However, this structure has the disadvantages of short transmission distance, high cost compared with other structures, and complicated network planning. In a grid-type network, the hub of the central node is located on a building in a central area and connected to the optical fiber ring. The transmission rate is 10 Mbps to 1 Gbps. Each central node in this structure can be used as both a user access point and a relay node for the next central node.

The use of FSO systems for point-to-point connections has been fully proven, for example, for backhaul links in local area networks (LANs), emergency backup, and backup during disaster recovery. However, it has not been fully proven for last-mile access. In recent years, with the continuous development of semiconductor laser and photodetector technology, FSO communication has attracted increasing attention, but it also faces challenges, as described in the literature [1, 2]. The main challenges are as follows. (1) Atmospheric absorption and scattering of the light signal results in atmospheric attenuation, even in a clear sky. (2) The turbulent motion of the atmosphere causes the light spot to drift and flicker. In particular, under strong turbulence, the optical signal undergoes serious interference or even misses the target, resulting in a high bit error rate and short-term communication interruption, which seriously affect the stability and reliability of laser communication. (3) In actual use, the optical carrier frequency is very high, and the signal bandwidth of the optical communication

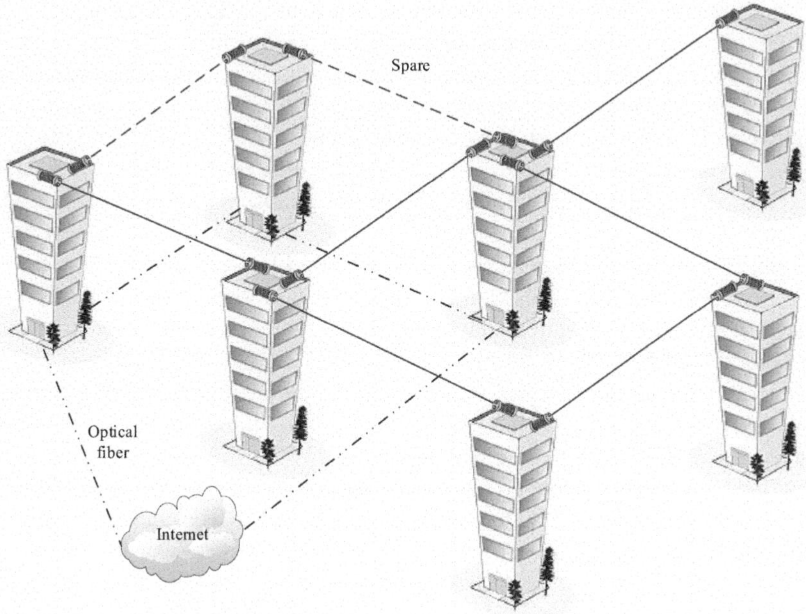

Fig. 1.7 Schematic diagram of mesh topology

system can exceed 1 THz. However, because the optical dispersion and the speed of electronic devices are limited, the transmission rate is often limited to 10 Gbps or lower. (4) Difficulties arise in the implementation of high-precision ATP technology. In particular, it is more difficult to achieve high-precision ATP using narrow beams. (5) To protect human eyes, the optical power provided by the transmitting antenna is limited. (6) Modulation methods such as OOK and M-ary pulse position modulation have typically been used in optical communication systems. For single-carrier systems, when the data transmission rate is high, the delay spread of light transmission in the atmosphere causes mutual interaction between symbols. Overlap causes intersymbol interference, necessitating stringent channel equalization. When the system code rate is high, channel equalization becomes difficult.

1.1.2 UV Communication

Ultraviolet (UV) communication is a type of wireless optical communication. It represents a new communication mode that uses the scattering of UV light in the atmosphere to transmit information. UV light has wavelengths of 400–10 nm and is typically divided into A, B, and C rays (referred to as UVA, UVB, and UVC respectively), with wavelength ranges of 400–315, 315–280, and 280–10 nm, respectively

[3]. Atmospheric ozone has two absorption bands in the UV: the strong Hartley band (200–320 nm), with a maximum absorption coefficient of 255.3 nm, and the Huggins band (320–360 nm). UV communication uses mainly UVC light with wavelengths of 200–280 nm as the transmission medium. Because solar radiation at these wavelengths is strongly absorbed by ozone molecules in the stratosphere, the solar background in the near-surface troposphere is lower than 10–13 W/m^2. In particular, in the near-Earth solar spectrum, this UV band is almost absent [4] and is usually called the solar-blind zone. Therefore, the background noise is low when solar-blind UV light is used for communication, and solar-blind UV communication exhibits little interference and can be used at any time. Because many particles are present in the atmosphere, UV radiation undergoes considerable scattering during transmission. This scattering enables non-line-of-sight (NLOS) signal transmission by UV optical communication systems. Consequently, these systems can adapt to complex terrain and overcome the disadvantage of other FSO communication systems, which must use line-of-sight (LOS) communication [5].

UV light has short wavelengths and is strongly scattered by the atmosphere during transmission. Solar-blind UV light is affected mainly by Rayleigh and Mie scattering during atmospheric propagation. It is rapidly attenuated, but the scattering propagation path can avoid artificial and Mie scattering by natural obstacles, enabling NLOS communication [6]. Sunlight in the UVC band is fully absorbed by atmospheric ozone. This frequency band has little background interference in the near-Earth range. NLOS communication is realized because of scattering by aerosol particles and atmospheric molecules. The divergence angle between the emitted beam and receiving field of view determines the effective communication range.

UV communication is a wireless optical communication technology based on atmospheric scattering and absorption. It uses UV light to send and receive information. When light propagates in the atmosphere, an electromagnetic field is generated that vibrates the charges carried by atmospheric particles. These molecules and particles that are forced to vibrate will become new point light sources and emit secondary waves. These secondary waves are located in a homogeneous medium. They are coherent, but the fixed phase relationship between them is destroyed in the low-altitude atmosphere. Thus, UV light propagates in all directions with the frequency of the original light and a fixed phase difference with respect to the secondary wave source. Therefore, the emitted UV signal is scattered in the atmosphere, and the scattered signals can retain the original information. As long as the scattered signal can reach the field of view of the light-receiving device, two parties can communicate. Therefore, NLOS communication is very suitable for environments with many obstacles, complex combat environments, and strong concealment during combat, and thus has great strategic significance. UV communication systems include modulators/demodulators, UV light sources/detectors, and optical antennas (some receiving antennas also contain optical filters). A schematic diagram is shown in Fig. 1.8 [7].

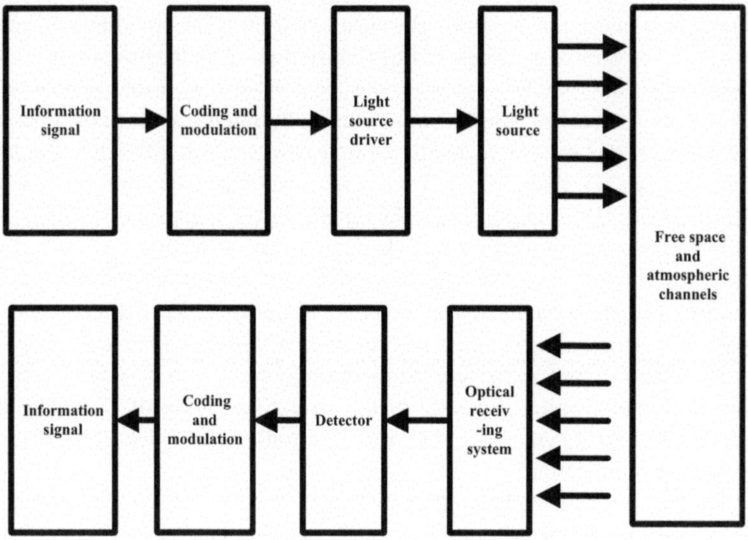

Fig. 1.8 Schematic illustration of UV communication

1.1.3 Characteristics of UV Communication

Electronic communication must be fast, accurate, confidential, and uninterrupted. The characteristics of UV communication systems makes them ideal for covert communication. The signal radiated by a light source diffuses into the low-altitude atmosphere and is absorbed by the atmosphere, and the signal intensity attenuates exponentially. Most of the UV light is scattered to the field of view of the receiver by tiny particles in the atmosphere and received by an antenna in the receiver. With the continuous development of communication technology, the public expects the communication transmission rate, anti-interference ability, confidentiality, and other features to improve. To improve the security of information, we must comprehensively use and develop various communication methods and ensure that communication systems have strong anti-reconnaissance, anti-jamming, and anti-destruction capabilities [8]. Compared with conventional communication methods, UV communication has the following advantages.

(1) Low eavesdropping rate. The energy decays very quickly because of the absorption and scattering of UV light by atmospheric molecules and suspended particles during transmission; therefore, the range of this type of wireless communication is limited. Eavesdropping is impossible outside of the communication range, even if a high-sensitivity UV detector is used.
(2) Low resolution. Because UV light is invisible to human eyes, it is difficult to identify the position of the signal source with the naked eye. In addition, because UV light carries signals mainly by scattering, it is difficult to determine the location of the signal source from the scattered signals.

(3) Strong anti-interference ability. Because UV light is absorbed by atmospheric ozone and oxygen, it experiences little interference near the ground. In addition, because of scattering, UV light near the ground is evenly distributed and can be filtered at the receiving end to remove the background signal. At the same time, the optical signal is not affected by radio waves, and it is difficult to implement long-distance UV interference. Therefore, it is not convenient for an enemy to interfere with UV communication by an army from a distance.
(4) Omnidirectional communication. There are many molecules and aerosol particles in the atmosphere, which have a strong scattering effect on UV light. After multiple scattering events, UV photons can be dispersed in all directions in local space. Therefore, information can be received within the effective coverage area; unlike a laser, it will not have a strong orientation.
(5) NLOS communication. Because of the strong scattering of UV light, it can propagate in a NLOS manner. Thus, it can naturally bypass obstacles and provide NLOS communication.
(6) There is no need for ATP equipment.
(7) Around-the-clock operation. UV communication generally uses a wavelength range in the solar-blind area (200–280 nm), where near-Earth solar radiation is weak, and interference is low even in the daytime. In addition, although the transmittance of UV light varies with the season, altitude, climate, visibility, etc., these factors have the same effect on solar radiation at a particular place and time. This variation is regarded as a low-frequency background signal that can be easily removed by a filter. Therefore, UV communication can be performed smoothly even in complex and changeable environments, regardless of weather conditions or the presence of smoke and dust. Thus, UV communication is an all-weather communication method [9–11]. To interfere with UV communication by an enemy, methods such as deceptive interference, suppressive interference, and decreasing the signal transmission efficiency are generally used to interfere with reception by the enemy's receiver [12].

UV communication systems are very broadly applied in military situations. For example, a fleet of ultralow-flying helicopters in formation can use UV light for internal security communications around the clock. Helicopter pilots can transmit speech or data to their ground crew, knowing that their communications will not be detected by the enemy. Small units patrolling urban areas or areas with complex terrain can also use a UV communication system to transmit secret information to coordinate ground operations if LOS communication is impossible. Conventional tactical operations can also be supported by short-range data communication between communication shelters, which currently use cable communication. The use of UV communication will reduce the cable burden on the logistics department and save the time required to retract the cable. A UV communication line to the shelter will greatly reduce the time needed to set up and dismantle communication equipment and lines. When a fleet must maintain radio silence, UV communication systems can be used to provide short-range communication between ships [12, 13].

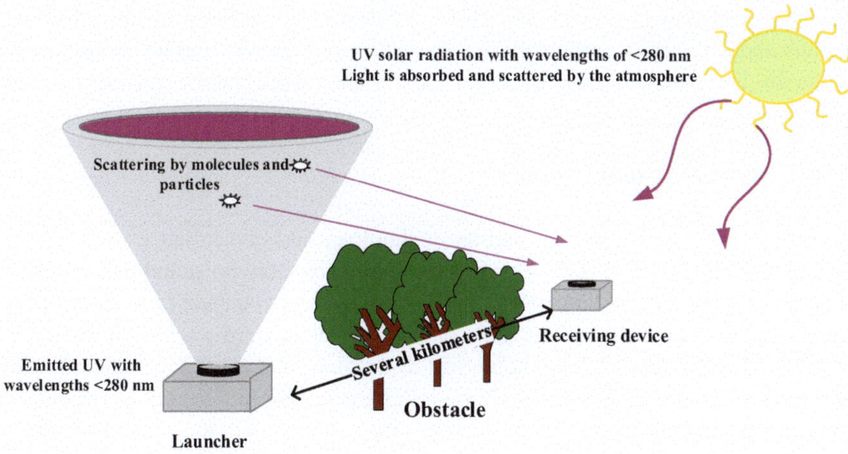

Fig. 1.9 Schematic diagram of UV communication [9]

A schematic diagram of UV communication is shown in Fig. 1.9. Owing to the advantages of UV communication, it can meet the special requirements of communication technology for modern warfare; it is an ideal means of tactical communication and has important research value.

1.1.4 Research Progress in UV Optical Communication

Because of the special technical demands, device development, and application fields of UV communication, it is still being explored and developed worldwide. UV detection technology was developed and applied as early as the 1950s and 1960s as another military/civilian dual-use photoelectric detection technology, after infrared and laser detection technology [14]. In modern warfare, information and electronic warfare have entered the military field as new forms of combat. To improve the reliability of information transmission in the communication system, block various forms of interference, and strive to take the initiative in war, UV confrontation and anti-resistance technology is gaining increasing attention from the military [15]. In the late 1980s, the first foreign military research on UV light was conducted, and some progress has been achieved. Examples include the detection of fingerprints, bodily fluids, etc., in policing [16]; the detection and location of corona discharge in high-voltage substation systems in electric power research; the observation of plasma and arc discharges in scientific research; the detection and reporting of forest fires and oil and gas field fires; measurement analysis in biological and medical research; microelectronics and ultrafine detection; astronomical research; and meteorology and environmental protection detection [17, 18]. After considerable research, experimentation, and development, UV technology is widely used in military and civilian applications such as space detection, UV guidance, UV interference, early warning systems and military communications, fire detection, disinfection technology, and corona discharge detection [19].

1.1.4.1 Research Status Abroad

As early as 1939, the US discussed the technology and performance of UV light sources, detectors, and filters for naval communications, but no actual research was conducted. In the mid- to late 1960s, the US military showed great interest in short-range communication that meets the requirements of modern warfare. In the mid- to late 1970s, Ross et al. began theoretical discussions and experimental studies on the transmission characteristics of UV radiation in the atmosphere and the feasibility and technical details of scattering atmospheric communication. In addition, Neer et al. [20], funded by the US Department of the Navy and Naval Electronic Systems Command, conducted a feasibility study of UV audio communication, developed an experimental prototype, and conducted many experiments using the prototype [20].

The low-transmittance series developed by the US Air Force Geophysics Laboratory is recognized as an effective and convenient software package. The low-resolution atmospheric model it proposes has a spectral resolution of 20 cm^{-1}. The model has been widely used to study atmospheric transmission from the UV to microwave regimes; the algorithm is relatively simple, and its accuracy is approximately 10–15%. Seven versions have been published since it was proposed in 1970. In 1979, Reilly et al. established a single-scattering model of UV communication based on an elliptical coordinate system and studied the time-spreading characteristics and transmittance of single scattering for NLOS UV communication [21].

In 1985 and 1986, Geller et al. at the US Naval Ocean Systems Center developed a set of UV solar-blind short-range communication systems. The reflected light source consists of two 25 W mercury discharge lamps filled with argon gas, and the detector is a photomultiplier tube. The system also uses a solar-blind UV filter with a transmission wavelength of approximately 240–280 nm. This communication system can operate in both LOS and NLOS modes. The communication rate was 1200 bps in 1985 and increased to 2400 bps in 1986, and the bit error rate was better than 10^{-5}. At the average ozone concentration, the maximum communication distances of the LOS and NLOS communication systems are 3 and 1 km, respectively. Under normal conditions, the communication distance is 0.75 km, and the system can operate normally for one year [22].

In 1988, Stamnes et al. obtained the eigenvalues and eigenvectors of the radiative transfer equation in the matrix form of the discrete coordinate method and obtained the integral constant. In this method, the radiative transfer equation is obtained under N-stream parallel plane atmospheric conditions. Only the atmospheric transmission of UV light can be simulated under LOS conditions.

In 1992, Yen et al. in the US military further discussed and compared various short-range communication systems for use in the battlefield environment. In the same year, Miller further improved the performance of solar-blind UV flash lamps and comprehensively studied the emission light source for UV communication [22].

In 2000, GTE Corporation in the US developed a practical new concealed UV wireless communication system for the US military [18]. The communication rate of the system was increased to 4800 bps, and the bit error rate reached 10^{-6}. The system is not easily detected or intercepted, and it is suitable for use in various close-range anti-jamming communication environments, especially for special operations and low-level conflicts; thus, it meets the requirements for tactical communication.

In 2002, the US Defense Advanced Research Projects Agency began to study semiconductor UV optical sources [23], which are smaller and use less power than low-pressure mercury lamps. Two generations of deep-UV light-emitting diodes (LEDs) have been developed. The central wavelengths of the first- and second-generation deep-UV LEDs are 274 and 254 nm, respectively. Although deep-UV LEDs are available on the market, their electro-optical conversion efficiencies are very low.

In 2007, DARPA began to study avalanche photodiodes (APDs) [24]. APDs provide larger amplification and have a lower dark current and smaller volume than photomultiplier tubes. This project is still in the research stage, and there is no commercial product yet. Products in development for military early warning include the AN/AAR-54 UV imaging detector (Northrop Grumman), the AN/AAR-57 UV imaging detector (Lockheed Martin and the Litton Applied Technology Division), the Miltz UV imaging detector (German Aerospace Center and Dems Benz Missile Company), and MILDS UV warning system (France). These systems have a target arrival angle measurement accuracy of 1°, an alarm time of 0.5–1 s, and a detection distance of 5 km. A nonmilitary UV image survey camera system has been developed by ATP (France). To detect corona discharge in power transmission, CSIR (South Africa) has developed the COROCAM UV detector, and the DAYCOR series of UV detectors has been developed by the Electrical Power Research Institute (Israel) [25].

The MIT Lincoln Laboratory [26] designed a hardware test platform for UV NLOS communication and measured the bit error rate on a test bench to verify the results of a numerical analysis model. An atmospheric transmission model of an NLOS light scattering communication system assuming single scattering has also been developed [27]. The relationship between system geometric parameters such as the divergence angle of the light source, receiving field of view, and receiving and transmitting elevation angles and the received scattered light energy was analyzed using the model. The results showed that the received light energy originates mainly from atmospheric molecular scattering when the transceiver elevation angle is large; otherwise, aerosol scattering contributes most of the received light energy. For NLOS communication systems operating in the solar-blind UV spectral region, increasing the receiving field of view can effectively increase the signal-to-noise ratio. The variation in received scattered light energy with the divergence angle of the light source is opposite for two typical sending and receiving elevation angles" or "for the typical case of different sending and receiving elevation angles., indicating that the divergence angle of the light source should be determined according to the application design. An experimental method has been introduced to study the transmission characteristics of UV NLOS communication channels [28]. The results showed that the UV communication distance can be extended by using highly sensitive detectors and large-diameter antennas. An experimental system for atmospheric scattering communication based on UV LEDs was designed and constructed, and an NLOS single-scattering model was used to simulate the performance of atmospheric scattering transmission of UV light [29]. Because UV signals are strongly scattered and absorbed in the solar-blind region, the transmission loss of UV light in the atmosphere is very large; therefore, scattering communication is generally suitable for short-distance applications. Different path loss models have been established for solar-blind UV NLOS communication; however, they assume that the coplanar light source and receiver point to the axis of the beam. The single-scattering coplanar analysis model was further expanded to non-coplanar geometries, where the model is a function of the geometry and atmospheric properties. The performance of the model was numerically analyzed in different non-coplanar geometries [30].

1.1.4.2 Domestic Research Status

Domestic research on UV communication has just begun. The main topics of this research are summarized below.

In 1999, Beijing Institute of Technology conducted an NLOS UV communication experiment using a low-pressure gas-filled mercury lamp as a light source and achieved good results within a linear distance of 500 m. Chongqing University has also established a UV communication system. Xi'an University of Technology has established an experimental system for UV communication and has realized voice transmission in the laboratory.

The National Defense University of Science and Technology has studied modulation methods for UV communication using mainly low-pressure mercury and iodine lamps as light sources. From 1998 to 2004, Beijing Institute of Technology conducted research and equipment development for UV communication in the battlefield using a low-pressure mercury lamp as the UV source and the analog frequency modulation method for the emission system. Voice communication was realized [31].

The Communication Technology Room of the Space Center of the Chinese Academy of Sciences has systematically studied the system composition and key technologies of UV communication since 2002 and conducted prototype trial production. The light source, transmitter structure, light source driving and modulation method, filter system, detector design, and modeling and simulation of UV atmospheric transmission characteristics have been studied in detail [22].

In 2006, the National University of Defense Technology proposed the use of a low-voltage iodine lamp as the UV light source [32], and a low-voltage frequency modulation circuit realized frequency modulation of the light source. A series of experimental systems was constructed. The maximum modulation frequency is 100 kHz, but the system is also limited to short-distance LOS communication.

As a new means of communication, UV communication systems can be connected to the modern communication network architecture as independent network units and interconnected with wireless and wired communication networks for resource information sharing. UV communication has broad application prospects in battlefield environments, urban mesh backbone network access, and short-range communication by underwater equipment. Figure 1.10 illustrates a military vehicle escort formation in an urban environment using UV NLOS communication to form a chain network for communication [33, 34]. The transmission of information between vehicles and the use of radio frequency communication will cause serious signal interference. Figure 1.11 shows an unmanned ground-based wireless sensor network [34] based on UV NLOS communication. Nodes communicate with the gateway node by UV NLOS communication.

Fig. 1.10 Scenario based on blind UV NLOS communication [34]

Fig. 1.11 Unattended ground-based sensor network using UV NLOS communication [34]

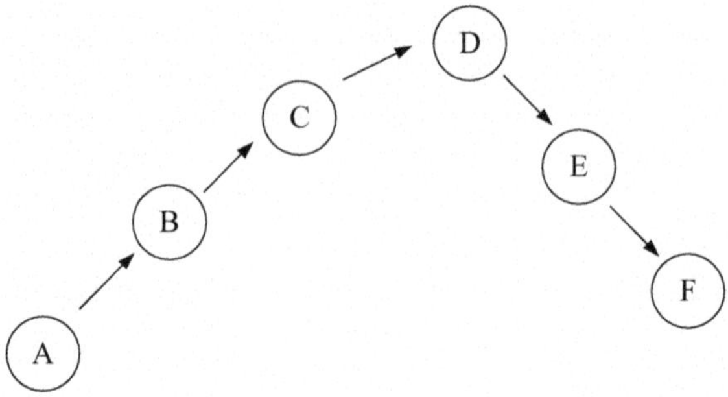

Fig. 1.12 Ad hoc network

1.2 Wireless Ad Hoc Network

1.2.1 Concept of Wireless Ad Hoc Network

An ad hoc network is a wireless mobile communication network composed of a group of mobile nodes with wireless transceivers. It is temporary and does not rely on a preset infrastructure. Mobile nodes in the network use their own wireless transceivers to exchange information. When they are not within communication range of each other, they perform multihop communication using intermediate node relays. In this case, the intermediate node receives a packet sent by the previous node and forwards it to the next node; therefore, the self-organizing network is also called a packet wireless network or multihop network [35].

Figure 1.12 shows a schematic diagram of an ad hoc network, in which nodes A and F cannot communicate directly, but node A can transmit data to node F through the multihop path A → B → C → D → E → F. The architecture of the mobile ad hoc network is shown in Fig. 1.13. The routing protocol is the core of the network layer and has two main functions. (1) When there is no available route between the source and destination nodes, it selects a transmission path that can forward packets for this pair of nodes. (2) After the route is selected, the packet data from the source node are sent to the destination node. Because the topology of the mobile ad hoc network changes constantly, and because the channel depends on the network configuration, the channel also changes constantly. Therefore, a routing algorithm that can consider the current network situation for path selection will greatly improve the communication efficiency of the network and can effectively reduce the routing overhead.

1.2.2 Development of Wireless Ad Hoc Networks

The concept of self-organizing networks can be traced back to the ALOHA network developed in 1968. The ALOHA network was built by the University of Hawaii to

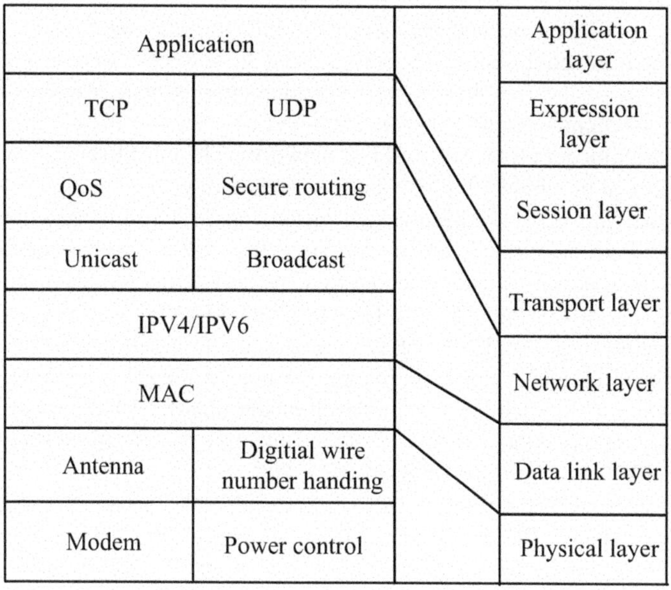

Fig. 1.13 Architecture of mobile ad hoc network

connect seven educational facilities on four islands. Its greatest contribution was the first study of the multisite access channel problem of shared media, which provides a foundation for future research on distributed channel access technology. The ALOHA protocol is a single-hop protocol and does not support routing functions.

Following the success of the ALOHA network and the development of early fixed packet-switched networks, the US DARPA developed the Packet Radio Network (PRNET) in 1972 [36, 37], which played a fundamental role in the development of the Internet. PRNET is a multihop network that provides centralized and distributed operation management mechanisms. By the 1980s, the feasibility of PRNET was proven. The success of PRNET demonstrated that multihop technology can improve network capacity and also proved the feasibility of mobile ad hoc networks.

However, because of the limitations of early wireless communication devices and controllers, such as large size, limited power, limited time, and weak anti-electronic interference capabilities, the US Department of Defense believed that it was still too early to fully develop wireless mobile ad hoc networks (MANETs). To solve these problems, DARPA launched the Survivable Adaptive Network (SURAN) research program in 1983. Its research goals were to develop a low-cost, low-power small radio that can support complex packet wireless network protocols; to develop a networking algorithm suitable for tens of thousands of nodes; and to develop a packet radio that can withstand complex electronic interference caused by web technology y. One of the research achievements of SURAN is the low-cost packet radio, which was developed in 1987.

In 1994, DARPA launched the Global Mobile Information Systems (GloMo) project, which covered almost all fields related to wireless communication. The research goals of the GloMo project were to support global seamless Ethernet-like multimedia connections between unlimited devices and to solve the mobile, multihop, multimedia problem for ad hoc networks. The GloMo project resulted in the development of wireless Internet gateway technology.

Self-organizing networks have a long history of military use, and commercial applications have been rapidly developed since the 1990s. The commercial application of ad hoc networks is represented by wireless LANs (WLANs) and wireless personal area networks (WPANs). In June 1997, the MANET working group of the Internet Engineering Task Force (IETF) released the first-generation WLAN standard, IEEE 802.11, which supports only the one-hop ad hoc mode. In 1994, the Ericsson Company of Sweden launched a development plan for Bluetooth technology, and in 1999 it released the first version of the Bluetooth specification. Bluetooth has become the IEEE standard for WPANs, IEEE 802.15.1.

Ad hoc network technology is still being developed, and the application field continues to expand. The expansion of the main application fields has two directions; one is to support the self-organization mode in public mobile communication systems. The 3GPP TR25.924 recommendation specification for opportunity-driven multiple access (ODMA) was released in December 1999. ODMA is an attempt to introduce ad hoc networks in the third generation (3G) of mobile communications. Although ODMA has not been adopted for 3G for technological and market-based reasons, it has resulted in the development of many technologies that combine public mobile communication systems and ad hoc networks, such as A-GSM [38], integrated cellular and ad hoc relaying systems [39], and self-organizing packet radio ad hoc networks with overlay [40]. The other direction for expansion is sensor networks. A sensor network has a data-centric layered architecture and is composed of a group of randomly distributed sensors integrated with modules for self-organized data collection, processing, and communication. Sensor networks represent a special type of self-organization related to communication.

1.2.3 Research Status of Wireless Sensor Networks Worldwide

Wireless sensor networks were developed to meet the operational needs of the US military. In 1978, DARPA established the Distributed Sensor Networks Working Group at Carnegie Mellon University [32]. The working group studied communication and computing problems in sensor networks in the context of the need for military reconnaissance systems [41]. DARPA and the US National Science Foundation (NSF) have since established a number of research projects related to sensor networks. Since the mid-1990s, wireless sensor networks have attracted considerable attention in academia, the military, and industry. The US has invested heavily in research on wireless sensor network technology through the Department of Defense and the NSFC, and other developed countries have followed suit. Many research programs on wireless sensor networks have been established [42]. For example, in 2003, the NSF formulated a sensor network research plan that allocated 34 million USD to support research on basic theories related to sensor networks. The US Department of Defense and various military departments have also attached great importance to sensor networks. On the basis of C4ISR, the C4KISR plan has been proposed; sensor networks are regarded as an important research field, and a series of military sensor network research projects have been established. Well-known American universities such as University of California, Los Angeles; Cornell University; Harvard University; Massachusetts Institute of Technology; and University of California, Berkeley have also conducted research on sensor networks. The main research projects include SensorWebs, NEST [42], SensorIT [43], SeaWeb, WINS [44], Smart Dust [45], and Hourglass [46], which cover almost all aspects of wireless sensor networks, from signal processing to network protocols. Other developed countries have also launched many research projects on wireless sensor networks. For example, in 2002 the European Union began to implement the three-year EYES program for self-organization and cooperation in an effective energy sensor network. In March 2004, Japan's Ministry of Internal Affairs and Communications established the Ubiquitous Sensor Network survey and program.

Research on wireless sensor networks and their modern applications in my country[China] is on a very similar timeline to that in developed countries. The theory and application of wireless sensor networks have been studied at domestic research institutes and universities including Shanghai Institute of Microsystems, Chinese Academy of Sciences, Shenyang Institute of Automation, Chinese Academy of Sciences Institute of Automation, Hefei Institute of Intelligence, Harbin Institute of Technology, Tsinghua University, Zhejiang University, Beijing University of Posts and Telecommunications, and Northwestern Polytechnical University. This research involves hardware design, operating system design, routing technology, network self-organization and management, energy management, data query and processing, node positioning, time synchronization, adaptive technology, network security, etc. "Research on basic software of sensor network systems and key data management technology" is one of the second-phase joint funding projects officially established by the Information Science Department of the National Natural Science Foundation of China and Microsoft Research Asia. The research project lists sensor networks as a major research topic. During the Eleventh Five-Year Plan period, according to the National Medium- and Long-Term Science and Technology Development Plan Outline, National "Eleventh Five-Year Science and Technology Development Plan, and 863 Plan Eleventh Five-Year Plan Development Outline, the information technology field in the 863 Plan includes sensor network technology under the special topic of self-organizing network and communication technology. The goal is to conduct frontier exploration and research and strive to develop breakthrough technologies that can compete with those of developed countries.

Wireless sensor networks have broad application prospects in both national security and national economic efforts. The in-depth study of this technology will promote the informatization construction of our country and has important strategic significance for national defense science and technology and national economic construction.

1.2.4 Research Status of Wireless Mesh Networks

With the rapid development of wireless mesh networks, many international standardization organizations have added support for mesh networking in various wireless network standards, each of which supports some functions of wireless mesh networks.

The IEEE 802.16 standard group designed support for point-to-multipoint and mesh topologies in the IEEE 802.16a broadband wireless metropolitan area network standard promulgated in April 2003. The IEEE 802.15 WPAN working group established Task Group 5 in November 2003 to study the use of short-distance, low-cost equipment to cover a larger environment, such as a living space, campus, or hospital, by a mesh network. The IEEE 802.11 Wireless LAN Working Group also established a Mesh Research Group (IEEE 802.11s) in early 2004, mainly to study protocols that support wireless distribution systems and thus realize networking among multiple access points in a WLAN by automatically configuring the topology. In addition, organizations such as the International Telecommunication Union, 3GPP, and the IETF have also incorporated wireless mesh networks into their work plans [47].

Many foreign companies have developed systems based on wireless mesh network technology. The SkyPilot System developed by SkyPilot Company in the US has a user data rate greater than 3 Mbps and a frequency band of 5.8 GHz, and nodes can realize LOS/NLOS communication within a range of 32/6.4 km. The Tropos 5110/3110 products developed by Tropos Company in the US offer a user data rate of 0.512–1 Mbps and a frequency band of 2.4 GHz, and the node can realize outdoor/indoor communication within range of 4.15 km/290 m. The MEA product developed by MeshNetworks in the US has a user data rate of 1.5–6 Mbps and a frequency band of 2.4 GHz. Nodes can realize NLOS communication within a range of 1.6 km. The BelAir200/100 products developed by BelAir Company in Canada offer a user rate exceeding 1 Mbps and a frequency band of 2.4/5.8 GHz; nodes can perform backbone/user communication. The Wireless AP7220 product developed by Nortel Company in Canada has a user rate exceeding 1 Mbps and a frequency band of 2.4/5.8 GHz. Nodes can realize backbone/user communication within a range of 1 km/100 m. The HotPoint 1000 product developed by Firetide in the US has a user rate exceeding 1 Mbps and a frequency band of 2.4 GHz; nodes can realize outdoor/indoor communication within a range of 3 km/200 m. The MESHWORKS product developed by Radiant Network Company in the UK has a user rate exceeding 4 Mbps and a frequency band of 5.8 GHz. Nodes can realize outdoor communication within a 2 km range. The development of these products will further the research, application, and promotion of wireless mesh network technology.

1.2.5 Mobile Ad Hoc Network and Its Development

1.2.5.1 Mobile Ad Hoc Network

A MANET is a temporary network without a fixed central node. The nodes in the network not only have the functions required for ordinary mobile terminals, but can also forward messages. Each node has the functions of both a host and a router. It does not require the support of any existing information infrastructure and can be used at any time. Thus, a mobile communication network can be quickly built anywhere. Nodes in the network can join or leave the network at any time; thus, the network topology changes very rapidly, causing problems such as unreachable communication nodes [48–50]. A MANET is also called a multihop wireless network.

The predecessor of the MANET is PRNET, which was launched in 1972 by the US DARPA [51–53]. In addition, DARPA launched the SURAN project in 1983 to study how PRNET could be expanded to support larger-scale networks. To study and develop adaptive network protocols that can adapt to rapidly changing battlefield environments [54], DARPA launched the GloMo project in 1994. More in-depth research is underway to develop highly invulnerable mobile information systems that meet the requirements for military application and can be quickly deployed [55].

The IEEE 802.11 standard committee adopted "ad hoc network" to describe this special self-organizing, peer-to-peer, multihop wireless mobile communication network. "Ad hoc" is a Latin phrase meaning "established according to need". Figure 1.14 shows a logical network diagram of a MANET.

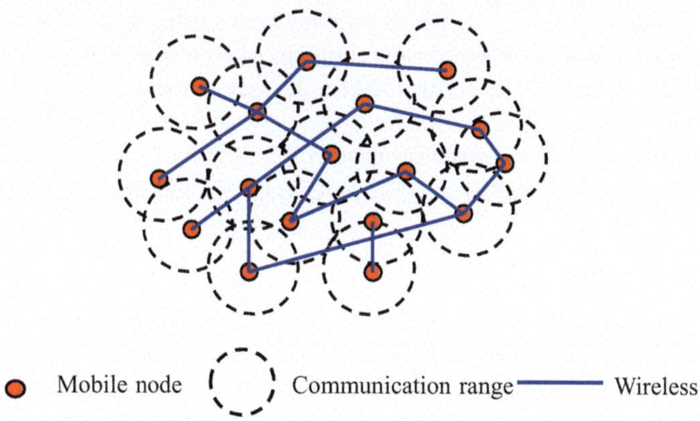

Fig. 1.14 MANET logical network diagram

1.2.5.2 Research Status of MANETs

(1) IEEE 802.11 and 802.15 working groups

The IEEE 802.11 working group is responsible mainly for writing WLAN standards. These standards describe primarily two modes: MANET mode (ad hoc mode) and access point mode. The standards specify the underlying physical layer and MAC layer standards. The IEEE 802.15 working group is responsible mainly for formulating WPAN standards. This committee has specified several different WPAN standards such as Bluetooth, ZigBee, and WPAN based on multimedia applications.

(2) MANET working group of IETF

The goal of the IETF working group is to standardize technologies related to the routing layer in MANETs to realize a routing mechanism based on the Internet Protocol (IP), and ultimately to make the IP truly network-independent regardless of the underlying hardware facilities and protocols. Research by the working group focuses mainly on the routing algorithm, network simulation and model, network security, quality of service (QoS), and multimedia transmission in MANETs. Most proposed standards focus on the routing algorithm of the routing layer; the most common algorithms include dynamic source routing (DSR) [56], ad-hoc on-demand distance vector (AODV) [57], temporally ordered routing algorithm (TORA) [57], destination-sequenced distance vector (DSDV) [54], and zone routing protocol (ZRP) [57].

(3) Monarch project team at Carnegie Mellon University

The Monarch project is a research project initiated by Carnegie Mellon University in the 1990s to provide wireless network access and interconnection support for the wireless mid-segment. It studies mainly network protocols and interfaces; the research scope includes protocol design, implementation, performance evaluation, and practicality-based validity verification for the second to sixth layers of the Open Systems Interconnection (OSI) seven-layer protocol structure. The Monarch project currently focuses on research on routing protocols for MANETs and the Transmission Control Protocol and support for high-level protocols and application layers for mobile networks.

(4) Routing protocol

Because mobile self-organizing networks have no central node, their rapid configuration and strong anti-destruction ability are intensively studied in the fields of military and civil communication in various countries. However, many problems with MANETs have not been completely solved, for example, problems with routing, channel access, topology maintenance, and security, among which routing technology is particularly important. In contrast to the nodes of conventional wired or cellular networks, those of MANETs exchange data by a multihop routing packet-forwarding mechanism. As the network topology changes, the routing process is initiated, selected, and maintained, and data are forwarded along the selected route.

However, changes in network topology caused by the mobility of nodes, the unreliability of wireless channels, the complexity of channel access, etc., complicate research on MANET routing protocols, which faces great challenges. Therefore, routing protocols have become a very important research topic for MANETs.

The main routing protocols used in wireless ad hoc networks include active, on-demand, hybrid, and location-based routing protocols [58]. Active routing protocols are also known as table-driven or proactive routing protocols. DSDV (Destination-Sequenced Distance-Vector, DSDV)is a typical table-driven routing protocol and is based on the Bellman–Ford routing mechanism. The main improvement is that the routing table entry includes the sequence specified by the destination node number to distinguish old and new routes and avoid routing loops. However, DSDV does not adapt to rapidly changing ad hoc networks, nor does it support unidirectional channels [59]. Passive routing protocols are also called on-demand or reactive routing protocols. DSR is a typical passive routing protocol, and it was also the earliest protocol to use the concept of on-demand routing. The header of the data packet contains complete routing information to avoid loops. Intermediate nodes use route caching technology to reduce resource consumption for route discovery, but expired routes affect the accuracy of route selection. DSR supports a single link and does not need to broadcast link status information periodically, which can reduce overhead. However, because the packet header contains complete routing information, link utilization is low [60]. AODV and TORA are two other widely used on-demand routing protocols. In contrast to DSR, AODV and TORA do not support unidirectional links; in addition, AODV cannot adapt well to rapid changes in network topology, and TORA must be supported by a global synchronization clock [60, 61]. As the scale of an ad hoc network increases, the plane routing protocol cannot be applied because of increased link and processing overhead. In this case, scalable and effective routing can be obtained using a hierarchical (layer, group) routing scheme, but the clustering process will consume more energy [58]. Commonly used hierarchical routing protocols suitable for single channels include the ZRP [62], fisheye state routing [63], core-extraction distributed ad hoc routing [64], optimized link state routing [65], and cluster-based routing [66]. Cluster head gateway switch routing [67] is a hierarchical routing protocol suitable for multiple channels. The location-aided routing protocol obtains information on the location of the mobile host using GPS to control the routing query range. Although the use of geolocation information by this routing protocol can provide high performance, delays result in inaccurate location information [68].

As a key technology for wireless self-organizing networks, routing technology will inevitably develop in the directions of practicability, adaptive behavior, and security. Increasing numbers of MANET researchers combine positioning technology with wireless ad hoc networks and make full use of location information to optimize routing [58]. Other major feature of wireless ad hoc networks are the vulnerability of routing protocols and diversity of intrusion behaviors. Various secure routing protocols have been proposed [69]. However, these protocols still have many problems, and more research on the security of routing protocols is needed.

1.2.5.3 Wireless Sensor Network Routing Protocol Algorithm

There are many types of routing algorithms for wireless sensor network protocols. Depending on the logical structure of the network, they can be divided into plane and hierarchical routing algorithms [70, 71]. Representative energy-saving plane routing algorithms are directed diffusion (DD) [72], routing algorithm with a negotiation mechanism [sensor protocol for information via negotiation (SPIN)] [73], and the multipath routing algorithm (highly resilient energy-efficient multipath routing) [74]. DD is an information dissemination protocol centered on data bits. Its routing strategy was designed by Estrin et al. for sensor networks. The SPIN protocol group includes the protocols SPIN-1 and SPIN-2. The sending decision process of these two protocols depends only on the local neighbor information, so it is suitable for use in environments where the sensors move. The multipath routing algorithm is based on the DD routing mechanism. The purpose is to improve the reliability of routing by maintaining multiple available paths, which not only realizes effective energy recovery, but also improves the robustness of the DD protocol by solving the energy waste problem caused by the use of periodic low-rate data diffusion. Representative energy-saving hierarchical routing protocols are the low-power adaptive cluster (LEACH) routing algorithm [75], power-efficient gathering in sensor information systems (PEGASIS) [76], the threshold-sensitive energy-efficient sensor network protocol [77], and the energy-aware sensor clustering routing protocol [78]. The LEACH algorithm is suitable only for applications with short transmission distances and small network coverage areas; it does not support network expansion and thus has limited applicability in wireless sensor networks. In addition, there is no mechanism to ensure the timeliness and necessity of cluster reconstruction. The PEGASIS algorithm improves the lifetime of sensor networks because all nodes have the global information of the network. When nodes are large, it is very difficult for a single node to save and update the global information of the network. The layered PEGASIS protocol proposed by Lindsey et al. [79] is an extension of PEGASIS and is designed to reduce the delay caused by the transmission of data packets to sink nodes.

Because of the characteristics of wireless sensor networks, we should consider mainly the following factors in routing protocol design: energy saving, fault tolerance, scalability, transmission delay, data fusion, and QoS. In addition, wireless sensor networks with many widely distributed nodes will be subject to many potential network attacks, and security should be considered. Therefore, protocols for wide use in wireless sensor networks must be associated with a secure routing protocol [80].

1.3 Characteristics and Key Technologies of Wireless Ad Hoc Networks

1.3.1 Characteristics of Wireless Ad Hoc Networks

Compared with conventional communication networks, wireless ad hoc networks have the following characteristics [81, 82].

(1) Centerlessness and self-organization

Ad hoc networks have a distributed structure in which all nodes have equal status. They are peer-to-peer networks with strong scalability. The layout of the network does not necessarily rely on any pre-installed network facilities. Nodes can join or leave the network at any time and can quickly and automatically form an independent network after starting up. Compared with networks having a strong central node, they have strong survivability.

(2) Dynamically changing network topology

The topology of ad hoc networks may change at any time because of the randomness of the speed and mode in which nodes move, the variable transmission power, and the effects of channel interference. In addition, the changes and their timing are difficult to predict.

(3) Multihop routing

The communication distance of nodes is limited by their transmission power. Internode communication in ad hoc networks often must be relayed and forwarded through intermediate nodes. In contrast to general multihop networks, ad hoc networks do not require dedicated routing equipment, and all ordinary nodes in the network perform multihop routing.

(4) Distributed control

Because there is no central control node, the control and management of the network are distributed to each node. Even if one or some nodes in an ad hoc network fail, the entire network will not be paralyzed, and the rest of the nodes will operate normally. This characteristic enhances the robustness of the network.

(5) Temporary nature

Ad hoc communication networks are usually created temporarily for a specific purpose without planning or establishing infrastructure. This temporary nature is reflected in the membership, topology, and survival time of the network.

(6) Unidirectionality of channels

Because of the characteristics of wireless channels, unidirectional channels may exist between nodes in an ad hoc network. These channels can be fully utilized through the network routing algorithm, which can improve the network performance.

(7) Limited bandwidth

Owing to the physical characteristics of the wireless channel itself, the bandwidth of ad hoc networks is much lower than that of wired networks. In addition, factors such as signal fading, collision, blocking, and noise interference caused by wireless channel competition must be considered; these factors reduce the bandwidth used well below the maximum transmission rate provided by the physical layer.

(8) Limited energy

Nodes in ad hoc networks generally rely on batteries to provide energy. Until battery capacity is greatly improved, power saving should be highly valued in ad hoc network technology.

(9) Poor security

Ad hoc networks have a very open wireless distributed structure. Because of the limited number of wireless channels and power supply, they are more vulnerable to passive eavesdropping, active intrusion, denial-of-service, deprivation, denial-of-sleep, and other cyber-attacks.

1.3.2 Key Technologies of Wireless Ad Hoc Networks

Since self-organizing networks were introduced, they have been extensively studied, and excellent progress has been made. However, many problems with self-organizing networks remain. These problems are associated with factors such as centerlessness, self-organization, node mobility, wireless communication technology, limited bandwidth, and the limited battery energy of nodes. They involve, for example, the MAC protocol, routing protocol, QoS assurance, network security, and network management of self-organizing networks. They have also become hot topics of research on wireless ad hoc networks.

(1) MAC protocol

Channel access technology is the basis of self-organizing network protocols. It controls how nodes access the wireless channel and plays a decisive role in the performance of self-organizing networks. An ad hoc network is a special type of peer-to-peer network. Conventional fixed or centralized network channel access protocols cannot meet the needs of ad hoc networks. A good MAC protocol should offer high spatial multiplexing and hardware independence, avoid packet conflict, and provide conflict resolution methods. It should also have high throughput, fairness, low latency, low energy consumption, support for multicasting and broadcasting, multiservice capability, and so on. In addition, ad hoc networks have problems with unique hidden terminals and exposed terminals that must be solved. These problems need to be resolved by designing a targeted MAC protocol.

(2) Routing protocol

Conventional distance vector and link-state routing protocols are not suitable for self-organizing networks with highly dynamic topologies. An ideal self-organizing network routing protocol should be loop-free and offer distributed operation, on-demand operation, support for unidirectional links, the maintenance of multiple routes, security, and energy saving. Therefore, it is necessary to study new routing protocols that meet the requirements of ad hoc networks, which have become a hot topic in ad hoc network research. The most mature self-organizing network routing protocols currently studied include DSDV [83], DSR [84], and AODV [85].

(3) QoS assurance

Self-organizing networks were initially used mainly to communicate small amounts of data. With the continuous development of applications, it has become necessary to transmit multimedia information such as voice and images via ad hoc networks. Because multimedia information has high technical requirements for bandwidth, delay, and delay jitter, it is necessary to provide a certain QoS guarantee. QoS assurance in ad hoc networks is a systemic problem involving every layer of ad hoc network protocols. For example, the link layer should provide a resource reservation strategy, the network layer should provide QoS routing, and the application layer should provide adaptive source coding and compression technology.

(4) Network security

Security is inevitably an issue for self-organizing networks. Because transmission occurs over a wireless channel, self-organizing networks are vulnerable to eavesdropping and interference and are subject to active intrusion, identity forgery, and denial-of-service and other attacks. In addition, the limited energy available to distributed structured networks and nodes also threatens network security. Proposed security strategies include key exchange based on password authentication, the resurrecting duckling security model, and asynchronous distributed key management.

(5) Network management

The management of self-organizing networks is more complicated than that of wired networks. Network management requires the effective and timely collection of network-related information and reasonable processing of dynamic network configurations based on this information.

1.4 UV Self-organizing Network

Most research on solar-blind UV communication focuses on point-to-point communication systems, and we can use the solar-blind band to form a network using UV communication technology. Self-organizing networks are a special type of peer-to-peer network that have the characteristics of centerlessness, self-organization, rapid deployment, mobility, and multihop capability. The combination of solar-blind UV communication and self-organizing network technology in a UV self-organizing network can compensate for the inherent shortcomings of UV scattering communication, such as low transmit power, severe atmospheric attenuation, and a large number of one-way channels.

1.4.1 Potential Applications of UV Self-organizing Networks

UV self-organizing networks show promise for applications in military and civil communication owing to their excellent characteristics. First, they can be rapidly deployed. Second, the use of multihop routing technology can reduce the transmit power of the terminal without reducing the network coverage, thus compensating for the limited UV communication distance. In addition, the high survivability and robustness of the network can also meet the requirements of specific applications [81, 86, 87].

(1) Military communication

Mobile self-organizing networks were developed to meet the ever-changing needs of the battlefield, and battlefield communication is expected to be the most direct military application. UV self-organizing networks are ideal carriers for self-organizing networks because of their solar blindness and NLOS transmission. In addition, the unique security features of UV light also make it the technology of choice for digital battlefield communications. MANET technology has become the core technology of the tactical Internet of the US military. UV self-organizing networks are expected to become a hot topic of research on military communication in various countries.

(2) Emergency rescue and disaster relief

When natural or other disasters disrupt the normal operation of the network infrastructure or make it impossible to establish communication base stations, temporary networks can be quickly established using UV self-organizing network technology, and the network infrastructure can be extended. UV self-organizing networks support communication in harsh and unusual environments. In addition, their independent networking and self-organizing characteristics can be useful for establishing communication between teams in remote locations.

(3) Sensor networks

Sensor networks are another important application of UV self-organizing network technology. This type of network has very broad application prospects for automatic monitoring and road traffic, industrial manufacturing, biomedical, and security monitoring.

(4) Personal communication

Personal area networks are another application field of UV self-organizing network technology. By using MANET technology, these mobile terminals can be quickly organized into a wireless network to realize communication between them; this approach can also meet special requirements such as mobile conferences. However, because high-power UV light is harmful to humans, application prospects in personal communication require further research.

(5) Environmental science

As the environment receives increasing attention, the scope of environmental science is extended, and it is difficult to collect raw data by conventional methods. Wireless sensor networks enable the convenient acquisition of random research data in the field, for example, by tracking the movement of migratory birds and insects, studying the impact of environmental changes on crops, and monitoring the composition of the ocean, atmosphere, and soil. The ALERT system uses multiple sensors to monitor rainfall, river water levels, and soil moisture and predict the likelihood of flash floods [88]. Similarly, wireless sensor networks are helpful for the accurate and timely forecasting of forest fires. In addition, they can be applied in precision agriculture to monitor crop pests, soil pH and fertilization status, etc.

(6) Medicine and health

The use of special-purpose sensor nodes on inpatients, such as heart rate and blood pressure monitoring equipment, in conjunction with wireless sensor networks, enables doctors to know the condition of patients at any time and provide timely treatment [89]. Wireless sensor networks can also be used for long-term monitoring. These data are very useful for the development of new drugs, and the tiny sensors will not inconvenience patients greatly or interfere with their normal activities. In addition, this type of monitoring has novel and unique applications in, for example, drug management.

(7) Space exploration

Humans have always dreamt of exploring other planets. Wireless sensor networks scattered by spacecraft should be an economical and feasible way to realize the long-term monitoring of planetary surfaces. NASA's Jet Propulsion Laboratory has developed sensor webs [90] for future Mars exploration and has tested and improved them in environmental monitoring projects around the Kennedy Space Center in Florida.

(8) Other commercial applications

Wireless sensor networks have excellent potential for commercial application owing to their three major characteristics of self-organization, miniaturization, and the ability to perceive the external world. For example, wireless networks of sensors and actuators embedded in furniture and home appliances can be connected to the Internet to provide a more comfortable, convenient, and user-friendly smart home environment. In addition, urban vehicle monitoring has been proposed [91]. Wireless sensor networks have been successfully applied in tracking systems; a German research institution is using this technology to develop an auxiliary system for football referees to reduce the misjudgment rate of offsides and goals in football matches. In addition, in many fields such as disaster rescue, warehouse management, and automated production lines, wireless sensor networks will enable new design and application models.

1.4.2 Key Technologies of UV Self-organizing Networks

(1) Research on atmospheric channel characteristics

The scattering characteristics of UV radiation are exploited in UV NLOS communication, and atmospheric particles with diameters similar to the wavelength of UV light have greater scattering intensity. When the diameter of the scattering particles is much smaller than the wavelength, the scattering is called Rayleigh scattering. The scattering of UV light by atmospheric molecules can be analyzed using Rayleigh scattering theory, but Rayleigh scattering is dominant only in clear weather (a visibility of 20 km), and it makes the smallest contribution to atmospheric scattering. Scattering by aerosol particles makes a much larger contribution than Rayleigh scattering; Mie scattering theory must be used to analyze aerosol scattering.

(2) Solar-blind UV self-organizing network access technology

The data transmission rate and quality of UV scattering communication depend strongly on the azimuth and angle of the transmitter and receiver in both LOS and NLOS modes. Nodes in the network first determine the relative node positions using the positioning algorithm of the network and then adjust the azimuth angles of transmission and reception. They attempt to operate in LOS communication mode and adjust the transmission and reception elevation angles when LOS communication is not possible. In that case, they attempt to enable the optimal NLOS communication so as to provide reliable network access and realize the rapid transmission of information in the self-organizing network.

In addition, it is necessary to study the access technology that can be effectively applied to solar-blind UV self-organizing networks and improve the fairness to ensure throughput. The fairness of network access can be improved by reducing the gap between the backoff window between nodes that fail and succeed in the competition. The window function used to ensure fairness must have the following characteristics. When the competition fails, the backoff window first grows slowly, and with increasing number of failures, the backoff window grows more rapidly until it reaches the maximum value. When the competition succeeds, the backoff window is slowly reduced until it reaches a minimum value.

Unidirectional links in self-organizing networks are caused mainly by factors such as hidden terminal problems and differences in the transmission capabilities, access channels, and energy control methods of wireless node devices. Asynchronous battery consumption caused by different usage rates and environmental impact will also cause nonuniform node coverage. Consequently, unidirectional links are ubiquitous. In atmospheric optical communication, many unidirectional optical links or asymmetric links often exist because of the presence of LOS and NLOS channel types and the effects of atmospheric turbulence, as well as hidden terminals or interference. In this case, only one-way communication is possible between two nodes, and ring network technology is a good choice for communication. First, the ring network discovery mechanism is implemented in the solar-blind UV self-organizing network, and effective communication nodes are selected to form a one-way ring circuit. To achieve point-to-point two-way communication, the ring network can carry out unicast, multicast, broadcast, and other types of communications; in addition, it can implement priority access technology to solve the duplex communication problem of one-way optical paths.

(3) Solar-blind UV self-organizing network routing technology

Channel-quality-based routing algorithms in conventional wireless ad hoc networks consider mainly parameters such as bandwidth and delay. Unlike the radio band, the solar-blind UV band does not have a bandwidth limitation; however, UV light in the solar-blind zone is rapidly attenuated, and the channel varies greatly with time because of scattering by atmospheric molecules and suspended particles. UV light propagation conditions change continuously in time and space; the channel bit error rate is high, and user terminals move frequently. These factors result in unreliable channels, easy data loss, and large fluctuations in network resources. A fixed routing algorithm clearly cannot meet the requirements of UV atmospheric communication in this volatile environment. Therefore, it is necessary to study adaptive routing algorithms for the solar-blind UV band channel. Channel quality should be evaluated by collecting data on some parameters in real time, such as packet loss rate, delay, and packet error rate, to determine which nodes provide a better channel and which are good next-hop nodes.

LOS communication is very easily realized when there is no obstacle between communication nodes. LOS communication using UV light has not only good real-time performance, but also a relatively long communication distance. The communication rate is relatively high. When LOS communication is impossible, the NLOS scattering channel can provide multiple scattering paths, and the wide field-of-view reception can improve the sensitivity of photoelectric detection and achieve reliable transmission of low-speed information. Therefore, voice and image services can be allocated to LOS channels with good real-time performance, and data services can be allocated to NLOS channels with high reliability. In addition, multicast technology realizes point-to-multipoint and multipoint-to-multipoint communication; multicast routing technology can be used in solar-blind UV self-organizing networks to realize multicast communication, which can greatly improve the communication efficiency of the network.

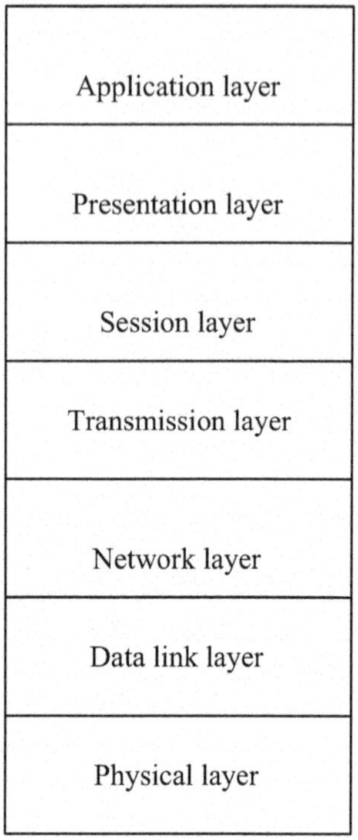

Fig. 1.15 OSI reference model

1.5 Structure of the Book

Figure 1.15 shows a conceptual diagram of the OSI reference model. This book essentially follows this model to introduce the theory of UV self-organizing networks. Chapter 2 discusses the physical mechanism of UV NLOS communication, including its range. Chapter 3 introduces a new node localization algorithm, and Chap. 4 describes multiuser detection in the self-organizing network algorithm. Chapter 5 analyzes the fairness of the MAC layer protocol of UV self-organizing networks, with a focus on a fairness algorithm that we propose. Chapter 6 introduces the UV wireless optical mesh network access protocol, and Chap. 7 focuses on a UV self-organizing network routing protocol based on node position and speed information. Chapter 8 introduces a UV communication network routing protocol based on the ant colony algorithm. Considering the relationship between UV sensor networks and UV self-organizing networks, we discuss UV sensor networks in Chap. 9.

This book focuses on our research on UV self-organizing networks. To ensure a systematic and complete presentation, existing protocols are analyzed in detail. To facilitate understanding of the information in this book, the necessary simulation programs and simulation scripts are presented.

References

1. Ke X (2004) Introduction to wireless laser communication. Beijing University of Posts and Telecommunications Press, Beijing
2. Ke X, Yin Z (2009) Coding theory in wireless laser communication system. Science Press, Beijing
3. Huffman RE (1992) Atmospheric ultraviolet remote sensing. Academic Press, Boston
4. Yen J (1992) Intentionally short-range communications exploratory development plan. National Technical Information Service
5. Li J, Koni Q (2005) Application of ultraviolet optical communication in military communication system. Opt Photoelectr Technol 3(4):19–21
6. Zachor AS (1978) Aureole radiance field about a source in a scattering—absorbing medium. Appl Opt 7(12):1911–1922
7. Neer ME, Schlupf JM, Fishburne ES et al (1979) The development and testing of an UV voice communication system. Nav Electron Syst Command 4(26):125–126
8. Li Y, Li W (2006) Research on short-range communication of atmospheric ultraviolet light. J Atmos Environ Opt 1(2):136–139
9. Gary AS, Melissa N, Mrinal I et al (2000) NLOS uv communication for distributed sensor systems. Proc SPIE 7(4126):83–96
10. Wang K (2006) Research on solar-blind ultraviolet voice communication system. Chongqing University Master's thesis, Chongqing
11. Xianmei Q, Wenyue Z et al (2008) Simulation of atmospheric transmission characteristics for ultraviolet communication. Acta Photonica Sinica 4(10):789–790
12. Liu X, Ju M (2002) A preliminary study on ultraviolet light communication and its countermeasures. Appl Optoelectron Technol 5(20):8–9
13. Xu G (2000) Ultraviolet communication. Mod Commun 4:6–7
14. Kaifu C (2001) New UV camera devices and their applications. Foreign Electron Compon 2(2):4–10
15. Hu Q (1999) Ultraviolet optical communication system. MilY Commun Technol 4(13):67–68
16. Guiping J, Qichang P (2003) Development of UV fingerprint detector. Opt Precis Eng 2(11):198–202
17. Yu X (2009) Design of real-time dual-spectral image processing system based on DSP. Master's thesis of Jinan University, Guangzhou
18. He Q, Sadler BM, Xu Z (2009) Modulation and coding tradeoffs for non-line-of-sight ultraviolet communications. Proc SPIE 7464(74640H):1–12
19. Arnon S, Kedar D (2008) UV solar-blind FSO sub-sea video communication: link budget study. Proc SPIE 10(7):1–8
20. Liu J, Jia H et al (2006) The development of military ultraviolet optical technology. Opt Optoelectron Technol (12):61–64
21. http://www.ecchn.com/20070727ecnews14264447.html
22. Ni G, Zhong S, Liu L (2000) Research on free atmosphere ultraviolet optical communication. Opt Technol 4(26):297–298
23. Shaw GA, Siegel AM, Model J (2006) Extending the range and performance of non-line-of-sight ultraviolet communication links. In: Proceedings of SPIE, vol 4, no 17, pp 1–12
24. Xu Z, Sadler BM (2008) Ultraviolet communications: potential and state- of-the-art. Proceedings of IEEE 5(46):67–68
25. Yu X (2009) Design of real-time dual-spectral image processing system based on DSP. Jinan University Master's Thesis, Guangzhou
26. Shaw GA, Nischan M (2001) Short-range NLOS ultraviolet communication test bed and measurements. Int Soc Opt Eng, Orlando 4396:31–40
27. Tao F, Gang C, Zujie F (2006) Atmospheric transmission model for non-line-of-sight light scattering communication. China Laser 11(33):1522–1526

28. Jia HH, Zhang HL, Yin HW et al (2007) The experimental research of NLOS UV propagation channel in the atmosphere based on LIA technology. Proc SPIE, Wuhan 6783(2):67833B.1–67833B.8
29. Xu Z, Shen L, Wang J et al (2009) Ultraviolet scattering communication experimental system and its performance analysis. J Southeast Univ 6(39):1087–1092
30. Wang LJ, Xu ZY, Sadler BM (2010) Non-line-of-sight ultraviolet link loss in noncoplanar geometry. Opt Lett 8(35):1263–1265
31. Lan L, Ni G (2003) Simulation of atmospheric transmission characteristics for ultraviolet communication. J Beijing Inst Technol 4(23):19–23
32. Honghui J, Jiankun Y, Shengli C et al (2005) Study and design on high data rate UV communication system. Int Soc Opt Eng 6021:440–446
33. Tang Y, Wu ZL, Ni GQ et al (2008) NLOS single scattering model in digital UV communication. Int Soc Opt Eng 7136(2):713615–1–713615–10
34. Moriarty D, Hombs B (2009) System design of tactical communications with solar blind ultraviolet non line-of-sight systems. In: Military communications conference, Boston, pp 1–7
35. Ning X, Wang H, Wang Z (2009) Research on wireless multi-hop self-organizing networks in small areas. Communication theory and technology. In: New development of communication theory and technology—proceedings of the 14th national youth communication academic conference, China Dalian, Liaoning, pp 641–645
36. Leiner BM, Nielson DL, Tobagi FA (1987) Issues in packet radio network design. Proc IEEE 75(1):6–20
37. Jubin J, Tornow JD (1987) The DARPA packet radio network protocols. Proc IEEE 75(1):21–32
38. Aggelou GN, Tafazolli R (2001) On the relaying capability of next-generation GSM cellular networks. IEEE Pers Commun 8(1):40–47
39. Wu H, Qiao C, De S et al (2001) Integrated cellular and ad hoc relaying systems: iCAR. IEEE J Sel Areas Commun 19(10):2105–2211
40. Zadeh AN, Jabbari B, Pickholtz R et al (2002) Self-organizing packet radio ad hoc with overlay (SOPRANO). IEEE Commun Mag 40(6):149–157
41. Meng X, Song B, Xu L (2007) Pyroelectric infrared sensor and its typical application. Instrum User 4(14):42–43
42. Shu W et al (2007) Theory and application of wireless sensor networks. Beijing University of Aeronautics and Astronautics Press, Beijing
43. Callaway EH (2004) Wireless sensor nework: architecture and protocols. CRC Press LLC
44. Pottie G, Kaiser W (2000) Wireless integrated network sensors. Commun ACM 43(5):551–558
45. Pister K (2010) Smart dust: autonomous sensing and communication in a cubic milimeter. http://robotics.Eecs.Berkeley.edu/-pister/SmartDust
46. Aynor M, Moulton S, Welsh M et al (2004) Integrating wireless sensor networks with the grid. IEEE Internet Comput 8(4):32–39
47. Hongqi J, Kai K, Xiaokang L (2005) Wireless mesh network technology to expand broadband access. Telecommun Sci 21(1):24–31
48. Sun B, Chen H, Li L (2004) A reliable multicast routing protocol in mobile ad hoc networks. In: Proceedings of the 16th international conference on computer communication, vol 1, no 1, pp 1123–1129
49. Chen N, Li L (2004) Research on the basis of QoS routing protocol of ad hoc network. DCABES 2004 Proc 1(1):206–210
50. Das SK, Manoj BS, Murthy CS (2002) A dynamic core based multicast routing protocol for ad hoc wireless networks. In: Proceedings of ACM/MOBIHOC, vol 1, no 1, pp 24–35
51. Jubin J, Tornow JD (1987) The DARPA packet radio network protocols. Proc IEEE 75(l):21–32
52. Davies BH, Davies TR (1987) The application of packet switching techniques to combat net radio. Proc IEEE 75(l):43–45
53. Shaeham N, Westeott J (1987) Future directions in packet radio architectures and protocols. Proc IEEE 75(l):83–99
54. Pei G, Gerla M, Chen TW (2000) Fisheye state routing: a routing seheme for ad hoc wireless networks. In: Proceedings of IEEE/ICC'00, vol 1, no 1, pp 70–74

55. Leiner BM, Ruth R, Ambatipudi RS (1996) Goals and challenges of the DARPA GloMo program. IEEE Pers Commun 3(6):34–43
56. Xiping Z (2008) Simulation of TORA based on NS2 mobile ad hoc network routing protocol. Simul Technol 24(3–1):185–187
57. Linxing C, Xi Z, Yi C (2004) Mobile ad hoc network—ad hoc wireless network technology. Electronic Industry Press, Beijing
58. Ma M (2007) Research on wireless ad hoc network routing protocol. Beijing University of Posts and Telecommunications, Ph.D, Beijing
59. Liang T, Dai G, Xin X, Li L (2008) Implementation of embedded wireless ad hoc network device based on DSDV protocol. Comput Eng Des 29(6):1370–1372
60. Peichao L, Xi Z, Hao Y (2009) Simulation and performance analysis of DSR routing protocol in wireless ad hoc networks. Microcomput Technol 25(25):169–171
61. Nadeem Q, Naeem I (2005) TORA stabilization via dynamic surface control technique. In: IEEE 2005 international conference on emerging technologies, Islamabad, pp 488–493
62. Liang B, Haas (2006) Hybrid routing in ad hoc network with a dynamic virtual backbone. IEEE Trans Wirel Commun 5(6):4–8
63. Pyssysalo T, Raatikainen P, Zidbeck J (1995) FSR-a fair switching architecture. In: Proceedings seventh euromicro workshop on real-time systems, Odense, pp 116–123
64. Qi W, Zhang X, Yu H (2004) An Improved CEDAR routing protocol. In: The fourth international conference on computer and information technology, Wuhan, China, pp 621–626
65. Dang-Quan N, Pascale M (2005) Interference-Aware QoS OLSR for mobile ad-hoc network routing. In: Proceedings of the sixth international conference on software engineering, Artificial Intelligence, Networking and Parallel, Maryland, pp 428–435
66. Cui Y, Cao J (2007) An improved directed diffusion for wireless sensor networks. In: Proceedings of the international conference on wireless communications, networking and mobile computing, Shanghai, China, pp 2380–2383
67. Khan UR, Zaman RU, Reddy AV (2008) A three-tier architecture for integrating mobile ad hoc network and the internet using a hierarchical integrated routing protocol. In: Proceedings of the international conference on advanced computer theory and engineering, Phuket, pp 518–522
68. Soumendra N, Gray RS (2006) Multipath location aided routing in 2D and 3D. In: Proceedings of the conference on wireless communications and networking, Las Vegas, pp 311–317
69. Li J (2006) Research on key security issues of mobile self-organizing networks. Doctoral Dissertation of PLA University of Information Engineering, Zhengzhou, Henan
70. Xiangquan Z (2004) A practical tutorial for wireless autonomous network. Tsinghua University Press, Beijing
71. Royer EM, Toh CK (1996) A review of current routing protocols for ad-hoc mobile network. IEEE Pers Commun 6(2):45–55
72. Intanagonwiwat C, Govindan R, Estrin D (2000) Directed diffusion: a scalable and robust communication paradigm for sensor networks. In: Proceedings of the annual international conference on mobile computing and networking, Boston, pp 56–67
73. Heinzelman W, Kulik J, Balakrishnan H (1999) Adaptive protocols for information dissemination in wireless sensor networks. In: Proceedings of the 5th ACM/IEEE mobicom conference, Seattle, pp 174–85
74. Ganesan D, Govindan R, Shenker S et al (2001) Highly-resilient energy-efficient multi-path routing in wireless sensor network. Mob Comput Commun Rev 5(4):3–5
75. Shah RC, Rabaey J (2002) Energy aware routing for low energy ad hoc sensor networks. In: IEEE wireless communications and networking conference, pp 350–354
76. Lindsey S, Raghavendra C (2002) Pegasis: power-efficient gathering in sensor information systems. In: Proceedings of the IEEE aerospace conference, Montana, pp 1–6
77. Manjeshwar A, Agrawal DP (2001) TEEN: a routing protocol for enhanced efficiency in wirelesssensor networks. In: Proceedings of the 15th parallel and distributed processing symposium, San Francisco, pp 2009–2015
78. YOunis M, YOussef M, Arisha K (2002) Energy-Aware routing in cluster-based sensor network. IEEE Computer Society, Fort Worth, pp 129–136

79. Lindsey S, Raghavendra CS, Sivalingam K (2001) Data gathering in sensor networks using the energy delay metric. In: Proceedings of the IPDPS workshop on issues in wireless network and mobile computing, Sanfrancisco, pp 2001–2008
80. Karlof C, Wagner D (2003) Secure routing in wireless sensor networks: attacks and countermeasures. In: First IEEE international workshop on sensor network on sensor network protocol and applications
81. Wang J, Wang C, Wu Q et al (2004) Ad hoc mobile wireless network. National Defense Industry Press, Beijing
82. Linxing C, Xi Z, Yi C (2006) Mobile ad hoc network—self-organizing packet wireless network technology. Electronic Industry Press, Beijing
83. Perkins CE, Bhagwat P (1994) Highly destination sequenced distance vertor routing (DSDV) for mobile computers. Proceeding Conf Commun Arch Protoc Appl 24(4):234–244
84. Hashim R, Nasir Q, Harous S (2006) Adaptive multi-path QoS aware dynamic source routing protocol for mobile ad-hoc network. Innovations in Information Technology, Dubai, pp 1–5
85. Perkins CE, Royer EM (1999) Ad-hoc on-demand distance vertor routing. Mobile computing systems and applications, New Orleans, pp 9–100
86. Xiangquan Z et al (2004) A practical tutorial on wireless ad hoc network technology. Tsinghua University Press, Beijing, p 6
87. Haas J et al (1999) Guest editorial: wireless ad hoc network. IEEE J Sel Areas Commun 17(8):1329–1332
88. ALERT (2010). http://www.altersystem.org
89. Noury N, Herve T, Rialle V et al (2000) Monitoring behavior in home using a smart fall sensor. In: Proceedings of the IEEE-EMBS special topic conference on microtechnologies in medicine and biology, Lyon, pp 607~610
90. Slijepcevic S, Tsiatsis V, Zimbeck S (2002) On communication security in wireless ad-hoc sensor networks. In: Proceedings of the lth IEEE Int'1 workshops on enabling technologies: infrastructure for collaborative enterprises, Pittsburgh, pp 1080–1383
91. Shih E, Cho S, Ickes N et al (2001) Physical layer driven protocol and algorithm design for energy-efficient wireless sensor networks. In: Proceedings of the ACM MobiCom, Rome, pp 272–286

Chapter 2
Atmospheric Channel of Ultraviolet Communication

Ultraviolet (UV) communication can be operated in non-line-of-sight (NLOS) mode. In NLOS mode, UV communication systems are more adaptable to complex terrain, and they avoid the need for line-of-sight (LOS) wireless laser communication. Compared with radio frequency communication, UV communication also has the advantages of resistance to eavesdropping, low position discrimination, omnidirectionality, and strong anti-interference ability; it meets the requirements of tactical communication. Therefore, in recent years, research on battlefield UV communication technology has attracted the attention of military powers worldwide. Owing to advances in science and technology and the need for commercial development, this technology has entered a new development period. It shows great application potential and value for military applications [1–3].

2.1 UV Transmission Characteristics

UV light is electromagnetic radiation with wavelengths of 10–400 nm. Because its characteristics and effects vary with wavelength, UV light is usually divided into near-UV (NUV, 315–400 nm), middle-UV (MUV, 280–315 nm), far-ultraviolet (200–280 nm), and vacuum ultraviolet (10–200 nm) light [4, 5]. O_3 in the upper atmosphere exhibits strong absorption of light with wavelengths of less than 200 nm; thus, UV transmission in this wavelength range is greatly limited and cannot be used for atmospheric communication. The solar background at 200–280 nm is less than 10^{-13} W/m^2, and this part of the spectrum is often referred to as the solar-blind region. In this region, there is minimal interference from background light in UV communication [6]. Unless otherwise indicated, here "UV communication" refers to the use of the solar-blind region for communication.

X. Ke, *UV Light Self Organizing Network Theory*, Optical Wireless Communication Theory and Technology, https://doi.org/10.1007/978-981-97-7854-6_2

2.1.1 Atmospheric Absorption and Scattering of UV Light

UV communication uses the free atmosphere as the transmission medium, and its transmission characteristics are determined mainly by absorption and scattering by atmospheric molecules. Matter generally absorbs optical radiation or light energy. When light passes through matter, the electric vector of the light wave causes charged particles in the material structure to vibrate, and some of its energy is used to supply the energy required for this forced vibration. The energy may be converted into translational kinetic energy, which increases the thermal motion of the molecules, increasing the object's temperature. Part of the light energy is converted to heat energy and disappears [7]. In atmospheric absorption during transmission, atmospheric molecules consume UV energy in this way. Various components of the atmosphere absorb light of different wavelengths to different degrees, and O_3 has the strongest ability to absorb UV light. With increasing O_3 concentration, more energy is absorbed by the atmosphere, and the transmission loss increases. Consequently, UV communication can be used only for short-distance communication.

In a medium with uniform optical properties or at the interface between two uniform media with different refractive indices, whether light is refracted or reflected, it is transmitted only in specific directions, and the light intensity is zero in other directions. No light can be seen outside of the beam. However, when light passes through a material with non-uniform optical properties, we can see scattered light in the lateral direction [7, 8]. With increasing optical inhomogeneity of the medium, scattering becomes stronger. UV communication is based on the scattering properties of UV radiation. Atmospheric scattering is caused mainly by atmospheric molecules and suspended particles, and their characteristics, such as concentration, size, uniformity, and geometric size, affect the transmission characteristics of UV light. The scattering of light by the atmosphere can be divided into Rayleigh and Mie scattering.

2.1.2 Transmission Characteristics of UV Communication

Solar-blind UV communication systems can be divided into LOS, collimated LOS (quasi-LOS, QLOS), and NLOS systems according to the operation mode. NLOS is further subdivided into NLOS(A), NLOS(B), and NLOS(C), as shown in Fig. 2.1.

In LOS communication, the transmitter is within the field of view of the receiver, and the optical communication path is not blocked by any obstacles. The communication distance is relatively long in LOS mode and is similar to that of typical lasers, but this mode lacks the advantage of omnidirectionality. In NLOS communication, the transmitter is in the field of view of the receiver, but the optical communication path is blocked by obstacles (for example, in the woods, by leaves). However, a communication link can still be established owing to the strong scattering of the UV signal, which enables omnidirectional communication. When the transmitter is not

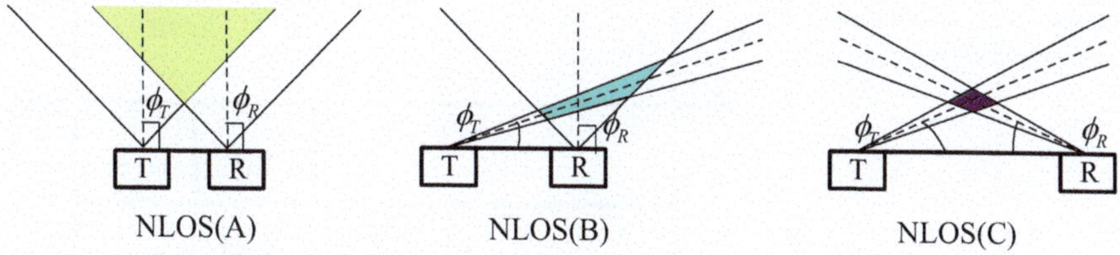

Fig. 2.1 Typical configurations of NLOS communication

Table 2.1 Performance comparison of different UV operation modes

Mode	ϕ_T	ϕ_R	Omnidir-ectionality	Communication distance (km)	FOV overlap	Communication band width
LOS	/	/	None	2–10	Limited	Widest
QLOS	/	/	None	< LOS	Limited	Wide
NLOS(A)	90°	90°	Good	0–1	Unlimited	Narrowest
NLOS(B)	≤ 90°	90°	Generally	1.5–2	Limited	Wider
NLOS(C)	< 90°	< 90°	Varies	2–5	Limited	Wide

within the receiver's field of view, an indirect transmission path can be established if the fields of view of the transmitter and receiver overlap because of the strong scattering of solar-blind UV light.

The overlap serves as a relay station for communication between the transmitter and receiver, which enables the transmitted signal to reach the receiver under NLOS conditions. The three types of NLOS communication are defined in terms of the angles ϕ_T and ϕ_R between the optical axis of the transmitter and receiver, respectively, and the horizontal plane (Table 2.1). For NLOS(B), ϕ_T and ϕ_R cannot both be 90° simultaneously, as shown in Fig. 2.1. UV communication systems can be easily switched between the three modes by changing ϕ_T and ϕ_R as needed. Table 2.1 compares the performance of UV communication systems in each operation mode.

2.2 Characteristics of Atmospheric Channel of UV Light

UV communication is a new means of information transmission. When the UV carrier signal passes through the atmosphere, it is affected by various components of the atmosphere and by weather conditions. The quality of communication and system performance are directly related to these effects. We used the atmospheric transmission software LOWTRAN to comprehensively analyze the spectral components of the atmosphere, atmospheric absorption, atmospheric scattering, and mathematical models of UV transmission, and discuss the atmospheric transmission of UV light.

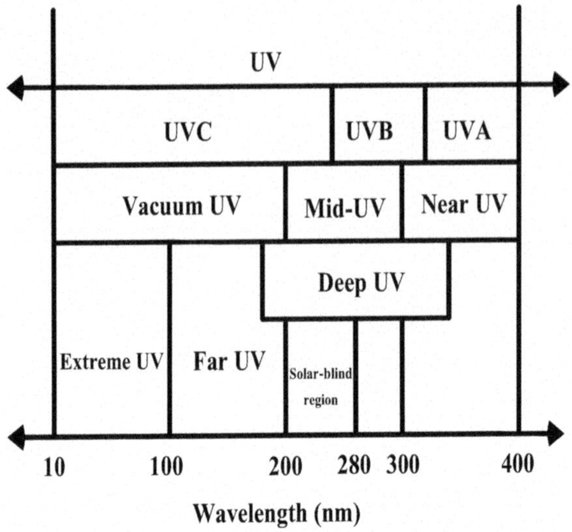

Fig. 2.2 UV spectral distribution [9]

2.2.1 *Solar-Blind UV Light*

More than 99% of solar radiation has wavelengths of 150–4000 nm. In this wavelength range, approximately 50% of solar energy is in the visible region (wavelengths of 400–760 nm), 7% is in the UV (wavelengths of < 400 nm), and 43% is in the infrared (wavelengths of > 760 nm). The maximum energy of the solar spectrum appears at a wavelength of 475 nm [9].

Figure 2.2 shows the regions of the UV spectrum. The strong absorption of 20–280 nm UV light by the ozone layer in the upper troposphere (10–50 km) decreases the solar background in the troposphere (especially the near-Earth background) in this band to below 10^{-13} W/m^2; that is, almost no sunlight in that band reaches the surface of the earth. This region in the UV spectrum is called the solar-blind region [10, 11, 58]. In the LOWTRAN software, we set the atmospheric environment to obtain the transmittance in the horizontal and vertical directions as shown in Fig. 2.3. UV light around 250 nm is greatly attenuated.

2.2.2 *Characteristics of the Atmosphere*

The main components of the atmosphere are nitrogen (78.1%); oxygen (20.9%); hydrogen (0.93%); and small amounts of carbon dioxide, rare gases (helium, neon, argon, krypton, xenon, and radon), and water vapor. Impurity particles are generally classified as solid or liquid particles. Solid particles include dust, smoke, and various industrial pollutants. Liquid particles include cloud droplets, fog droplets, raindrops,

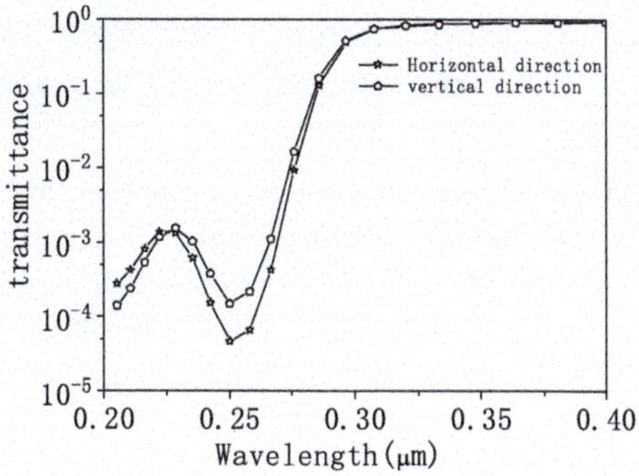

Fig. 2.3 Transmittance of UV light

ice crystals, snowflakes, and hail. However, because of differences in temperature and wind, the atmosphere is often in motion [9].

According to the distribution of temperature, composition, and ionization state in the vertical direction, the atmosphere can be divided into five layers: troposphere (< 10 km), stratosphere (10–50 km), mesosphere (50–80 km), thermosphere (80–500 km), and dissipative layer (above 500 km), as shown in Fig. 2.4. The troposphere, which makes up 80% of the atmosphere, has the strongest effect on the transmission of UV light, and most weather occurs in the troposphere. The stratosphere is less dense and stable, and it has little effect on the transmission of UV light.

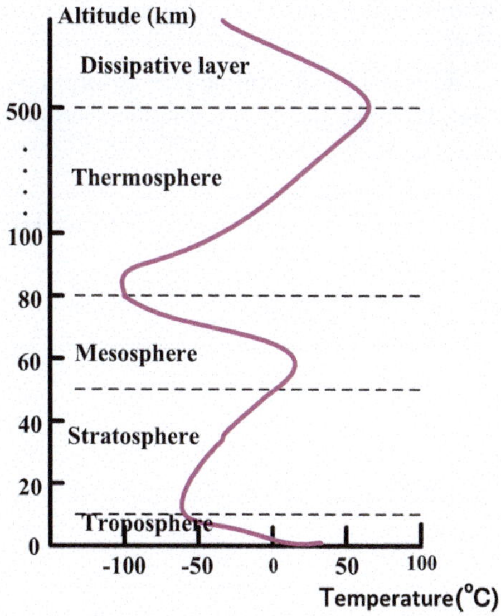

Fig. 2.4 Vertical structure of the atmosphere [12]

2.2.3 *Main Factors Affecting UV Optical Communication in the Atmospheric Channel*

NLOS UV optical communication uses the atmosphere as the transmission medium. When UV light carrying a signal is transmitted in space, the quality of its transmission, performance of the communication system, and transmission range are affected by the O_3 concentration in the atmosphere; the concentration, size, uniformity, and geometric dimensions of scattering particles; and the operating wavelength.

2.2.3.1 Atmospheric Absorption

When UV light passes through the atmosphere, various components of the atmosphere absorb it to different degrees. In the UV region (0.2–0.4 μm), ozone has a strong absorption band between 0.2 and 0.264 μm and a weak band at 0.3–0.36 μm. Sulfur dioxide and ozone have a strong ability to absorb UV light. Although ozone accounts for only 0.01–0.1% of the atmosphere, it strongly absorbs solar radiation [12].

We used LOWTRAN to simulate the atmospheric transmission attenuation for a horizontal transmission distance of 1 km and a visibility of 23 km, as shown in Fig. 2.5. Ozone has the strongest UV light absorption ability, accounting for approximately 75% of the total attenuation. In addition, the ozone concentration strongly affects the atmospheric transmittance, and thus the communication range [13].

Inn et al. [13] determined the absorption coefficient of ozone in the UV absorption band. For example, at 266 nm, the absorption coefficient of ozone K_a (km^{-1}) can be expressed as follows:

$$K_a = 0.025 \times d \tag{2.1}$$

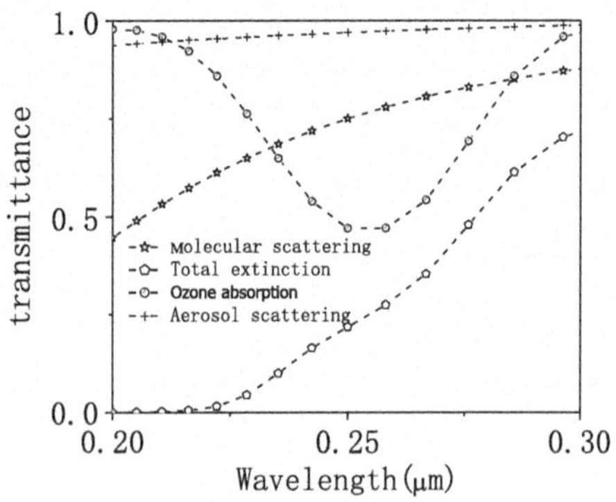

Fig. 2.5 UV atmospheric transmittance

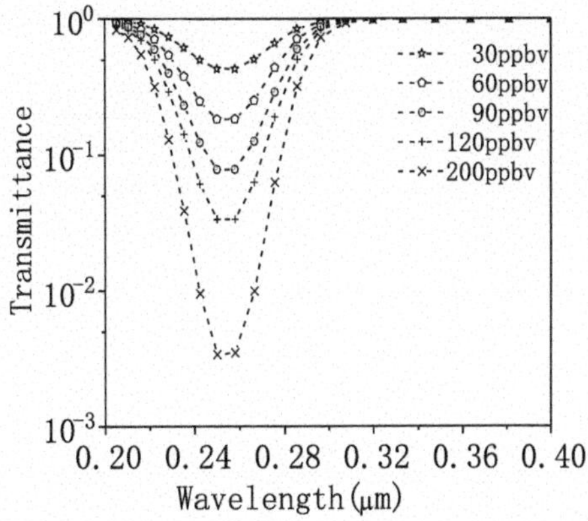

Fig. 2.6 Effect of ozone concentration on transmittance

where *d* represents the ozone concentration in the atmosphere in units of parts per billion by volume (ppbv). Figure 2.6 shows the effect of ozone concentration on the UV atmospheric transmittance. Absorption by ozone is strongest around 250 nm, and above 300 nm ozone has almost no effect on atmospheric attenuation. With increasing ozone concentration, the transmittance gradually decreases. The atmospheric transmittance is lowest at a concentration of 200 ppbv, indicating that the absorption capacity of ozone is strongest at that concentration. Absorption by ozone reduces the number of signals that can be transmitted; it is precisely because of this absorption that UV light has a large attenuation factor in atmospheric transmission, making signals outside the transmission range difficult to intercept.

2.2.3.2 Atmospheric Scattering

The basis of UV communication is the scattering characteristics of UV light. The scattering of UV light by atmospheric molecules and aerosol particles causes the optical signal to change direction during transmission; thus, it can bypass obstacles and be received by the detector. The main scatterers in the atmosphere are atmospheric molecules and aerosol particles. The scattering properties of scatterers depend on their size. Atmospheric particles with diameters closer to the wavelength of UV light scatter UV light more strongly. Molecules are much smaller than the wavelength of UV light and thus cause Rayleigh scattering; by contrast, aerosol particles are much larger than the wavelength of UV light and cause Mie scattering. Studies have shown that Rayleigh scattering is dominant when the sky is clear, that is, when the atmospheric concentration of aerosols is very low. Therefore, in theoretical calculations, only Rayleigh scattering is usually considered in clear weather, and the scattering effect of suspended particles is ignored, which can be considered a reasonable approximation. However, in severe weather, mainly Mie scattering is considered [14]. Table 2.2 gives the radii and concentrations of several important scattering particles in the atmosphere.

Table 2.2 Radius and concentration of atmospheric scattering particles [59]

Type	Radius/μm	Concentration/cm^{-3}
Air molecules	10^{-4}	10^{19}
Aitken nuclei	10^{-3}–10^{-2}	10^{-4}-10^{2}
Haze particles	10^{-2}–1	10–10^{3}
Droplets	1–10	10–10^{2}
Cloud drops	1–10	$10^{-3} \times 10^{2}$
Raindrop	10^{2}–10^{4}	10^{-5}–10^{-2}

(1) Rayleigh scattering

The transmission of UV light in a clear atmosphere occurs mainly by Rayleigh scattering, which is also known as molecular scattering. The Rayleigh scattering coefficient is determined using (2.2) [15–17]:

$$K_{SR} = \frac{8\pi^3[n(\lambda)^2 - 1]^2}{3N_A\lambda^4} \times \frac{6+3\delta}{6-7\delta} \times F_{King}(x) \tag{2.2}$$

where δ is a depolarization term, and $n(\lambda)$ is the particle number density of the scatterers:

$$n(\lambda) = \frac{0.05791817}{238.0185 - \lambda^{-2}} + \frac{0.00167909}{57.362 - \lambda^{-2}} + 1$$

N_A is the refractive index of the atmosphere: $N_A = 2.686763 \times 10^{19}\ \mathrm{mol/cm^3}$.
F_{King} is a correction factor: $F_{king}(x) = \frac{6+3d(x)}{6-7d(x)}$, $d(x) = 0.035$.
The unit of measurement is micrometers. The classical formula [16, 22] for calculating the Rayleigh scattering is

$$K_{SR} = 2.677 \times 10^{-17}\frac{P\gamma^4}{T} \tag{2.3}$$

where P is the atmospheric pressure, T is the absolute temperature (typically 300 K under standard atmospheric conditions), and γ is the UV wave number. The Rayleigh scattered light intensity is inversely proportional to the fourth power of the light wavelength. The forward and backward scattering energies are equal, and the scattered light intensity is [15]

$$I_0 \propto \frac{1}{\lambda^4} \tag{2.4}$$

Figure 2.7 shows the intensity distribution of Rayleigh scattering. Which reveals that Rayleigh scattering is symmetrical. The phase function of Rayleigh scattering is given by [13]

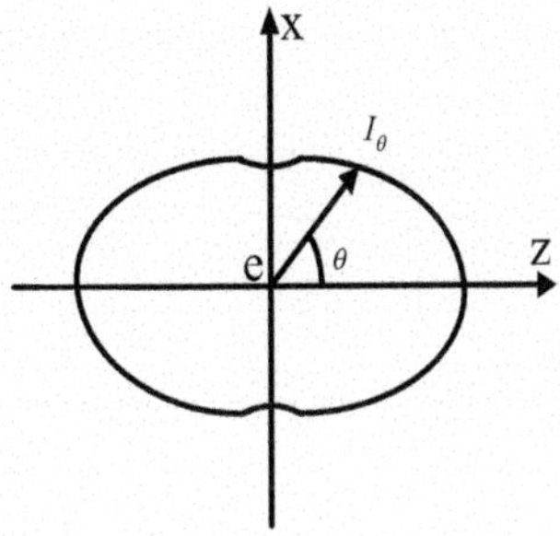

Fig. 2.7 Angular distribution of Rayleigh scattering light intensity

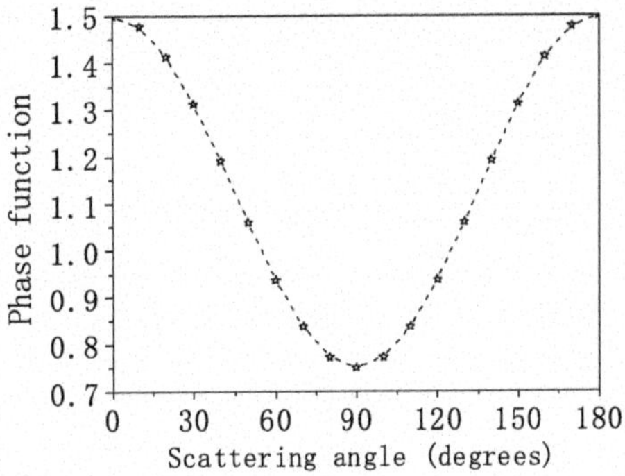

Fig. 2.8 Rayleigh scattering phase function

$$P_R(\theta_s) = \frac{3}{4}(1 + \cos^2\theta_s) \tag{2.5}$$

Figure 2.8 shows the Rayleigh scattering phase function, which is almost symmetric about 90° between 0° and 180°; the forward and backward scattering are strongest and the phase function is minimum at 90° and 270°.

(2) Mie scattering

Mie scattering occurs when the size of the scatterer is equal to the wavelength of incident light. As shown in Table 2.2, most atmospheric particles are aerosols. Atmospheric aerosol particles are too large for Rayleigh scattering of UV light and must be analyzed using Mie scattering theory. A simple model is often used to estimate the atmospheric aerosol scattering of UV light; it is based on the relationship between the magnitude of the scattering and atmospheric visibility. The scattering coefficient can be expressed as [16, 18]

$$K_{SM} = \frac{3.91}{R_V} \times (\frac{\lambda_0}{\lambda})q \tag{2.6}$$

where R_v is the visibility (km^{-1}), and q is a correction factor (nm); $\lambda_0 = 550$ nm, q is the correction factor determined by R_v. Table 2.3 shows the values under different weather conditions.

Table 2.3 Correction factors corresponding to different meteorological distances [16]

Visibility R_v	Visibility class	Weather conditions	Q (nm)
$R_v > 50$ km	9	Very sunny	1.6
6 km $< R_v <$ 50 km	6–8	Sunny	1.3
1 km $< R_v <$ 6 km	4–6	Frost	$0.16R_v + 0.34$
500 m $< R_v <$ 1 km	3	Mist	$R_v - 0.5$
$R_v <$ 500 m	< 3	Fog	0

The Mie scattered light intensity is given by [17, 19]

$$I_\theta \propto \frac{1}{\lambda^N} \tag{2.7}$$

where N is an integer in$(1,\infty)$.

Figure 2.9 shows the distribution of the Mie scattering light intensity, which reveals its asymmetry. Because forward scattering is much stronger than backward scattering, in UV communication, the effects of back scattering and polarized light are negligible. As shown in Fig. 2.10, for scattering angles of less than 60°, with increasing asymmetry factor g, the phase function increases. Between 60° and 180°, the opposite is true.

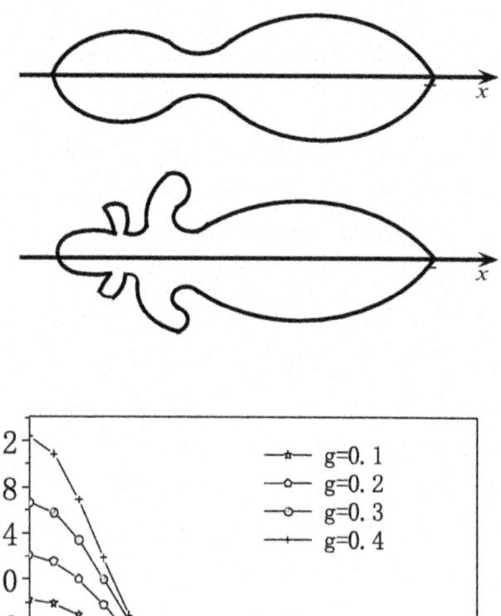

Fig. 2.9 Angular distribution of Mie scattering light intensity

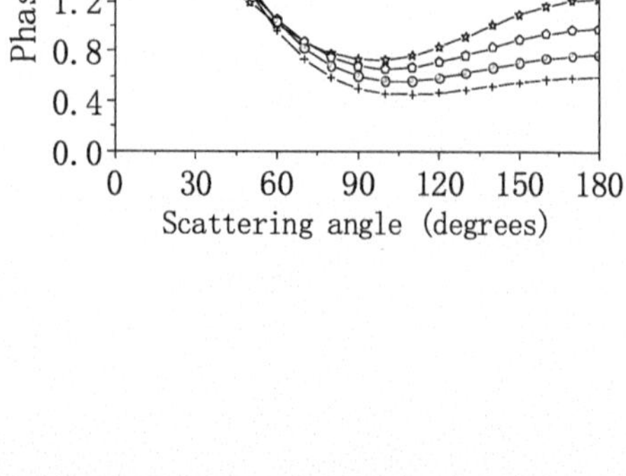

Fig. 2.10 Mie scattering phase function

The scattering phase function is used to study UV scattering communication. The Mie scattering phase functions commonly used in atmospheric transport simulations are the empirical Henyey–Greenstein function, the modified Henyey–Greenstein function, and the Henyey–Greenstein phase function defined by Cronette and Shanks, where the adjustable asymmetry factor g is added [20, 21]. The Henyey–Greenstein phase function can be expressed as [22]

$$p_{HG}(\theta, g) = \frac{1 - g^2}{(1 + g^2 - 2g\cos\theta)^{3/2}} \tag{2.8}$$

The Henyey–Greenstein phase function defined by Cronette and Shanks is [23]

$$p_{HG1}(\theta, g) = \frac{3}{2} \times \frac{1 - g^2}{2 + g^2} \frac{1 + \cos^2\theta}{(1 + g^2 - 2g\cos\theta)^{3/2}} \tag{2.9}$$

Equations (2.8) and (2.9) consider only forward scattering. By contrast, the modified Henyey-Greenstein function in (2.10) is a mixed-phase function that considers both forward and backward scattering [24]:

$$p_{HG2}(\theta, g) = \frac{1 - g^2}{4\pi}\left[\frac{1}{(1 + g^2 - 2g\cos\theta)^{3/2}} + f\frac{0.5(3\cos^2\theta - 1)}{(1 + g^2)^{3/2}}\right] \tag{2.10}$$

where θ is the scattering angle, and f is the scattering factor.

2.2.3.3 Atmospheric Turbulence

In the field of atmospheric optics, turbulence is defined as random changes in local atmospheric temperature, pressure, and density. Atmospheric turbulence produces many vortex elements with different refractive indices. These elements move rapidly according to wind speed and are continuously generated and annihilated. The frequency at which they change can reach hundreds of hertz, and the spatial scale of the changes may be as small as a few millimeters or as large as tens of meters. When a beam passes through vortex elements with different refractive indices, atmospheric turbulence effects such as beam bending, drift, and expansion distortion occur, resulting in scintillation and jitter in the received light intensity. Because turbulence is a stochastic process, it is appropriate to use statistics for its thorough analysis [25].

(1) Physical model of turbulence

Kolmogorov studied time-dependent changes in the velocity difference between two points separated by a displacement Y. He found that over a wide range of spatial motion scales, the root-mean-square velocity difference can be described in a general form, $D_{i,j}(r, r_1)$, which is called the structure tensor and is defined as [25]

$$D_{ij}(\overline{r},\overline{r_1}) = \{[V_i(\overline{r_i}-\overline{r}) - V_i(\overline{r_1})] \times \left[V_j(\overline{r}_1-\overline{r}) - V_j(\overline{r}_1)\right]\} \tag{2.11}$$

Here V_i, V_j, are the components of velocity. Two assumptions are made about the atmosphere. First, the atmosphere is assumed to be locally uniform; that is, the distribution function of the velocity difference is assumed to be constant under the translation of a point and to be related only to the statistics of the displacement vector. Second, the atmosphere is assumed to be locally isotropic; that is, the distribution function of the velocity difference is assumed to be constant under the rotation or mirror reflection of the vector and to be related only to the magnitude of the displacement vector. With these assumptions, $D_{ij}(r, r_1)$ can be simplified to

$$D_{ij}(\overline{r}) = \left[D_{rr}(\overline{r}) - D_{jj}(\overline{r})\right]\overline{r_i r_j} + D_{rr}(\overline{r})\delta_{ij} \tag{2.12}$$

Here $\delta_{ij} = 1, i = j$, and $\delta_{ij} = 0, i \neq j$; in addition, D_{rr} and D_{jj} are the structure functions of the wind speed components parallel and perpendicular to the displacement vector, respectively. We further assume that the turbulent field is incompressible $(V \times \upsilon = 0)$; then D_{rr} can be represented by D_{jj}:

$$D_{ij} = \frac{1}{2r}\frac{d}{dr}(r^2 D_{rr}) \tag{2.13}$$

The statistical structure function of turbulent flow can be represented by a single structure function D_{rr}:

$$D_{rr} = \left\{[V_r(\overline{r_1}+\overline{r}) - V(\overline{r_1})]^2\right\} \tag{2.14}$$

According to Kolmogorov spectral theory, if the distance between two points is within the so-called turbulent inertial subinterval, D_{rr} has the general form

$$D_{rr} = C_n^2 r^{2/3} \tag{2.15}$$

Here C_n^2 is the velocity structure constant, which depends on the turbulent kinetic energy dissipation rate ε, that is, the rate at which turbulent kinetic energy is converted into molecular thermal kinetic energy under the action of molecular viscosity. It is measured as the turbulent kinetic energy loss with time. The structure function D_{rr} holds only if the value of r lies between the turbulent inner scale l_0 and the turbulent outer scale L_0. The viscous effect becomes important when the inner scale corresponds to a vortex element scale smaller than it is, whereas the outer scale L_0 is the largest turbulent eddy scale when the turbulent field is considered to be isotropic. The inner scale l_0 of near-surface atmospheric turbulence is on the order of millimeters, and the outer scale of atmospheric turbulence L_0 is on the order of meters (generally 1–100 m).

The atmospheric refractive index structure function is a physical quantity that characterizes the strength of turbulence, and its changes are very complex. In general, turbulence is stronger in summer than in winter and stronger on sunny days than on cloudy days; it is stronger around noon (10:00–15:00) each day and weaker at other times. The refractive index structure constant exhibits the same variation trends. The change with height, C_n^2, is more complicated; it generally decreases with increasing height [26–28].

(2) Effect of atmospheric turbulence

The degree and type of the effects of atmospheric turbulence on beam characteristics are related to the beam diameter and turbulence scale. The effects can be roughly divided into three cases [26]:

(a) When the beam diameter is much smaller than the turbulence scale ($d \ll l$), the main effect of turbulence is random deflection of the beam, and the beam at the receiver end drifts.
(b) When the turbulence scale is comparable to the beam diameter ($d = 1$):, the main effect of turbulence is random deflection of the beam cross section, which results in fluctuations in the angle of arrival and image jitter on the focal plane at the receiving end.
(c) More commonly, the beam diameter is much larger than the turbulent scale ($d \gg l$). In this case, the beam cross section contains many small turbulent eddies, each of which diffracts a small part of the irradiated beam, and the intensity and phase of the beam vary in space and time. The beam is randomly distributed, and the coherence is degraded. The beam area also expands, causing fluctuations in the light intensity at the receiving end and attenuation of the overall received light intensity. In practical situations, disturbances associated with temperature differences will cause continuous mixing of the atmosphere, producing many unpredictable turbulent elements on various scales. These turbulent elements jointly strengthen the fluctuation in light intensity at the receiving end. In addition, the fluctuation in light intensity is related to the current wind speed and weather conditions. Therefore, it is difficult to detect and observe atmospheric turbulence. Atmospheric turbulence makes signal detection difficult to understand and poses a considerable obstacle to stable communication using optical communication systems [29, 30].

2.2.4 Simulation and Analysis of Atmospheric UV Transmission Characteristics

LOWTRAN is a software package for calculating atmospheric transmittance and background radiation; it was produced by the US Air Force Geophysical Laboratory. Its spectral resolution is 20 cm^{-1}, and the minimum sampling interval is 5 cm^{-1}. It can calculate the spectrum with wavelengths from 200 nm to infinity frequency band. The LOWTRAN transport model takes into account attenuation caused by atmospheric molecular scattering and molecular absorption; scattering and absorption by water molecules, carbon dioxide, ozone, nitrogen, ammonia, and aerosols; and single scattering and surface scattering by sunlight or moonlight. The effects of temperature, pressure, etc. are simulated by setting cards. Card 1 is used to select the atmospheric mode, path geometry type, program execution method, inclusion or omission of multiple scattering, boundary conditions, etc. Card 2 is used to select aerosol and cloud modes, and Card 3 is used to define the geometric path parameters for specific problems. Card Four covers the calculated spectral region and step size. Card Five uses IPRT to control the loop of the program to calculate a series of problems in one run.

We use the LOWTRAN transmission model to simulate and study atmospheric transmission in the UV band and analyze the effects of transmission distance, visibility, geographic location, season, propagation direction, altitude, and weather conditions on the atmospheric UV transmittance.

Figure 2.11 shows the atmospheric transmittance at visibilities of 0.2 and 0.5 km and transmission distances of 0.1 and 0.3 km on foggy days. The communication quality improves with increasing visibility; for certain visibility values, the transmittance decreases with increasing communication distance, degrading the communication quality. Figure 2.12 shows the atmospheric transmittance at a transmission distance of 1 km under the US standard atmosphere.

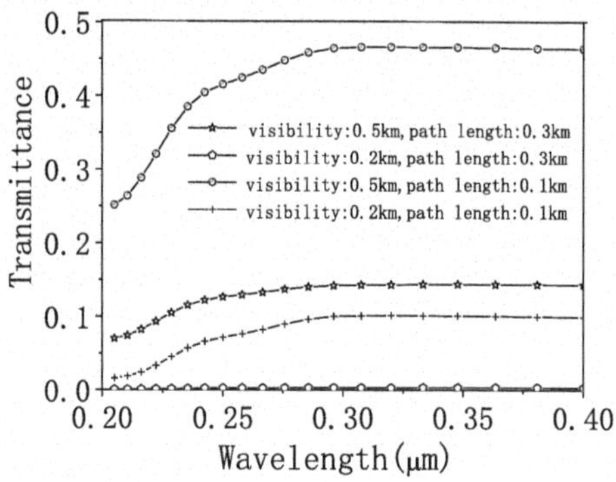

Fig. 2.11 Transmittance at different visibilities and transmission distances

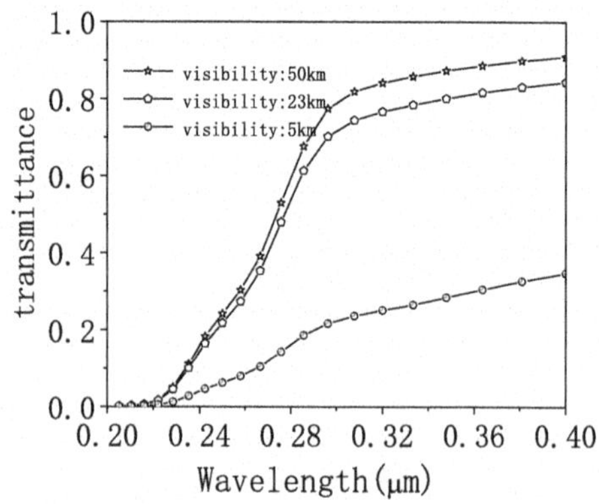

Fig. 2.12 Transmittance at different visibilities under US standard atmosphere

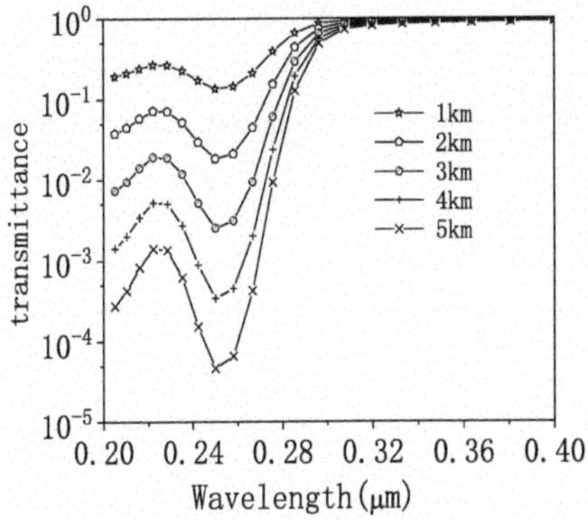

(a) Horizontal direction

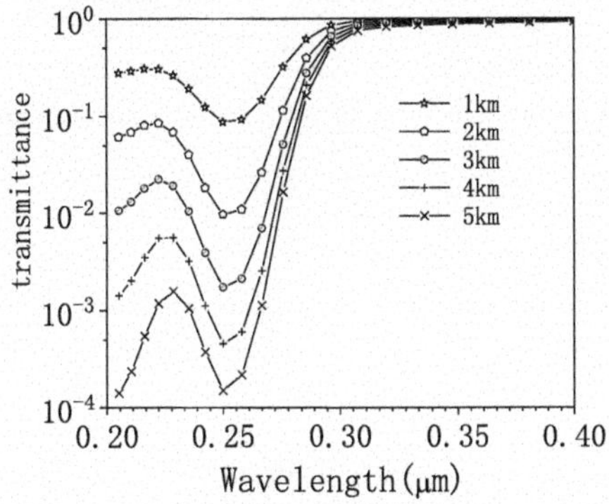

(b) Vertical direction

Fig. 2.13 UV transmittance at various transmission distances

Figure 2.13 shows the variation in atmospheric UV transmittance with transmission distance simulated by LOWTRAN software. The UV transmittance is generally attenuated with increasing transmission distance. The transmittance in the horizontal direction decays approximately exponentially with increasing transmission distance. The reason is that with the increase of the transmission distance, the vertical direction (zenith angle At the same altitude, the atmospheric UV transmittance travels the same distance along the horizontal path and the vertical path but has different transmittances is the layered structure of the atmosphere. That is, because the atmospheric density is uniformly distributed in the horizontal direction at sea level, the atmospheric UV transmittance distribution follows a negative exponential law with increasing horizontal transmission distance. By contrast, in the vertical direction (perpendicular to sea level), the atmospheric density varies with altitude. The density of scattering and absorption sources decreases with decreasing height; consequently, the atmospheric UV transmittance decreases with increasing transmission distance.

The absorption and scattering of UV signals by the atmosphere are closely related to the waveband and height of the signal source. The parameters are US standard atmosphere, a rural aerosol model (a visibility of 23 km), a signal height of 1–10 km above the ground, a transmission distance of 1 km, and wavelengths of 200–400 nm. Figure 2.14 shows the transmission curves in the horizontal and vertical directions for UV signal source and receiver system at different heights.

Because the surface atmosphere is dense and the ozone concentration is low, there is transmittance in the 200–400 nm band; with increasing height, the atmospheric molecular density decreases, and the ozone concentration increases. Consequently, the transmittance at 300–400 nm increases, and that at 200–300 nm decreases rapidly. The solar-blind region near 250 nm does not change with altitude. The UV transmittance in the horizontal direction does not change significantly within 1 km above sea level; with increasing altitude, the UV transmittance gradually increases. The gradient of UV transmittance in the solar-blind region is smaller than that of the

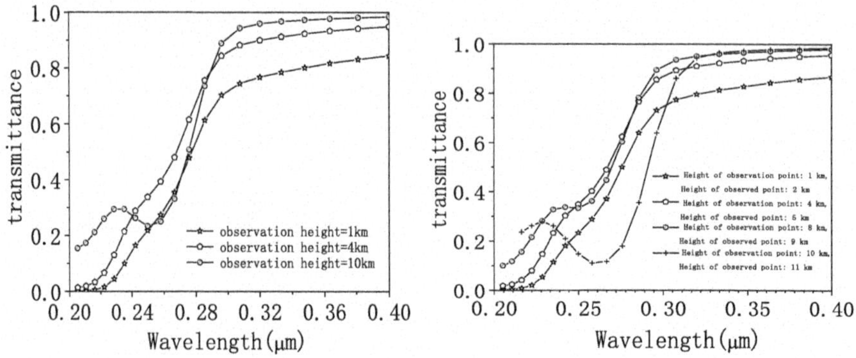

(a) Horizontal direction (b) Vertical direction

Fig. 2.14 UV transmittance at various altitudes

other bands and reaches a minimum at 10 km altitude under the selected conditions. Because of the limited installation height of actual systems, the installation range of the transceiver system is limited. Therefore, in terms of the UV transmittance, the installation height has little effect on system performance. The atmospheric UV transmittance drops sharply in the solar-blind region at an altitude of 10 km because the ozone concentration is higher in this layer. This change will decrease the maximum communication range.

Figure 2.15 shows the effects of different cloud layers on the UV transmittance. Because UV communication is relatively short-range and is generally performed in the lower atmosphere, low-altitude turbulent clouds have a greater impact, and high nimbus clouds have little effect. Figure 2.16 shows the effects of different atmospheric modes on the UV transmittance. The atmospheric transmittance decreases with decreasing visibility (decreasing meteorological distance), and the UV transmittance of the tropical atmosphere is slightly higher than that of the US standard atmosphere. The mid-latitude UV transmittance is slightly higher in summer than in winter. The UV transmittance does not change greatly with geographical location and season; the solar-blind region near 250 nm changes very little with location and season, and the transmittance in this region is high when the visibility is high. The over-rate attenuation is very high; therefore, it is appropriate to select the solar-blind region as the working band of UV communication systems.

Figure 2.17a, b shows the atmospheric UV transmittance in the horizontal direction at a sea level visibility of 23 km for five rain conditions in the US standard atmosphere in desert and ocean environments, respectively.

The atmospheric UV transmittance decreases with increasing rain rate, but the decrease in the solar-blind region is approximately 0.1, which is smaller than those in the NUV and MUV regions. In the absence of other factors, UV communication should be operable on rainy days; however, the transmittance is almost 0 during heavy rain, disrupting normal communication. Therefore, optical UV communication is not feasible on rainy days.

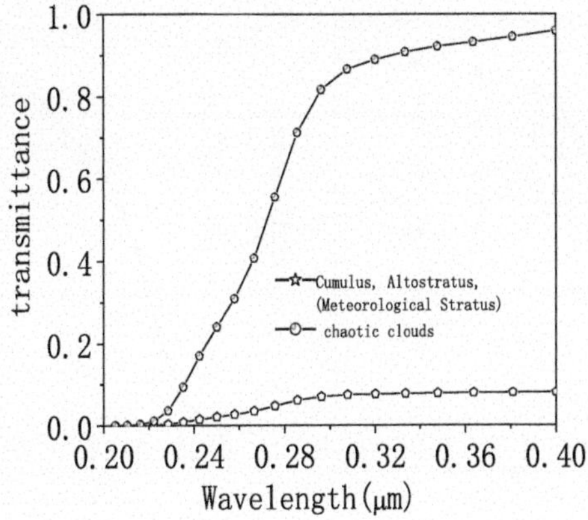

Fig. 2.15 Transmittance in various cloud layers

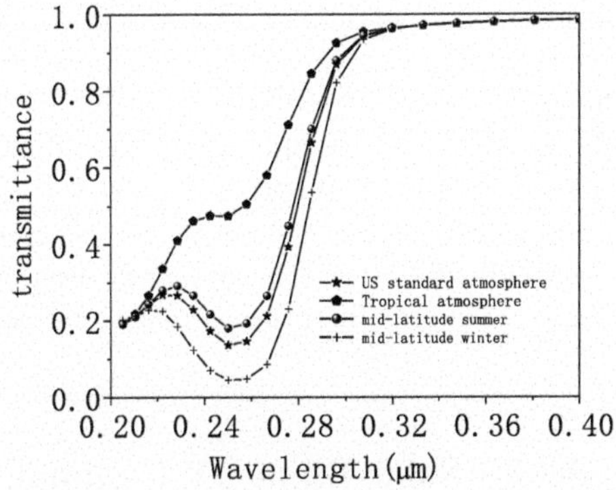

Fig. 2.16 Atmospheric model

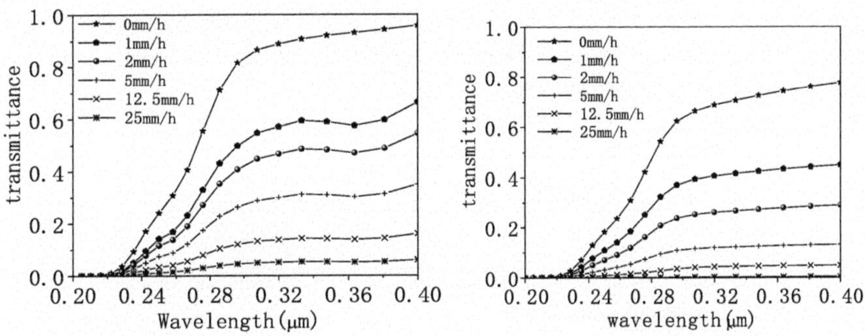

(a) Desert rainfall (b) Ocean rainfall

Fig. 2.17 UV transmittance under various rainfall conditions

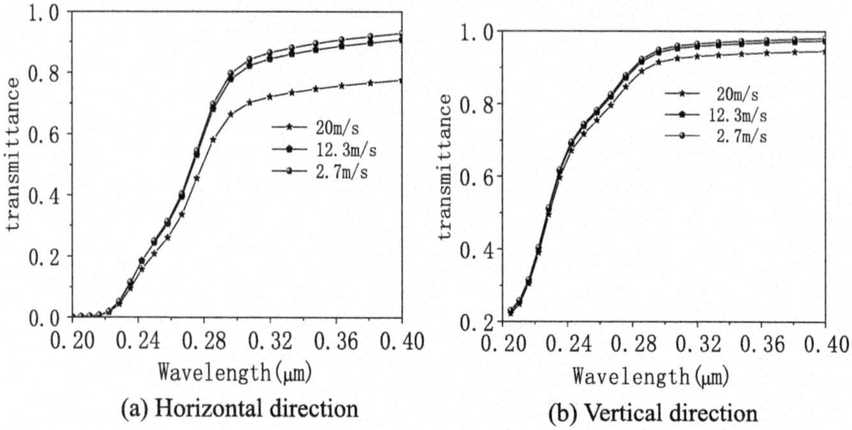

Fig. 2.18 Wind speed and transmittance

Figure 2.18a, b shows the effect of wind speed in the horizontal and vertical directions, respectively, on the UV transmittance at an altitude of 200 nm, a sea level visibility of 23 km, under the wind speed of 2, 6, and 9 in the US standard atmosphere. The transmittance decreases gradually with increasing wind speed, but the decrease is not large, indicating that UV communication systems can operate normally in windy weather, assuming there are no other interference factors.

2.3 Channel Model for Solar-Blind UV Communication

Overall, the analysis of the atmospheric channel model for UV optical communication adopts the NLOS single-scattering channel model proposed by Luettgen et al. [29] in 1990. The model analyzes the energy loss of light pulses scattered by scatterers in short-range NLOS UV systems.

2.3.1 Link Model for Solar-Blind UV Communication

Most UV communication systems operate in a pulsed manner, as described by the following mathematical model [30]:

$$C_{pp} = \sum_{\Delta\lambda} 1.33 \times 10^{18} [W_{in} \eta(\lambda) \Delta\lambda] \times N_{ATM} [K_{SCATT}(h), K_{ABS}(h), p(\Psi), \theta, \Omega, \beta] \times M(\lambda) \quad (2.16)$$

where C_{PP} is the number of received photons per pulse; W_{in} is the average input power of the UV source; $\eta(\lambda)$ is the efficiency density distribution of the UV source for converting electrical energy into light radiation in a wide band; N_{ATM} is the atmospheric transfer function; $K_{SCATT}(h)$ is the atmospheric scattering coefficient; $K_{ABS}(h)$ is the atmospheric absorption coefficient; $p(\Psi)$ is the primary scattering phase function; θ is a function of the field of view of the detector; Ω is a function of the field of view of the emission source; $M(\lambda)$ is the angle between the detector and the axis in the emission source quality factor. This model reveals that the performance of UV communication is affected mainly by three factors: the power of the UV radiation emitted by the light source, transmission characteristics of UV radiation in the atmosphere, and performance of the detection system.

When the characteristics of the equipment, background noise, and atmospheric channel are given, the geometric space represented by the positions of the transmitter and receiver determines the communication performance. Therefore, the geometry of the scattering channel of the UV optical link must be analyzed. The geometry of LOS transmission is shown in Fig. 2.19. In Fig. 2.19a, the transmitter and receiver perform LOS transmission. The light source at the transmitting end emits a signal to space with a divergence angle of ϕ_1. The receiver has a field of view of ϕ_2. The overlap between the two angles is the effective scattering area V. During operation, many photons are not scattered but reach the receiving end directly to realize communication; thus, the channel capacity of LOS communication is relatively large. The communication quality is relatively high, and the communication distance is relatively long (Fig. 2.19a). In NLOS communication (Fig. 2.19b), the light source at the transmitting end sends an optical signal to the space at the divergence angle ϕ_1 and transmitter elevation angle θ_1, and the receiver field of view ϕ_1 and elevation angle θ_2 are used to receive the optical signal. The receiver field of view forms an effective scatterer connected to transceivers in the atmosphere in the area of overlap. The transmitted optical signal reaches the scatterer via atmospheric absorption and scattering, and the receiving end receives the scattered light signal from the scatterer, thus completing the signal. The scattered light is typically very weak. Figure 2.19b shows that the effective communication range can be increased by increasing the receiver field of view to receive more scattered light.

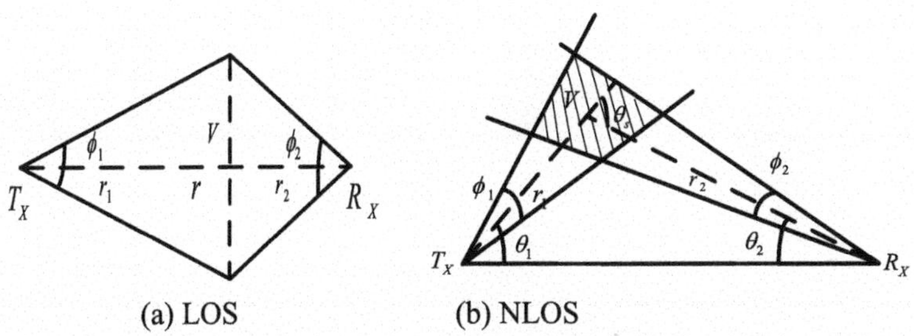

Fig. 2.19 Geometry of optical links in scattering communication [31]

2.3.1.1 Optical Link Model for UV LOS Mode

In wireless laser communication, there must be no obstacles between the transmitter and receiver that hinder the transmission of light. In general LOS communication, the transmission of light in the atmosphere is most important. Because of the large attenuation of solar-blind UV communication, for better communication performance, the LOS mode optical link model also considers the divergence angle of the emitted beam and the receiver field of view. According to the difference between these angles, LOS communication using solar-blind UV light can be divided into three cases: wide-divergence-angle transmission/wide-field-of-view reception, narrow-divergence-angle transmission/wide-field-of-view reception, and narrow-divergence-angle transmission/narrow-field-of-view reception, as shown in Fig. 2.20.

The UV power in the LOS link in UV communication is exponentially attenuated. In free space, the path loss is inversely proportional to the square of the transmission distance, which is expressed as $(\frac{\lambda}{4\pi r})^2$, and the atmospheric attenuation is expressed as $e^{-K_e r}$. In addition, the gain of the receiver detector is $\frac{4\pi A_r}{\lambda^2}$. Considering these parameters, the received optical power of the UV LOS link is expressed as follows [32]:

$$P_{r,LOS} = P_t(\frac{\lambda}{4\pi r})^2 e^{-K_e r} \frac{4\pi A_r}{\lambda^2} \tag{2.17}$$

This expression can be simplified to

$$P_{r,LOS} = \frac{P_t A_r}{4\pi r^2} e^{-K_e r} \tag{2.18}$$

where P_t is the transmit power, r is the communication distance, λ is the wavelength, and $K_e = K_A + K_s$ is the atmospheric channel attenuation coefficient, where K_A is

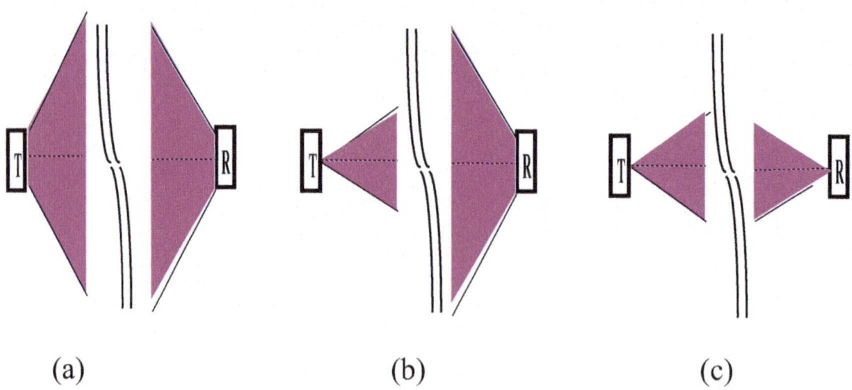

Fig. 2.20 Solar-blind UV LOS communication

the atmospheric absorption coefficient, K_s is the atmospheric scattering coefficient, and A_r is the receiver aperture area. After the receiver aperture of the UV optical communication system is selected, the atmospheric attenuation and communication distance are the main factors affecting the UV received power of the LOS link. Thus, the LOS link is very sensitive to changes in atmospheric conditions.

From (2.18), the received energy varies with K_e, which depends on weather conditions. The received energy is inversely proportional to r^2, and this attenuation is much more severe than that of a radio frequency link. The bandwidth of the detector is assumed to be twice the data transfer rate. The signal-to-noise ratio at the receiver is [33]

$$SNR_{r,LOS} = \frac{\eta_r G P_{r,LOS}}{2Rhc/\lambda} \tag{2.19}$$

Substituting (2.18) into (2.19) yields

$$SNR_{r,LOS} = \frac{\eta_r \lambda G P_t A_r}{8\pi r^2 hcR} e^{-K_e \cdot r} \tag{2.20}$$

The bit error rate (BER) after OOK modulation is

$$BER_{r,LOS} = Q(\frac{\sqrt{SNR_{r,LOS}}}{2}) = \frac{1}{2} erfc(\frac{\sqrt{SNR_{r,LOS}}}{2\sqrt{2}}) \tag{2.21}$$

where $Q(\cdot)$ is the Q function, and $erfc(\cdot)$ is the complementary error function.

The Q function is defined as

$$Q(a) = \int_a^\infty \frac{1}{\sqrt{2\pi}} e^{-\frac{y^2}{2}} dy \tag{2.22}$$

It has the following properties:

(1) $Q(0) = \frac{1}{2}$; (2) $Q(-a) = 1 - Q(a), a > 0$; (3) $Q(a) \approx \frac{1}{a\sqrt{2\pi}} e^{-\frac{a^2}{2}}, a \gg 1$

The complementary error function is defined as

$$erfc(\beta) = 1 - erf(\beta) = \frac{2}{\sqrt{\pi}} \int_\beta^\infty e^{-y^2} dy \tag{2.23}$$

It has the following properties:

(1) $erfc(-\beta) = 1 - erfc(\beta)$; (2) $erfc(\infty) = 0$; (3) $erfc(\beta) \approx \frac{1}{\sqrt{\pi}} e^{-\beta^2}, \beta >> 1$。

The Q function and error function are related as follows:

(1) $Q(a) = \frac{1}{2} erfc(\frac{a}{\sqrt{2}})$; (2) $erfc(a) = 2Q(\sqrt{2}a)$; (3) $erfc(a) = 1 - 2Q(\sqrt{2}a)$

When the received power is determined, we can roughly calculate the signal-to-noise ratio and BER of the LOS link according to the above theory to qualitatively describe the quality of communication.

2.3.1.2 Optical Link Model of UV NLOS Mode

In NLOS communication, the transmitted optical information can bypass obstacles between the transmit and receive ends. A solar-blind UV NLOS link is realized mainly using the characteristics of intense scattering of UV radiation during atmospheric transmission. UV NLOS communication can be divided into three communication modes according to the angle between the optical and horizontal axes: NLOS(A), NLOS(B), and NLOS(C), as shown in Fig. 2.1 [33]. Table 2.1 compares the performance of these communication methods. The communication mode can be switched by making adjustments. For example, the NLOS(A) mode can be transformed into the NLOS(B) and NLOS(C) modes by reducing θ_1 and θ_2, and finally converted into LOS communication [34]. In this transition, the working distance of the system changes from the minimum to the maximum, the communication bandwidth changes from the narrowest to the widest, and the omnidirectionality changes from the best to the worst.

Figure 2.19b shows a schematic diagram of an NLOS transmission link. It is assumed that the NLOS transmission energy is P_t and the energy per unit solid angle is $\frac{P_t}{\Omega_1}$. In single-scattering NLOS UV communication, the effects of path loss and attenuation are considered. The transmitter power is $(\frac{P_t}{\Omega_1})(\frac{e^{-K_e r_1}}{r_1^2})$ after transmission, and the power becomes $(\frac{P_t}{\Omega_1})(\frac{e^{-K_e r_1}}{r_1^2})(\frac{K_s}{4\pi}P_s V)$ after the signal is scattered by an effective scatterer. The arrival of the scattered beam at the receiving end can be regarded as LOS transmission. The influencing factors include the space link loss $(\frac{\lambda}{4\pi r_2})^2$, atmospheric attenuation $e^{-K_e r_2}$, and receiving detector. The gain is $\frac{4\pi A_r}{\lambda^2}$, and the single scattering of UV light in an NLOS link in atmospheric optical communication can be divided into three stages. The path r_1 from the transmitter to the scatterer can first be treated as a LOS link, and then the UV light is scattered. Finally, the path r_2 from the scatterer to the receiving end is also treated as a LOS link. By combining the expressions for the above processes, the received optical power of the UV NLOS link can be written as follows [32]:

$$P_{r,NLOS} = (\frac{P_t}{\Omega_1})(\frac{e^{-K_e r_1}}{r_1^2})(\frac{K_s}{4\pi}P_s V)(\frac{\lambda}{4\pi r_2})^2 e^{-K_e r_2}\frac{4\pi A_r}{\lambda^2} \tag{2.24}$$

where $\Omega_1 = 2\pi[1 - \cos(\phi_1/2)]$, $\theta_s = \theta_1 + \theta_2$, $r_1 = r\sin\theta_2/\sin\theta_s$, $r_2 = r\sin\theta_1/\sin\theta_s$, and $V \approx r_2 \cdot \phi_2 \cdot d^2$ By substituting (2.22) into (2.24), the expression can be simplified to [16]

$$P_{r,NLOS} = \frac{P_t \cdot A_r \cdot K_s \cdot P_s \cdot \phi_2 \cdot \phi_1^2 \cdot \sin(\theta_1 + \theta_2)}{32\pi^3 \cdot r \cdot \sin(\theta_1) \cdot (1 - \cos\frac{\phi_1}{2})} \cdot e^{-\frac{K_e \cdot r \cdot (\sin\theta_1 + \sin\theta_2)}{\sin(\theta_1 + \theta_2)}} \tag{2.25}$$

where r is the straight-line communication distance, Ω_1 is the transmission solid angle, V is the effective scattering region volume, and P_s is the phase function of the scattering angle θ_s. Equation (2.25) can be simplified according to the geometric configuration of the scattering channel, and the final expression is inversely proportional to the straight-line communication distance r. The characteristics of the UV NLOS link are affected by the factors that affect the LOS link and also by the channel scattering characteristics.

2.3.2 *Single-Scattering Model of UV Light*

NLOS solar-blind UV communication can be used as the basic link in short-range wireless data transmission and can also be used for communication between distributed sensor nodes. The UV signal from the transmitting end is scattered by the atmosphere and received at the receiving end within a certain distance. A signal collector with a large field of view is typically used at the receiving end to improve the receiving sensitivity. This signal collector can be placed anywhere within the effective scattering radius with the transmitting end as the center and can receive light-scattering signals. Factors such as altitude and terrain obstacles, which affect LOS communication, typically have a negative impact on sunlight. Blind UV communication can be avoided [35].

Assuming that the atmospheric channel is a linear time-invariant system, the transmitter emits a UV pulse with an energy of Q_t into the air at time $t = 0$. Let us examine the optical signal that the receiver can receive under the geometric relationship shown in Fig. 2.19b. To satisfy the single-scattering condition, we use an ellipsoidal channel model developed by Luettgen et al. [36] to analyze the NLOS transmission of solar-blind UV light. This model is established in an ellipsoidal coordinate system, as shown in Fig. 2.22. The detector is located at the focal point F1, and the light source is located at the focal point F2. The optical signal is scattered by the scatterer at the intersection of the transmitter and receiver elevation angles and reaches the receiver, thereby completing the communication. The area where the transmit and receive angles intersect is called the channel capacity. The number of optical signals reaching the receiver is determined by the geometry of the system and the atmospheric scattering factor (K_s) and attenuation factor (K_e). The geometric parameters include the aperture angle of the transmitted beam, field of view of the receiver, elevation angles of the transmitter and detection receiver, and transmission distance.

2.3.2.1 Ellipsoidal Coordinate System

The ellipsoidal coordinate system used to model the UV single-scattering link is shown in Fig. 2.21 [36]. An ellipsoidal surface is obtained by rotating the ellipse around its main axis once, and the coordinates of any point on the ellipsoid are uniquely determined by the radial, angular, and azimuthal coordinates ξ, η, and φ, respectively. The parameters for transforming the Cartesian coordinate system *x*-*y*-*z* to the ellipsoidal coordinate system are defined as follows [37]:

$$r_1 = [x^2 + y^2 + (z + r/2)^2]^{1/2} \tag{2.26}$$

$$r_2 = [x^2 + y^2 + (z - r/2)^2]^{1/2} \tag{2.27}$$

$$\xi = (r_1 + r_2)/r \quad (1 \le \xi \le \infty) \tag{2.28}$$

$$\eta = (r_1 - r_2)/r \quad (-1 \le \eta \le 1) \tag{2.29}$$

$$\varphi = \arctan(x, y) \quad (-\pi \le \varphi \le \pi) \tag{2.30}$$

$$\theta_s = \psi_1 + \psi_2 \tag{2.31}$$

From (2.26), (2.27), (2.28), (2.29), and (2.30), the following can be derived

$$\cos(\psi_1) = (1 + \xi\eta)/(\xi + \eta) \tag{2.32}$$

$$\sin(\psi_1) = [(\xi^2 - 1)(1 - \eta^2)]^{1/2}/(\xi + \eta) \tag{2.33}$$

$$\cos(\psi_2) = (1 - \xi\eta)/(\xi - \eta) \tag{2.34}$$

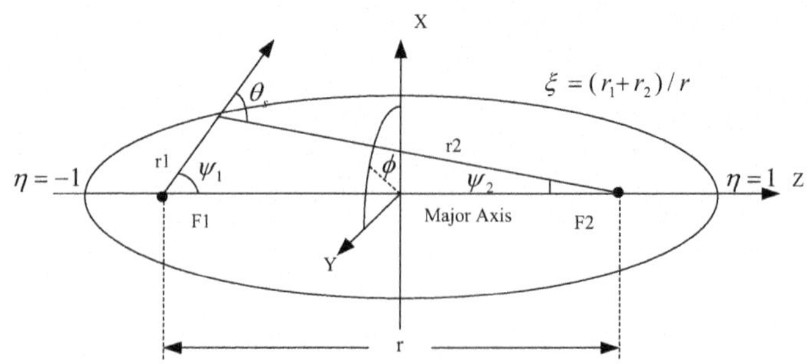

Fig. 2.21 Ellipsoidal coordinate system

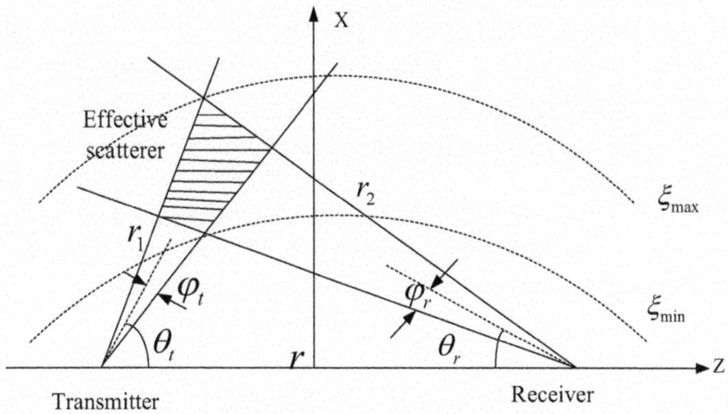

Fig. 2.22 UV single-scattering communication link model

$$\sin(\psi_2) = [(\xi^2 - 1)(1 - \eta^2)]^{1/2}/(\xi - \eta) \quad (2.35)$$

$$\cos(\theta_s) = (2 - \xi^2 - \eta^2)/(\xi^2 - \eta^2) \quad (2.36)$$

Equations (2.32) and (2.34) show that if we set $\cos(\psi_1) = f(\xi, \eta)$, then $\cos(\psi_2) = f(\xi, -\eta)$; similarly, from (2.33) and (2.35), if we set $\sin(\psi_1) = h(\xi, \eta)$, then $\sin(\psi_2) = h(\xi, -\eta)$. These relationships indicate the unique symmetry of the ellipsoidal coordinate system, which provides convenient conditions for modeling NLOS single scattering.

2.3.2.2 Analysis of UV Scattering Communication

The single-scattering UV communication link is represented in an ellipsoidal coordinate system, and the transmitting and receiving devices are placed at the two focal points. Many calculations of the UV communication process can then be made. A good formula greatly reduces the difficulty of analyzing and studying the communication process. Figure 2.22 shows the UV single-scattering link model based on the ellipsoidal coordinate system.

At time $t = 0$, UV light with a total energy of E_t leaves the emitting end at a divergence angle of φ_t [the emission energy per unit solid angle is $\frac{E_t}{\Omega}$ (unit: J/sr), solid angle $\Omega = 4\pi \sin^2(\theta_t/2)$] along the emission angle θ_t. At $t = r_1/c$, it reaches the scatterer at a distance r_1 from the emission end with an energy density of $\frac{E_t}{\Omega}\frac{e^{-kr_1}}{r_1^2}$, where e^{-kr_1} is the extinction coefficient, and r_1 is the ratio of UV transmission in the atmosphere to the energy after transmission in a vacuum.

The volume differential dv of the scatterer can be regarded as a secondary point light source, which is produced by the interaction between UV light and the transmission medium. It can be regarded as a spherical light source; because the solid angle on the entire spherical surface is 4π, the energy density of this secondary point light source is $\left[\frac{E_t}{\Omega}\frac{e^{-kr_1}}{r_1^2}\right]\frac{k_s}{4\pi}p(\cos\theta_s)dv$ [unit: J/sr), and $p(\cos\theta_s)$ is the single-scattering phase function. Note that $p(\cos\theta_s)$ is not always the same, and the out-of-phase functions discussed in the previous section should be chosen according to the prevailing atmospheric conditions.

Some of the energy in the secondary light source will be transmitted along a direction that enables it to reach the receiving end. The total energy density from the secondary point light source received by the receiving end at time $t = (r_1 + r_2)/c$ is given by

$$\delta E_r = \left[\frac{E_t}{\Omega}\frac{e^{-kr_1}}{r_1^2}\right]\left[\frac{k_s}{4\pi}p(\cos\theta_s)dv\right]\frac{e^{-kr_2}}{r_2^2} \tag{2.37}$$

The expression for the volume differential of the scatterer can be obtained as follows:

$$dv = (r/2)^2(\xi^2 - \eta^2)d\xi d\eta d\varphi \tag{2.38}$$

In addition, $r_1 = \frac{r}{2}(\xi + \eta)$, and $r_2 = \frac{r}{2}(\xi - \eta)$. The total volume-integrated energy density y at the receiving end is given by

$$\delta E_r = \frac{E_t k_s}{4\pi\Omega}\frac{e^{-kr\xi}p(\cos\theta_s)}{(r/2)(\xi^2 - \eta^2)}d\xi d\eta d\varphi \tag{2.39}$$

Because $\xi = (r_1 + r_2)/r$ (that is, $\xi = ct/r$), and thus $d\xi = cdt/r$ (that is, $\frac{d\xi}{dt} = \frac{c}{r}$), the instantaneous volume-integrated receiver power density at the receiving end can be written as

$$\delta P_r = \frac{cE_t k_s}{2\pi\Omega r^2}\frac{e^{-kr\xi}}{(\xi^2 - \eta^2)}p(\cos\theta_s)d\eta d\phi \tag{2.40}$$

Finally, by integrating each parameter, the total received power at the receiver is given by

$$P_r(\xi) = \begin{cases} 0 & \xi < \xi_{\min} \\ \int\limits_{\eta_1(\xi)}^{\eta_2(\xi)}\int\limits_{\varphi_1}^{\varphi_2}\frac{cE_t k_s}{2\pi\Omega r^2}\frac{e^{-kr\xi}}{(\xi^2+\eta^2)}p(\cos\theta_s)d\eta d\phi & \xi_{\max} < \xi > \xi_{\min} \\ 0 & \xi > \xi_{\max} \end{cases} \tag{2.41}$$

The process of UV scattering communication is divided into two subprocesses: transmission from the transmitting end to the scatterer and transmission from the scatterer to the receiving end. However, in the actual process, the loss of light energy must be determined for the upper and lower limits of three sets of integrals (η_1, η_2; $\xi_{\min}$, $\xi_{\max}$; φ_1, φ_2).

2.3.2.3 Analysis and Derivation of Three Sets of Integral Limits

(1) Upper and lower limits of angular coordinates: $\eta_1(\xi)$, $\eta_2(\xi)$.

In the ellipsoidal coordinate system, $\eta_1(\xi)$ is a function of ξ and ψ_1, and $\eta_2(\xi)$ is a function of ξ and ψ_2. Equations (2.42) and (2.43) can be obtained from (2.32) and (2.34), respectively:

$$\eta_1(\xi) = \frac{\xi\cos(\psi_1) - 1}{\xi - \cos(\psi_1)} \tag{2.42}$$

$$\eta_2(\xi) = \frac{1 - \xi\cos(\psi_2)}{\xi - \cos(\psi_2)} \tag{2.43}$$

Then, $\psi_1 = \theta_t + \varphi_t$, and $\psi_2 = \theta_r + \varphi_r$.

(2) Upper and lower limits of radial coordinates: $\xi_{\min}$, $\xi_{\max}$

When ξ takes the minimum value, $\xi_{\min}$, $\eta_1(\xi) = \eta_2(\xi)$. Thus, (2.42) and (2.43) are equal, where $\psi_1 = \theta_t - \varphi_t$, and $\psi_2 = \theta_r - \varphi_r$. $\xi_{\min}$ can be written as

$$\xi_{\min} = \left(\frac{1 + \cos(\theta_t - \varphi_t)\cdot\cos(\theta_r - \varphi_r)}{\cos(\theta_t - \varphi_t) + \cos(\theta_r - \varphi_r)}\right) + \sqrt{\left(\frac{1 + \cos(\theta_t - \varphi_t)\cdot\cos(\theta_r - \varphi_r)}{\cos(\theta_t - \varphi_t) + \cos(\theta_r - \varphi_r)}\right)^2 - 1} \tag{2.44}$$

When ξ takes the maximum value, $\xi_{\max}$, $\eta_1(\xi) = \eta_2(\xi)$. Consequently, (2.42) and (2.43) are equal, where $\psi_1 = \theta_t + \varphi_t$, and $\psi_2 = \theta_r + \varphi_r$. $\xi_{\max}$ can be written as

$$\xi_{\max} = \left(\frac{1 + \cos(\theta_t + \varphi_t)\cdot\cos(\theta_r + \varphi_r)}{\cos(\theta_t + \varphi_t) + \cos(\theta_r + \varphi_r)}\right) + \sqrt{\left(\frac{1 + \cos(\theta_t + \varphi_t)\cdot\cos(\theta_r + \varphi_r)}{\cos(\theta_t + \varphi_t) + \cos(\theta_r + \varphi_r)}\right)^2 - 1} \tag{2.45}$$

(3) Upper and lower limits of azimuthal coordinates: φ_1, φ_2

Because the emitted and received light cones are symmetric about the X–Z plane, the scatterers generated by their intersection are also symmetric about the X–Z plane, and (2.46) can be obtained:

$$\phi_2 = -\phi_1 \quad \text{(radians)} \tag{2.46}$$

Equation (2.30) defines the azimuthal coordinate ϕ as the arctangent of the Y-axis and X-axis coordinates of a point on the light cone, in radians. Figure 2.23 shows the emitted light cone from the emission point to the intersection with the receiving light cone in the three-dimensional coordinate system. A and D are the farthest intersection points of the light cone on the positive and negative sides of the Y axis, respectively, and AD is parallel to the Y axis. Therefore,

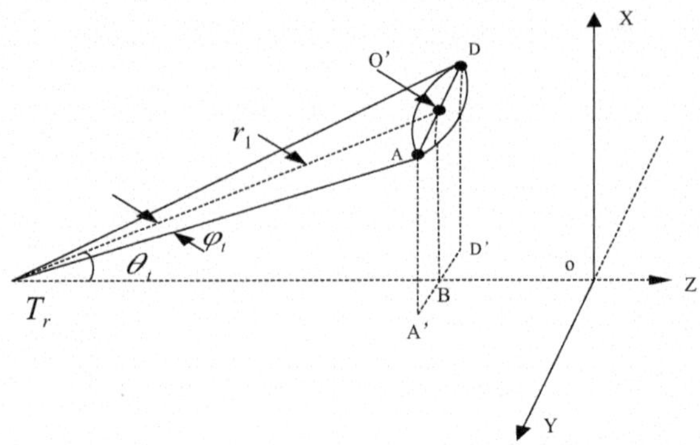

Fig. 2.23 Diagram of emission light cone

$$\varphi_2 = \arctan(x_A, y_A) \tag{2.47}$$

$$\varphi_1 = \arctan(x_D, y_D) \tag{2.48}$$

According to (2.46),

$$\varphi_2 = \arctan(x_A, y_A) = -\arctan(x_D, y_D) = -\varphi_1 \tag{2.49}$$

where (x_A, y_A) are the coordinates of point A, and (x_D, y_D) are the coordinates of point D. In Fig. 2.23,

$$x_A = AA' = O'B = r_1 \sin(\theta_t) \tag{2.50}$$

$$y_A = A'B = AO' = r_1 \tan(\varphi_t) \tag{2.51}$$

Therefore,

$$\varphi_2 = \arctan(r_1 \sin(\theta_t),\ r_1 \tan(\phi_t)) = -\varphi_1 \tag{2.52}$$

In the figure, Tr is the transmission end in UV communication. The above derivation process is based on the intersection length of the emitted and received light cones (AD in Fig. 2.23), where the intersection length of the emitted light cone is assumed to be less than or equal to the intersection length of the received light cone. In the opposite situation, the same calculation method can be used to analyze the received light cone to obtain the desired results.

2.3.3 Simulation Analysis of Single-Scattering Model

On the basis of the above theoretical analysis, to quantitatively analyze the performance of LOS and NLOS links, we assume that UV light undergoes isotropic scattering ($p_s = 1$) and ignore atmospheric turbulence. For OOK modulation, the other parameters are as follows: $\lambda = 265$ nm, $A_r = 1.8\,\text{cm}^2$, $\eta_r = 0.2$, $G = 100$, $K_e = 1.23\,\text{km}^{-1}$, $K_s = 0.157\,\text{km}^{-1}$, $\theta_1 = 45°$, $\theta_2 = 60°$, $\phi_1 = 1°$, and $\phi_2 = 60°$, where λ is the wavelength of UV light, η_r is the detection efficiency of the detector, G is the receiver gain, θ_1 is the transmitter elevation angle, θ_2 is the receiver elevation angle, ϕ_1 is the divergence angle, and ϕ_2 is the field angle.

We simulated the effects of transmission distance and BER on the data transmission rate of LOS and NLOS communication (Fig. 2.24). The transmission distance ranges from 1 to 1000 m, and the BER ranges from 10^{-10} to 10^{-1}.

At BERs of 10^{-3} and 10^{-6} for the LOS communication link, the data transmission rates are 6 and 2 Gbps, respectively, for a transmission distance of 20 m. When the transmission distance is increased to 1000 m, the rates decrease to 730 and 240 Kbps, respectively. From the above data, we can conclude that the BER increases gradually with increasing data transmission rate.

At BERs of 10^{-3} and 10^{-6} for the NLOS communication link, the data transmission rates are 6 and 2 Mbps, respectively, at a transmission distance of 20 m. At a distance of 200 m, the rates are 240 and 112 Kbps, and when the transmission distance increases to 1000 m, the rates decrease to 7 and 2 Kbps, respectively. The change in transmission rate is closely related to the receiver gain G. When $G = 1$, the data transmission rate decreases by a factor of 100. The loss of NLOS communication links is greater than that of LOS communication links.

The relationship between the transmitter elevation angle (0° to 90°) and data transmission rate is shown in Fig. 2.25 for a BER of 10^{-4}, $\theta_2 = 60°$, $\phi_1 = 1°$, and $\phi_2 = 60°$. The relationship between the receiver elevation angle (0° to 90°) and data transmission rate is shown in Fig. 2.26 for a BER of 10^{-4}, $\theta_1 = 45°$, $\phi_1 = 1°$, and $\phi_2 = 60°$.

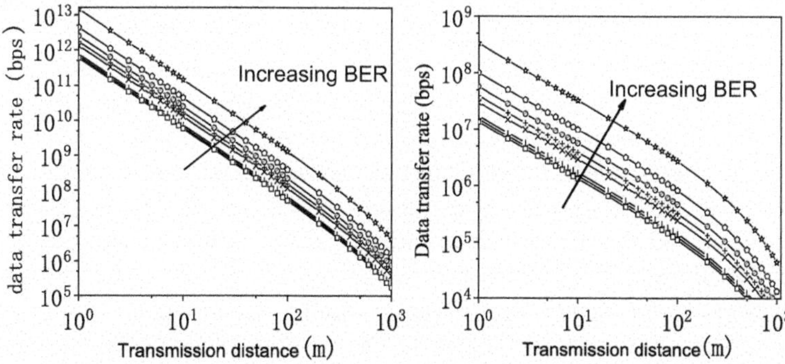

Fig. 2.24 Data rates for LOS and NLOS links

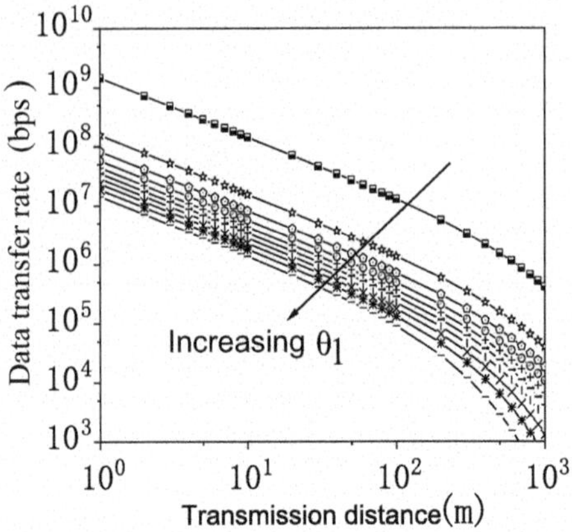

Fig. 2.25 Relationship between transmitter elevation angle θ_1 and data transfer rate for NLOS link

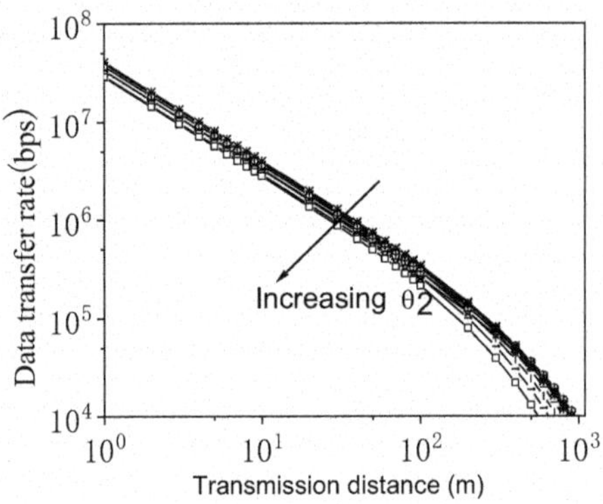

Fig. 2.26 Relationship between receiver elevation angle θ_2 and data transfer rate for NLOS link

The relationship between the divergence angle ϕ_1 (0° to 90°) of the light source at the transmitting end and the data transmission rate is shown in Fig. 2.27 for a BER of 10^{-4}, $\theta_1 = 45°$, $\theta_2 = 60°$, and $\phi_2 = 60°$. The relationship between the receiver field of view ϕ_2 (0° to 90°) and data transmission rate is shown in Fig. 2.28 for a BER of 10^{-4}, $\theta_1 = 45°$, $\theta_2 = 60°$, and $\phi_1 = 1°$.

Figures 2.25, 2.26, 2.27 and 2.28 show that the transmitter elevation angle θ_1 and receiver field of view ϕ_2 strongly affect the data transmission rate. They become sensitive to changes around 30°. As the receiver field of view increases, the data transmission rate increases, and the receiver field of view is more sensitive to changes above 30°. The effect of the receiver elevation angle on the transmission performance is clearly not as large as that of the transmitter elevation angle. The reason is the properties of the ellipsoid. With increasing θ_2, the emission solid cone angle decreases, whereas the scatterer volume V increases gradually, compensating for the decrease in solid cone angle somewhat. Therefore, the elevation angle of the receiving end changes from 0° to 90°, which has little effect on the data transmission rate. In UV networking, omnidirectional receiving antennas can be used for convenience. The

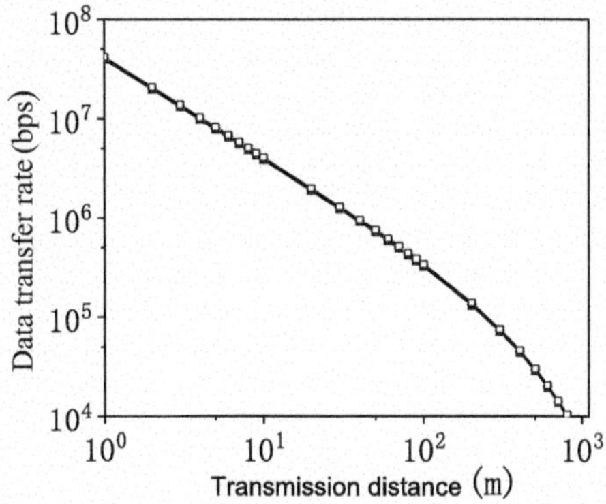

Fig. 2.27 Relationship between divergence angle ϕ_1 and data transfer rate for NLOS link

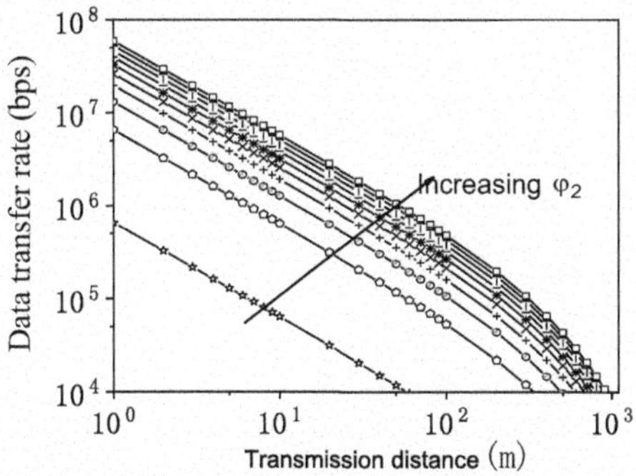

Fig. 2.28 Relationship between divergence angle ϕ_2 and data transfer rate for NLOS link

transmitter field of view has little effect on the data transfer rate. To concentrate the beam and reduce the loss, the transmitter field of view is generally reduced as much as possible. We compare the data transmission rates of the three NLOS communication modes below.

Figure 2.29 shows the data transmission rate of NLOS(A) communication at a BER of 10^{-4}, $\phi_1 = 1°$, $\phi_2 = 60°$, $\theta_1 \approx 90°$, and $\theta_2 \approx 90°$. The data transmission rate is below 1 Kbps at approximately 20 m, which has certain guiding significance experimentally but little significance in actual communication.

Figure 2.30 shows the data transmission rate of NLOS(B) communication at a BER of 10^{-4}, $\phi_1 = 1°$, $\phi_2 = 60°$, $\theta_1 = 15°$, and $\theta_2 \approx 90°$. NLOS(B) communication has a range of approximately 1000 m, which is important for actual communication.

Figure 2.31 shows the data transmission rate of NLOS(C) communication at a BER of 10^{-4}, $\phi_1 = 1°$, $\phi_2 = 60°$, $\theta_1 = 15°$, and $\theta_2 = 30°$. The data transfer rates of NLOS(C) and NLOS(A) communication differ by an order of magnitude, and the data transmission rate of NLOS(C) communication is better than that of NLOS(B) As shown in Fig. 2.32, in actual communication, NLOS(C) communication is preferable.

In a study [38], a 254 nm low-pressure mercury lamp was used to establish a UV digital communication system. For a transmitter elevation angle of 60°, a bit rate of

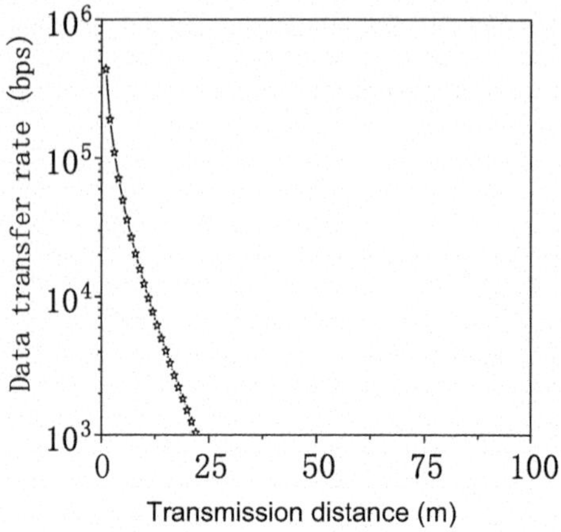

Fig. 2.29 Data transmission rate of NLOS(A) communication

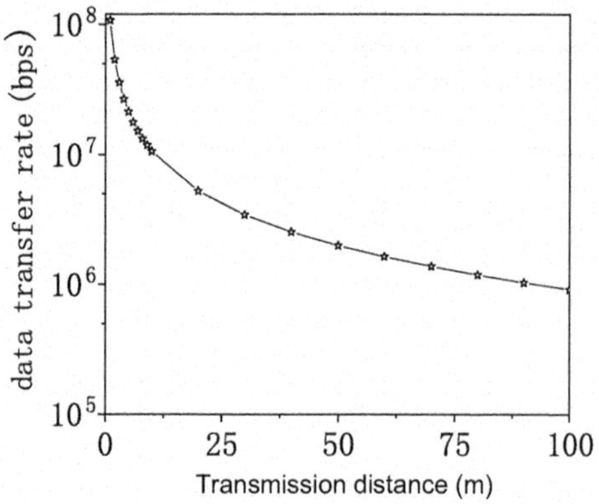

Fig. 2.30 Data transmission rate of NLOS(B) communication

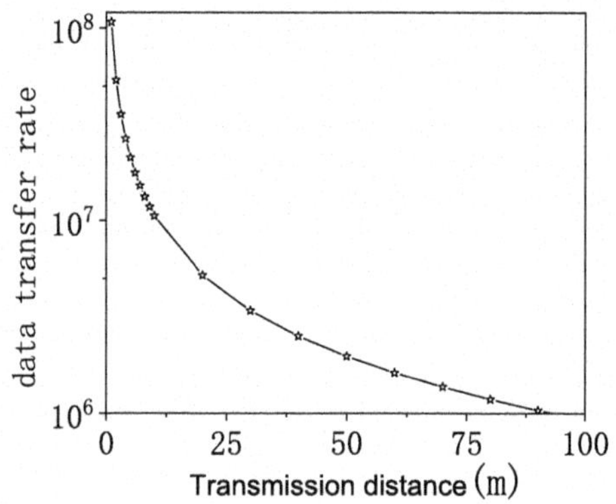

Fig. 2.31 Data transmission rate of NLOS(C) communication

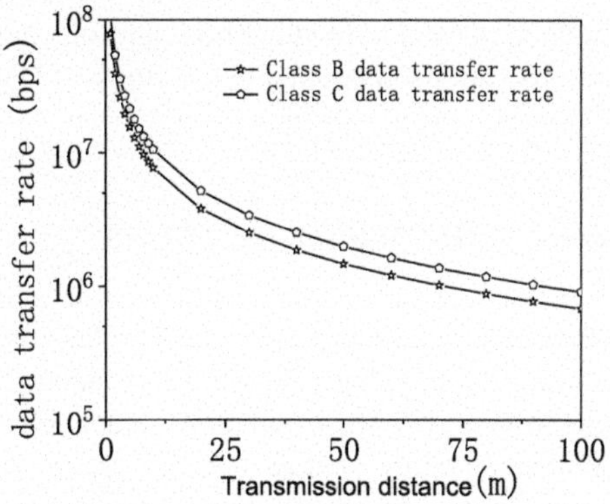

Fig. 2.32 Data transmission rates of NLOS(B) and NLOS(C) communication

1200 bps, and various transmission distances and receiver elevation angles, the BER of the system was up to 10^{-4} orders of magnitude.

2.4 Mie Scattering Mechanism in UV Communication

Light propagates in a straight line in a homogeneous medium, but when it encounters inhomogeneous obstacles in the atmosphere such as aerosol particles, it deviates from the initial propagation direction; that is, light is scattered. The scattering of UV light by atmospheric components enables it to bypass obstacles and communicate with another party, forming single-scattering and multiple-scattering links. The link model is shown in Fig. 2.33 [39].

The scattering phase function is central to the study of UV scattering communication links. Aerosol particles are present in the atmosphere at very high concentrations, and the scattering of UV light can be explained by Mie scattering theory. Therefore, it is necessary to study the Mie scattering phase function.

To calculate the Mie scattering phase function, the Mie scattering coefficient must first be calculated. The calculation of this coefficient requires multiple calculations of the Legendre polynomial and Bessel and Hankel functions, and this calculation method is very tedious and time-consuming. Researchers have proposed a recursive

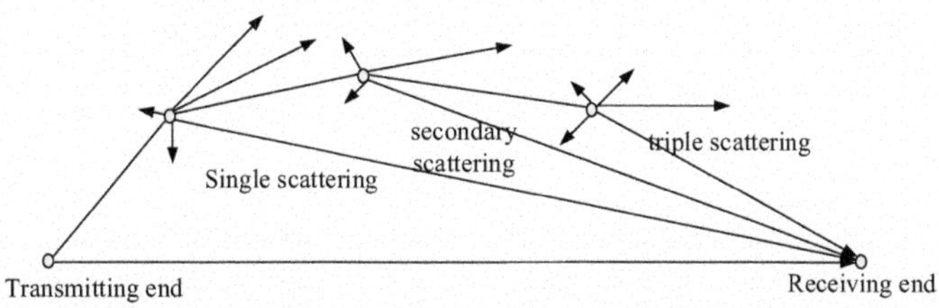

Fig. 2.33 UV scattering link model

method of calculating the Mie scattering, but it lacks a correct and detailed derivation and has not been applied to the study of UV scattering [40, 41, 60].

2.4.1 Mie Scattering Theory

Mie scattering theory provides a precise understanding of plane-wave scattering by homogeneous spherical particles in electromagnetic fields obtained from Maxwell's electromagnetic wave equations under certain boundary conditions through rigorous mathematical derivation. It is a rigorous theory describing light scattering. In Mie scattering theory, when natural light (completely unpolarized light) with a wavelength of λ irradiates a spherical particle, as shown in Fig. 2.34 [8], the initial light intensity is denoted as I_0, the scattering angle is θ, and the distance from the scatterer is r. The scattered light intensity at r, I_s, is defined as

$$I_s = \frac{\lambda^2 I_0}{8\pi^2 r^2}(i_1 + i_2) \tag{2.53}$$

$$\begin{cases} i_1 = |S_1(m, \theta, \alpha)|^2 \\ i_2 = |S_2(m, \theta, \alpha)|^2 \end{cases} \tag{2.54}$$

In (2.54) [42], i_1 is the intensity function of scattered light in the vertical direction, which is called the vertical component of the scattered light intensity function; i_2 is the intensity function of scattered light in the horizontal direction, which is called the parallel component of the scattered light intensity function weight. S_1 is the vertical component of the scattered light complex amplitude function, and S_2 is the parallel component of the scattered light complex amplitude function, which is an infinite series composed of the Bessel and Legendre functions. In addition, $\alpha = 2r/\lambda$ represents the particle's scale parameter, r represents the scattering particle radius,

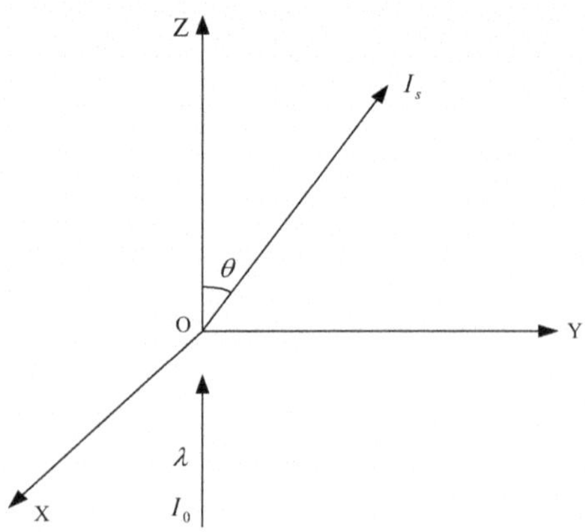

Fig. 2.34 Schematic diagram of scattering coordinate system

and $m = m_1 + im_2$ is the relative complex refractive index of the particle in the surrounding medium. A nonzero value of the imaginary part m_2 indicates that the particle absorbs light.

$$S_1 = \sum_{n=1}^{\infty} \frac{2n+1}{n(n+1)}(a_n \Pi_n + b_n \mathrm{T}_n) = \sum_{n=1}^{\infty} S_{1n} \tag{2.55}$$

$$S_2 = \sum_{n=1}^{\infty} \frac{2n+1}{n(n+1)}(a_n \mathrm{T}_n + b_n \Pi_n) = \sum_{n=1}^{\infty} S_{2n} \tag{2.56}$$

where a_n is the electric multipole coefficient, and b_n is the magnetic multipole coefficient; together they are called the Mie scattering coefficients.

The Mie scattering coefficients are defined as

$$\begin{cases} a_n = \dfrac{\psi_n(\alpha)\psi_n'(m\alpha) - m\psi_n'(\alpha)\psi_n(m\alpha)}{\varsigma_n(\alpha)\psi_n'(m\alpha) - m\varsigma_n'(\alpha)\psi_n(m\alpha)} \\ b_n = \dfrac{m\psi_n(\alpha)\psi_n'(m\alpha) - \psi_n'(\alpha)\psi_n(m\alpha)}{m\varsigma_n(\alpha)\psi_n'(m\alpha) - \varsigma_n'(\alpha)\psi_n(m\alpha)} \end{cases} \tag{2.57}$$

Equation (2.57) shows that the Mie scattering coefficient and complex refractive index m form Bessel and Hankel functions of various orders.

$$\psi_n(z) = (\frac{z\pi}{2}2)^{1/2} J^{(1)}_{n+\frac{1}{2}}(z) \tag{2.58}$$

$$\varsigma_n(z) = (\frac{z\pi}{2}2)^{1/2} H^{(2)}_{n+\frac{1}{2}}(z) \tag{2.59}$$

$\psi_n'(z)$ and $\varsigma_n'(z)$ are the derivatives of the respective variables.

$$\Pi_n = P_n^{(1)}(\cos\theta)/\sin\theta = \frac{dP_n(\cos\theta)}{d(\cos\theta)} \tag{2.60}$$

$$\mathrm{T}_n = \frac{d}{d\theta} P_n^{(1)}(\cos\theta) \tag{2.61}$$

The ratio of the scattered energy per unit solid angle of light in a given direction to the scattered energy per unit solid angle averaged in all directions is called the scattering phase function. For a monodisperse system, the scattering phase function of UV light, $P(\theta, \lambda)$, is defined as [4]

$$P(\theta, \lambda) = \frac{|S_1|^2 + |S_2|^2}{\sum_{n=1}^{\infty}(2n+1)(|a_n|^2 + |b_n|^2)} \tag{2.62}$$

In summary, the calculation of the scattering phase function is transformed into the calculation of the scattering coefficients a_n, b_n and angular scattering coefficients Π_n, T_n. A detailed derivation is given in the following two subsections.

2.4.2 Calculation of Mie Scattering Coefficients a_n, b_n

Both the Bessel and Hankel functions satisfy the following recurrence formulas [9]:

$$Y_{n+1}(z) = \frac{2n}{z} Y_n(z) - Y_{n-1}(z) \tag{2.63}$$

$$Y_n'(z) = \frac{1}{2}[Y_{n-1}(z) - Y_{n+1}(z)] \tag{2.64}$$

2.4.2.1 Calculation with $\psi_n(z)$ and $\varsigma_n(z)$

We substitute $n = n - \frac{1}{2}$ into (2.65) to obtain

$$Y_{n+\frac{1}{2}} = \frac{2n-1}{z} Y_{n-\frac{1}{2}} - Y_{n-\frac{3}{2}} \tag{2.65}$$

which can be written as

$$J^{(1)}_{n+\frac{1}{2}} = \frac{2n-1}{z} J^{(1)}_{n-\frac{1}{2}} - J^{(1)}_{n-\frac{3}{2}} \tag{2.66}$$

We substitute $n = n - 1$ and $n = n - 2$ into (2.58) to obtain

$$\psi_{n-1}(z) = (\frac{z\pi}{2})^{\frac{1}{2}} J^{(1)}_{n-\frac{1}{2}}(z) \tag{2.67}$$

$$\psi_{n-2}(z) = (\frac{z\pi}{2})^{\frac{1}{2}} J^{(1)}_{n-\frac{3}{2}}(z) \tag{2.68}$$

$$\psi_{n-1}(z) \times \frac{2n-1}{z} - \psi_{n-2}(z) = (\frac{z\pi}{2})^{\frac{1}{2}} [\frac{2n-1}{z} J^{(1)}_{n-\frac{1}{2}}(z) - J^{(1)}_{n-\frac{3}{2}}(z)] \tag{2.69}$$

From (2.66), we obtain

$$\psi_n(z) = \frac{2n-1}{z} \psi_{n-1}(z) - \psi_{n-2}(z) \tag{2.70}$$

Similarly,

$$\varsigma_n(z) = \frac{2n-1}{z}\varsigma_{n-1}(z) - \varsigma_{n-2}(z) \tag{2.71}$$

2.4.2.2 Calculation with $\psi_n'(z)$ and $\varsigma_n'(z)$

We substitute (2.63) into (2.64) to obtain

$$Y_{n+1}(z) = Y_{n-1}(z) - \frac{n}{z}Y_n(z) \tag{2.72}$$

which is rewritten as

$$J_{n+1}^{(1)}(z) = J_{n-1}^{(1)}(z) - \frac{n}{z}J_n^{(1)}(z) \tag{2.73}$$

Let $n = n + \frac{1}{2}$; then (2.73) becomes

$$J'_{n+\frac{1}{2}}{}^{(1)}(z) = J_{n-\frac{1}{2}}^{(1)}(z) - \frac{n+\frac{1}{2}}{z}J_{n+\frac{1}{2}}^{(1)}(z) \tag{2.74}$$

Then

$$\psi_n'(z) = \frac{\pi}{4}\left(\frac{z\pi}{2}\right)^{\frac{1}{2}}J_{n+\frac{1}{2}}^{(1)}(z) + \left(\frac{z\pi}{2}\right)^{\frac{1}{2}}J'_{n+\frac{1}{2}}{}^{(1)}(z) \tag{2.75}$$

We substitute (2.73) into (2.75) to obtain

$$\psi_n'(z) = \psi_{n-1}(z) - \frac{n}{z}\psi_n(z) \tag{2.76}$$

Similarly,

$$\varsigma_n'(z) = \varsigma_{n-1}(z) - \frac{n}{z}\varsigma_n(z) \tag{2.77}$$

2.4.3 *Calculation of Mie Angular Scattering Coefficients Π_n and T_n*

If we set $\xi = \cos\theta$, then

$$\Pi_n = \frac{dP_n(\xi)}{d(\xi)} = P_n'(\xi) \tag{2.78}$$

$$\mathrm{T}_n = \frac{d}{d\theta} P_n^{(1)}(\xi) \tag{2.79}$$

$$\because\ P_n^{(1)}(\xi) = \frac{dP_n(\xi)}{d\theta} = \frac{dP_n(\xi)}{d\xi}\frac{d\xi}{d\theta} \tag{2.80}$$

$$\therefore\ P_n^{(1)}(\xi) = \frac{d}{d\theta} P_n(\xi) \cdot (-\sin\theta) = -(1-\xi^2)^{\frac{1}{2}} \cdot \frac{d}{d\theta} P_n(\xi) \tag{2.81}$$

$$\begin{aligned}\frac{dP_n^{(1)}(\xi)}{d\theta} &= \frac{dP_n^{(1)}(\xi)}{d\xi} \cdot \frac{d\xi}{d\theta} \\ &= [-\frac{1}{2}(1-\xi^2)^{-\frac{1}{2}} \cdot \Pi_n(\xi) \cdot 2\xi - (1-\xi^2)^{\frac{1}{2}} \Pi_n'(\xi)][-\sin(\theta)]\end{aligned} \tag{2.82}$$

$$\mathrm{T}_n = \frac{dP_n^{(1)}(\xi)}{d\theta} = \xi \cdot \Pi_n + (\xi^2 - 1)\Pi_n' \tag{2.83}$$

Because the recurrence formula of the Legendre function satisfies (2.84) and (2.85),

$$(\xi^2 - 1)P_n'(\xi) = n\xi P_n(\xi) - nP_{n-1}(\xi) \tag{2.84}$$

$$nP_n(\xi) = \xi P_n'(\xi) - P_{n-1}'(\xi) \tag{2.85}$$

By differentiating both sides of (2.84), we obtain

$$(\xi^2 - 1)P_n''(\xi) = nP_n(\xi) + (n-2)\xi P_n'(\xi) - nP_{n-1}'(\xi) \tag{2.86}$$

We substitute (2.85) into (2.86) and obtain

$$\mathrm{T}_n = nP_n(\xi) + (n-1)\xi P_n'(\xi) - nP_{n-1}'(\xi) \tag{2.87}$$

We substitute (2.86) into (2.87) and obtain

$$\mathrm{T}_n = n\xi P_n'(\xi) - (n+1)P_{n-1}'(\xi) \tag{2.88}$$

Then

$$\mathrm{T}_n = n\xi \cdot \Pi_n - (n+1) \cdot \Pi_{n-1} \tag{2.89}$$

Equation (2.86) is rewritten as

$$P_{n-1}(\xi) = \frac{\xi}{n-1} P'_{n-1}(\xi) - \frac{1}{n-1} P'_{n-2}(\xi) \tag{2.90}$$

We substitute (2.85) and (2.90) into (2.84) and obtain

$$P'_n(\xi) = \frac{2n-1}{n-1} \xi \cdot P'_{n-1}(\xi) - \frac{n}{n-1} P'_{n-2}(\xi) \tag{2.91}$$

which can be written as

$$\Pi_n(\xi) = \frac{2n-1}{n-1} \xi \cdot \Pi_{n-1}(\xi) - \frac{n}{n-1} \Pi_{n-2}(\xi) \tag{2.92}$$

The following initial conditions are used in the simulation:

$$\begin{cases} \psi_0(z) = \sin z \\ \psi_1(z) = \frac{1}{z} \sin z - \cos z \\ \varsigma_0(z) = \sin z + i \cos z \\ \varsigma_0(z) = \frac{1}{z}(\sin z + i \cos z) - (\cos z - i \sin z) \\ \Pi_0(z) = 0 \\ \Pi_1(z) = 1 \\ \mathrm{T}_0(z) = 0 \end{cases} \tag{2.93}$$

2.4.4 Simulation and Calculation

The use of solar-blind UV light for communication is almost unaffected by background noise, and the receiver can receive a signal with a higher signal-to-noise ratio by receiving fewer UV photons [43, 44].

Figure 2.35 shows the the simulated phase function of solar-blind band UV (200–280 nm) scattering for haze particles with a radius of 0.1 μm and scattering angles of 0–π. The forward scattering of UV light in this band is smaller than the backward scattering. The difference between forward and backward scattering increases with decreasing wavelength of UV light. However, this difference is never large.

Figure 2.36 shows the simulated scattering phase function for fog drop particles with a radius of 2 μm. The results in Fig. 2.36a show the phase function of UV scattering in the solar-blind band for scattering angles of $0 \sim \pi$. Because the scattering phase function changes sharply in the middle of the scattering angle range, that portion the curve could not be obtained. Therefore, the results for scattering angles of $0.3\pi \sim 0.7\pi$ are shown in Fig. 2.36b.

Using the Mie scattering theory, we derived a recursive calculation formula for the Mie scattering and Mie angle scattering coefficients and used the results to simulate the scattering phase function of UV light in the solar-blind band. We found that as the

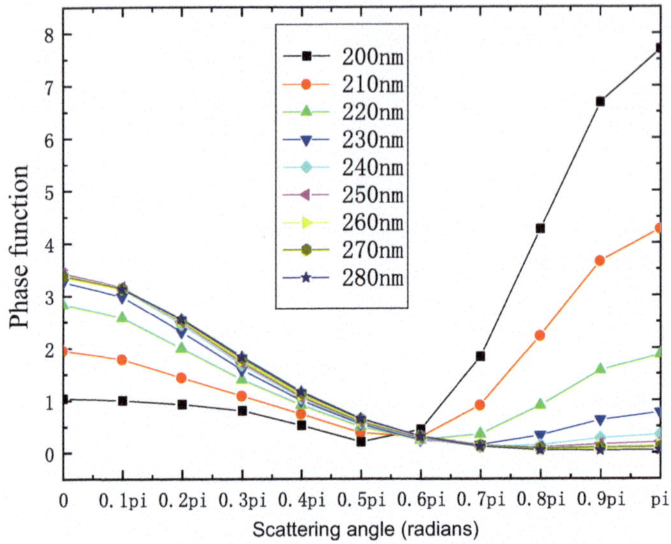

Fig. 2.35 Simulated scattering phase function for 0.1 μm particles

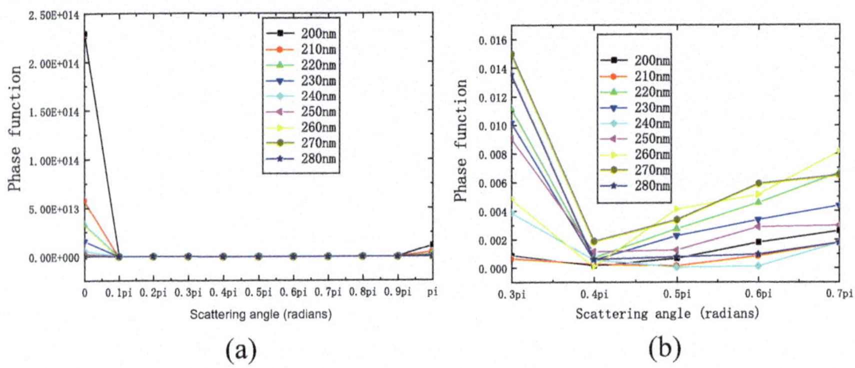

Fig. 2.36 Simulated scattering phase function for 2 μm particles

radius of the scattering particle increases, the forward and backward scattering of UV light are enhanced. When the scattering particles are small, the forward scattering of UV light is smaller than the backward scattering, and the scattering range does not vary greatly; for larger scattering particles, the forward scattering of UV light is greater than the backward scattering, and the scattering at zero scattering angle is much higher than that in other directions.

2.5 Calculation and Analysis of UV Communication Range

2.5.1 UV Communication Security

UV photons have high energy and thus strong chemical effects on objects. Modern medicine has found that UV light is harmful to human skin and eyes. The safe use of UV light is an important part of UV communication. Therefore, solar-blind UV communication must strictly control the emission power of the UV source and strictly limit the exposure time of human eyes and skin. The International Commission on Non-Ionizing Radiation Protection has formulated detailed regulations on the safe use of UVC radiation. In the absence of protective measures, human eyes and skin are continuously exposed to UVC light. Figure 2.37 shows the relationship between the safe light intensity for 8 h of exposure at various wavelengths. The minimum light intensities at wavelengths of 200, 270, and 280 nm are 100, 3, and 2.4 mJ/cm^2, respectively. Therefore, the transmit power of solar-blind UV communication systems must be controlled within a reasonable range [45].

The UV damage calculation formula is

$$E_{eff} = \sum_{200\,\text{nm}}^{300\,\text{nm}} E_\lambda S_\lambda \triangle \lambda \tag{2.94}$$

The expressions for wavelengths between 210 and 400 nm are as follows:

$$S_\lambda = 0.959^{(270-\lambda)}, \quad 210\,\text{nm} \le \lambda \le 270\,\text{nm};$$

$$S_\lambda = 1 - 0.36\left(\frac{\lambda - 270}{20}\right)^{1.64}, \quad 270\,\text{nm} < \lambda \le 300\,\text{nm};$$

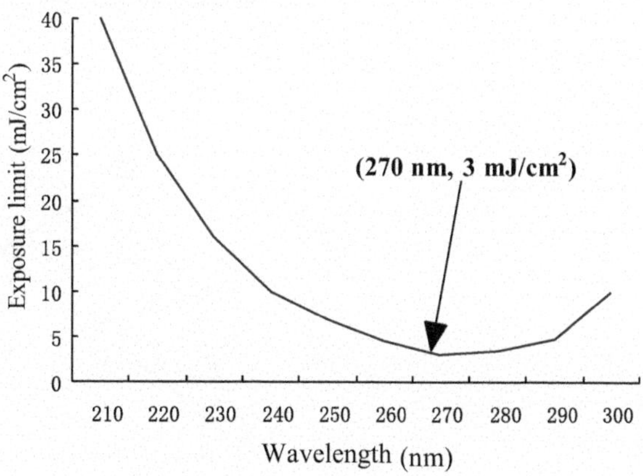

Fig. 2.37 Solar-blind UV exposure limits

$$S_\lambda = 0.3 \times 0.736^{(\lambda-300)} + 10^{(2-0.0163\lambda)}, \quad 300\,\text{nm} < \lambda \leq 400\,\text{nm}.$$

UV light safety should be considered in the design of UV communication systems. Through reasonable design, a lower average transmit power can be obtained, and the maximum instantaneous power can be limited.

2.5.2 Approximate Analysis of Effective Scattering Volume V

Figure 2.38 shows a schematic diagram of the geometric parameters of the transceiver for single-scattering UV communication. Using the parameters O, P, O', P', $AB \perp r_1$, $CD_1 \perp r$, $EF \perp r_2$, and $GH \perp r_2$, the volume is approximated by the cut-and-complement method. To improve UV scattering communication, the effective scatterer volume V is taken with a smaller intersecting volume.

First, the volume of the effective scatterer is found, as shown in Fig. 2.39; for each point, calculate $AB \perp TO'$ and $CD \perp TO'$.

As shown in Fig. 2.40, $VR = r_2$, and we set $VG = x$; then $GR = r_2 - x$, $O'G = tg(\theta_s) \cdot x$, and $O'G = (r_2 - x) \cdot tg(\frac{\phi_2}{2})$, namely:

$$x \cdot (tg(\frac{\phi_2}{2}) + tg(\theta_s)) = r_2 \cdot tg(\frac{\phi_2}{2})$$

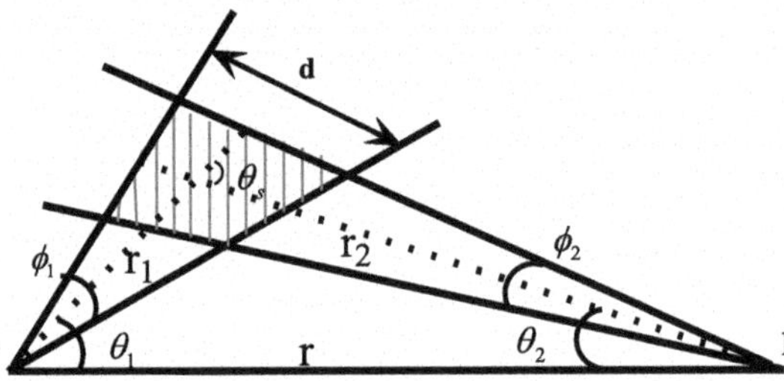

Fig. 2.38 Schematic diagram of transmission and reception of single-scattered communication

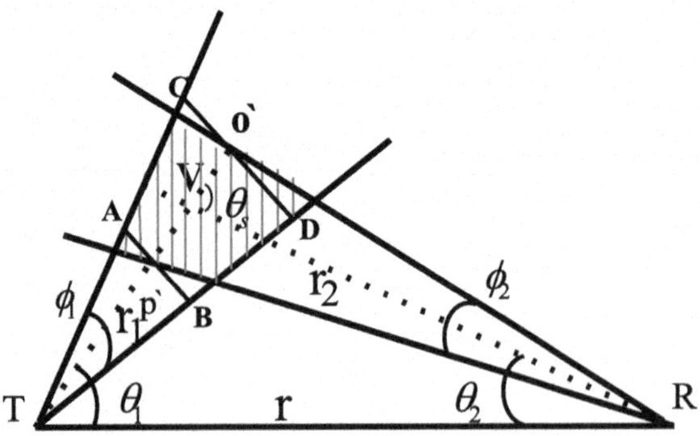

Fig. 2.39 Schematic diagram of reception and transmission of optical communication

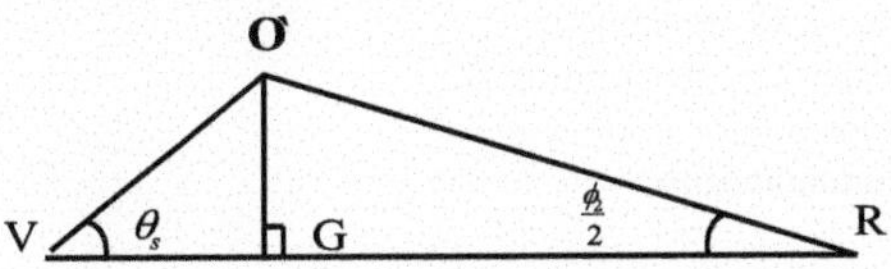

Fig. 2.40 Schematic diagram of geometric calculation

$$x = \frac{r_2 \cdot tg(\frac{\phi_2}{2})}{tg\theta_s + tg(\frac{\phi_2}{2})}$$

$$VO' = \frac{x}{\cos(\theta_s)} = \frac{r_2 \cdot tg(\frac{\phi_2}{2})}{(tg\theta_s + tg(\frac{\phi_2}{2})) \cdot \cos(\theta_s)}$$

From Fig. 2.41, $\angle O'P'R = 180° - [(180° - \theta_s) + \frac{\phi_2}{2}] = \theta_s - \frac{\phi_2}{2}$, and $VJ = \sin(\frac{\phi_2}{2}) \cdot r_2$; then $VP' = \frac{VJ}{\sin(\angle O'P'R)} = \frac{\sin(\frac{\phi_2}{2}) \cdot r_2}{\sin(\theta_s - \frac{\phi_2}{2})}$. In addition,

$$O'D = (r_1 + VO') \cdot tg(\frac{\phi_1}{2}) = (r_1 + \frac{r_2 \cdot tg(\frac{\phi_2}{2})}{(tg\theta_s + tg(\frac{\phi_2}{2})) \cdot \cos(\theta_s)}) \cdot tg(\frac{\phi_1}{2})$$

$$P'B = (r_1 - VP') \cdot tg(\frac{\phi_1}{2}) = (r_1 - \frac{\sin(\frac{\phi_2}{2}) \cdot r_2}{\sin(\theta_s - \frac{\phi_2}{2})}) \cdot tg(\frac{\phi_1}{2})$$

$$\begin{aligned} V_{ABCD} &= \frac{1}{3}\pi \cdot [O'D^2 \cdot (\mathrm{r}_1 + VO') - P'B^2 \cdot (\mathrm{r}_1 - VP')] \\ &= \frac{1}{3}\pi \cdot tg^2(\frac{\phi_1}{2}) \cdot [(r_1 + \frac{r_2 \cdot tg(\frac{\phi_2}{2})}{(tg\theta_s + tg(\frac{\phi_2}{2})) \cdot \cos(\theta_s)})^3 - (r_1 - \frac{r_2 \cdot \sin(\frac{\phi_2}{2})}{\sin(\theta_s - \frac{\phi_2}{2})})^3] \end{aligned} \tag{2.95}$$

Second, to find the volume of the effective scatterer, as shown in Fig. 2.42, for each point calculate $EF \perp RO$, $GH \perp RO$.

As shown in Fig. 2.43, pass point O to obtain $OI \perp TV$; it is known that $TV = r_1$, and we set $VI = x$. Then $TI = r_1 - x$, $OI = (r_1 - x) \cdot tg(\frac{\phi_1}{2})$, and $OI = x \cdot tg(\theta_s)$; thus,

$$x \cdot (tg(\frac{\phi_1}{2}) + tg(\theta_s)) = r_1 \cdot tg(\frac{\phi_1}{2})$$

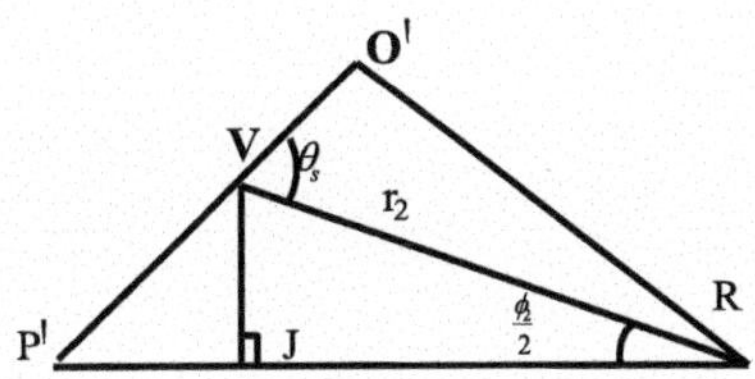

Fig. 2.41 Schematic diagram of geometric calculation

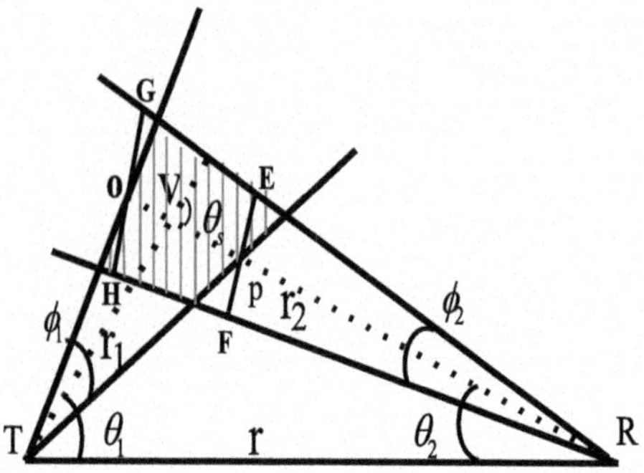

Fig. 2.42 Schematic diagram of reception and transmission of optical communication

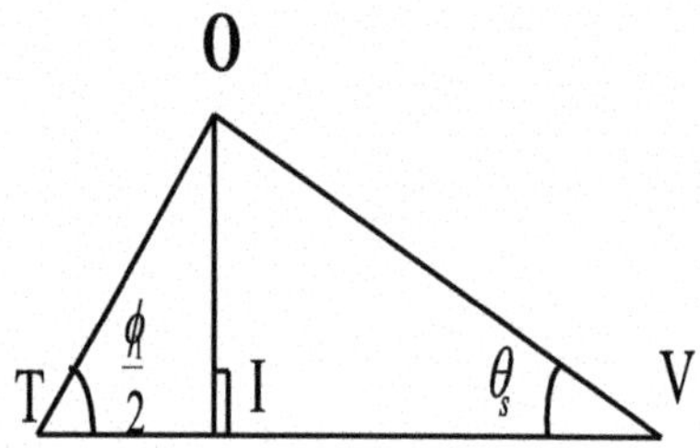

Fig. 2.43 Schematic diagram of geometric calculation

$$x = \frac{r_1 \cdot tg(\frac{\phi_1}{2})}{tg\theta_s + tg(\frac{\phi_1}{2})}$$

$$VO = \frac{x}{\cos(\theta_s)} = \frac{r_1 \cdot tg(\frac{\phi_1}{2})}{(tg\theta_s + tg(\frac{\phi_1}{2})) \cdot \cos(\theta_s)}$$

From point V, obtain $VM \perp TP$. From Fig. 2.44,

$$VM = r_1 \cdot \sin(\frac{\phi_1}{2})$$

$$\angle MVP = 180^\circ - \theta_s - (90^\circ - \frac{\phi_1}{2}) = 90^\circ - \theta_s + \frac{\phi_1}{2}$$

$$\angle VPM = 90^\circ - \angle MVP = \theta_s - \frac{\phi_1}{2}$$

$$\sin(\angle VPM) = \frac{VM}{VP}$$

$$VP = \frac{r_1 \cdot \sin(\frac{\phi_1}{2})}{\sin(\theta_s - \frac{\phi_1}{2})}$$

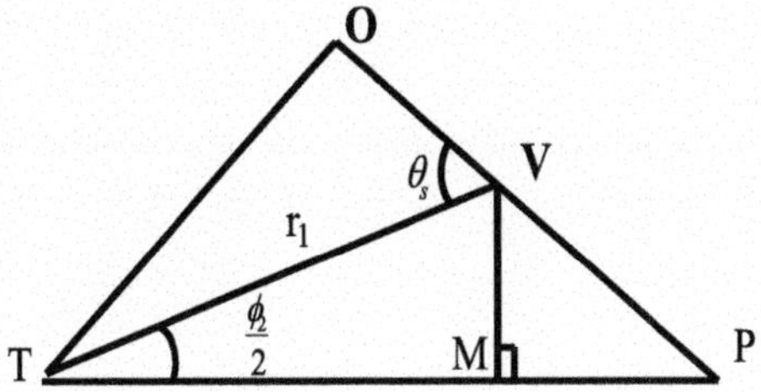

Fig. 2.44 Schematic diagram of geometric calculation

$$PR = r_2 - VP = r_2 - \frac{r_1 \cdot \sin(\frac{\phi_1}{2})}{\sin(\theta_s - \frac{\phi_1}{2})}$$

$$EP = tg(\frac{\phi_2}{2}) \cdot PR$$

$$OR = r_2 + VO = r_2 + \frac{r_1 \cdot tg(\frac{\phi_1}{2})}{(tg\theta_s + tg(\frac{\phi_1}{2})) \cdot \cos(\theta_s)}$$

$$GO = tg(\frac{\phi_2}{2}) \cdot OR$$

$$\begin{aligned} V_{EFGH} &= \frac{1}{3}\pi \cdot [GO^2 \cdot (\mathrm{r}_2 + VO) - EP^2 \cdot (\mathrm{r}_2 - VP)] \\ &= \frac{1}{3}\pi \cdot tg^2(\frac{\phi_2}{2}) \cdot [(r_2 + \frac{r_1 \cdot tg(\frac{\phi_1}{2})}{(tg\theta_s + tg(\frac{\phi_1}{2})) \cdot \cos(\theta_s)})^3 - (r_2 - \frac{r_1 \cdot \sin(\frac{\phi_1}{2})}{\sin(\theta_s - \frac{\phi_1}{2})})^3] \end{aligned} \quad (2.96)$$

$$V = \min(V_{ABCD}, V_{EFGH})$$

Using the observation plane in Fig. 2.45 as the cut plane, we obtain the three situations illustrated in Fig. 2.46. Among them, $d \approx r_1 \cdot \phi_1$ [46, 47] is usually relatively rare, and the spatial distribution of the effective scatterer is divided into the following three cases. (1) At $d \approx r_2 \cdot \phi_2$, as shown in Fig. 2.46a, the effective scatterer is a cylindrical body with a radius approximately that of the bottom surface and a height of d; therefore, $V \approx \pi(\frac{r_2 \cdot \phi_2}{2})^2 \cdot d$. The effective scatterer is small, and the noise interference is relatively low. (2) At $d \approx r_2 \cdot \phi_2$, as shown in Fig. 2.46(b), the effective scatterer is approximated as a square column with bottom side length d and height r. Thus, $V \approx r_2 \cdot \phi_2 \cdot d^2$, and the effective scatterer is large and noisy. (3) At $d \approx r_2 \cdot \phi_2$, as shown in Fig. 2.47c, the effective scatterer volume is $V \approx r_2 \cdot \phi_2 \cdot d^2$, and the noise is much smaller than that in Fig. 2.46b.

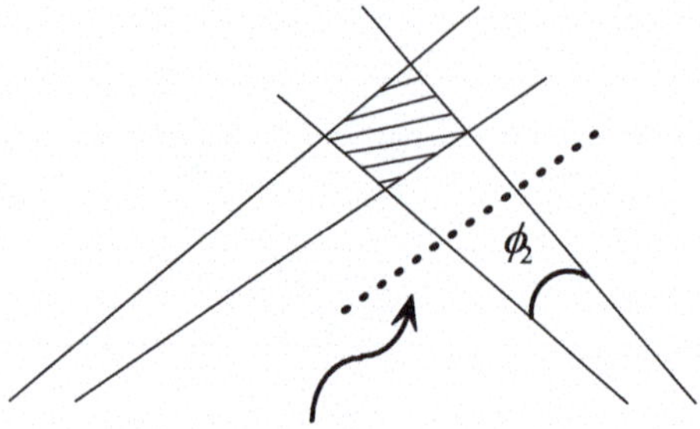

Fig. 2.45 Observation surface of effective scatterer

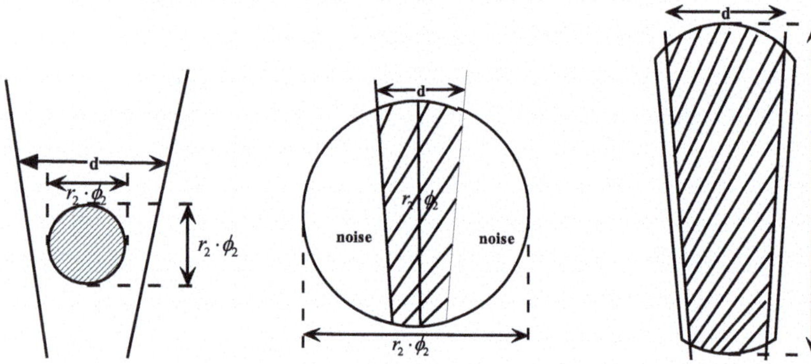

Fig. 2.46 Cutaway view of effective scatterer

Fig. 2.47 Stereo view of NLOS(A) coverage

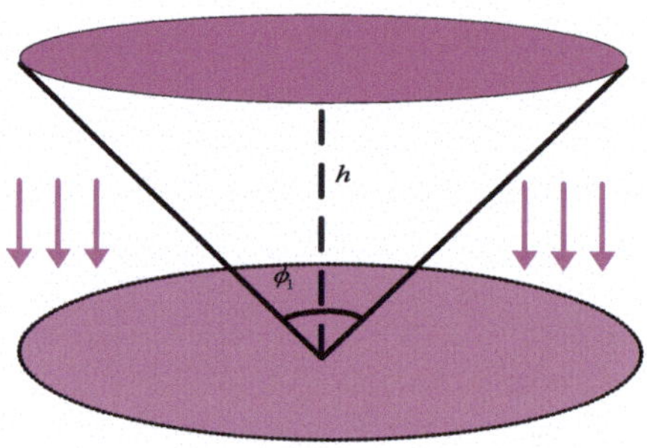

2.5.3 UV Communication Node Coverage Model

We analyzed the coverage of UV LOS and NLOS communication. Calculation formulas for the coverage of various communication methods were obtained.

2.5.3.1 Coverage in NLOS Method

The coverage of the three modes of UV NLOS communication was analyzed.

(1) NLOS(A) communication

In this type of communication, the elevation angles of the transmitter and receiver are both 90°. Assuming that the effective scatterer has a limited volume, the NLOS communication model is used, and the UV coverage after scattering has a circular area, as shown in Fig. 2.47. The communication distance is the same in all directions. The backscatter at the transmitting end is large, and the signal transmission capability is poor, which seriously limits the communication effect and transmission distance. Therefore, this communication method is rarely used.

(2) NLOS(B) communication

In this mode, the elevation angle of the transmitter is less than 90°, and the elevation angle of the receiver is 90°. UV communication occurs mainly by forward scattering, but some backscattering also occurs near the transmitting end. The coverage of UV communication is shown in Fig. 2.49. For communication to occur, the receiver must be within the range of forward scattering. The projection of the cone AGH on the ground is shown in Fig. 2.48.

According to the sine law and Fig. 2.48, $\frac{a}{\sin(A)} = \frac{b}{\sin(B)} = \frac{c}{\sin(C)}$, where a, b, and c are the sides of a triangle, and the angles of the triangle are A, B, and C. Then, in triangle ABO, $AB = r$, and $\angle AOB = 180° - \theta_1 - \theta_2 - \frac{\phi_2}{2}$, where $\theta_2 = \frac{\pi}{2}$, and $\frac{AO}{\sin(\theta_2+\frac{\phi_2}{2})} = \frac{AB}{\sin(\angle AOB)}$, which can be written as $\frac{AO}{\cos(\frac{\phi_2}{2})} = \frac{r}{\cos(\theta_1+\frac{\phi_2}{2})}$. Therefore, $AO = \frac{r\cdot\cos(\frac{\phi_2}{2})}{\cos(\theta_1+\frac{\phi_2}{2})}$. For $\frac{AB'}{AO} = \cos(\theta_1)$,

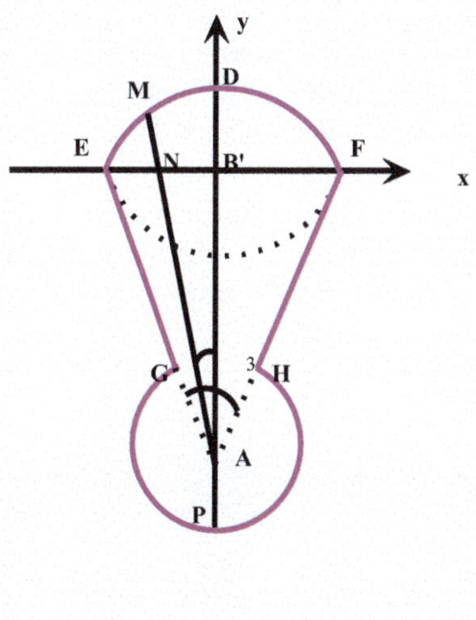

Fig. 2.48 Plan view of NLOS(B) coverage

Fig. 2.49 Stereo view of NLOS(B) coverage

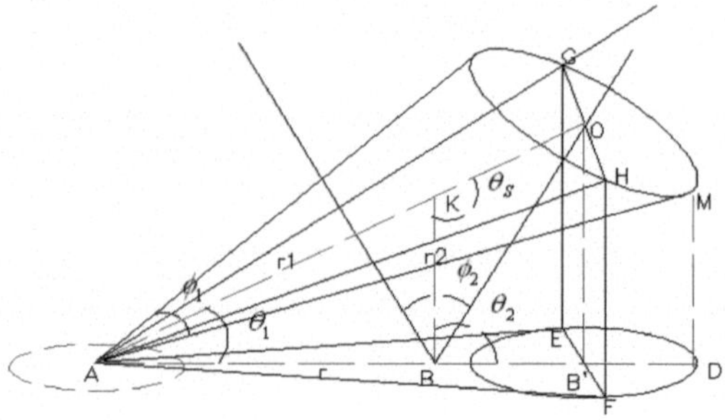

$$AB' = \frac{r \cdot \cos(\theta_1) \cdot \cos(\frac{\phi_2}{2})}{\cos(\theta_1 + \frac{\phi_2}{2})}.$$

$$\frac{AD}{AM} = \cos(\theta_1 - \frac{\phi_1}{2}),$$

$$AD = \frac{r \cdot \cos(\theta_1 - \frac{\phi_1}{2}) \cdot \cos(\frac{\phi_2}{2})}{\cos(\theta_1 + \frac{\phi_2}{2}) \cdot \cos(\frac{\phi_1}{2})},$$

$$B'D = AD - AB' = \frac{r \cdot \cos(\theta_1 - \frac{\phi_1}{2}) \cdot \cos(\frac{\phi_2}{2})}{\cos(\theta_1 + \frac{\phi_2}{2}) \cdot \cos(\frac{\phi_1}{2})} - \frac{r \cdot \cos(\theta_1) \cdot \cos(\frac{\phi_2}{2})}{\cos(\theta_1 + \frac{\phi_2}{2})}$$

$$= \frac{r \cdot \cos(\theta_1 - \frac{\phi_1}{2}) \cdot \cos(\frac{\phi_2}{2}) - r \cdot \cos(\theta_1) \cdot \cos(\frac{\phi_2}{2}) \cdot \cos(\frac{\phi_1}{2})}{\cos(\theta_1 + \frac{\phi_2}{2}) \cdot \cos(\frac{\phi_1}{2})}$$

$$B'D = \frac{r \cdot tg\frac{\phi_1}{2}}{ctg\theta_1 - tg\frac{\phi_2}{2}}. \tag{2.97}$$

Figure 2.50 shows the coverage on the ground centered on a cone with vertex A. Forward scattering is described by an electron-momentum distribution function (EMDF) forming an elliptical arc. For backward scattering, a circular arc is used for approximate correction. The radius of the circular arc is taken as the short semi-axis of the forward-scattering elliptical arc. The conditions for receiving the signal within the effective coverage area are as follows. For any point on the elliptic arc EMDF, the included angle with point A of the transmitting end is a. The equations for the elliptic arc EMDF and straight line are

$$\frac{x^2}{a^2} + \frac{y^2}{b^2} = 1,$$

$$a = EF/2 = tg(\frac{\phi_1}{2}).\frac{r \cdot \cos(\frac{\phi_2}{2})}{\cos(\theta_1 + \frac{\phi_2}{2})},$$

$$b = B'D = \frac{r \cdot tg(\frac{\phi_1}{2})}{ctg(\theta_1) - tg(\frac{\phi_2}{2})}.$$

$$\text{AM}: \quad k = -ctg\alpha \quad y = kx + d \quad d = -\frac{r \cdot \cos(\theta_1) \cdot \cos(\frac{\phi_2}{2})}{\cos(\theta_1 + \frac{\phi_2}{2})}$$

$$\frac{x^2}{a^2} + \frac{y^2}{b^2} = 1$$

$$\frac{x^2}{[tg(\frac{\phi_1}{2}) \cdot \frac{r \cdot \cos(\frac{\phi_2}{2})}{\cos(\theta_1 + \frac{\phi_2}{2})}]^2} + \frac{y^2}{[\frac{r \cdot tg(\frac{\phi_1}{2})}{ctg(\theta_1) - tg(\frac{\phi_2}{2})}]^2} = 1$$

$$\frac{x^2 \cdot \cos^2(\theta_1 + \frac{\phi_2}{2})}{r^2 \cdot tg^2(\frac{\phi_1}{2}) \cdot \cos^2(\frac{\phi_2}{2})} + \frac{y^2 \cdot [ctg(\theta_1) - tg(\frac{\phi_2}{2})]^2}{r^2 \cdot tg^2(\frac{\phi_1}{2})} = 1,$$

$$x^2 \cdot \cos^2(\theta_1 + \frac{\phi_2}{2}) + \cos^2(\frac{\phi_2}{2}) \cdot [ctg(\theta_1) - tg(\frac{\phi_2}{2})]^2 \cdot y^2 = r^2 \cdot tg^2(\frac{\phi_1}{2}) \cdot \cos^2(\frac{\phi_2}{2}),$$

$$y = -ctg(\alpha) \cdot x - \frac{r \cdot \cos(\theta_1) \cdot \cos(\frac{\phi_2}{2})}{\cos(\theta_1 + \frac{\phi_2}{2})},$$

$$x^2 \cdot \cos^2(\theta_1 + \frac{\phi_2}{2}) + \cos^2(\frac{\phi_2}{2}) \cdot [ctg(\theta_1) - tg(\frac{\phi_2}{2})]^2 \cdot (ctg(\alpha) \cdot x$$
$$+ \frac{r \cdot \cos(\theta_1) \cdot \cos(\frac{\phi_2}{2})}{\cos(\theta_1 + \frac{\phi_2}{2})})^2 = r^2 \cdot tg^2(\frac{\phi_1}{2}) \cdot \cos^2(\frac{\phi_2}{2})$$

$$x^2\{\cos^2(\theta_1 + \frac{\phi_2}{2}) + ctg^2(\alpha) \cdot \cos^2(\frac{\phi_2}{2}) \cdot [ctg(\theta_1) - tg(\frac{\phi_2}{2})]^2\}$$
$$+ x \cdot \frac{2r \cdot \cos(\theta_1) \cdot ctg(\alpha) \cdot \cos^3(\frac{\phi_2}{2}) \cdot [ctg(\theta_1) - tg(\frac{\phi_2}{2})]^2}{\cos(\theta_1 + \frac{\phi_2}{2})})$$
$$+ \frac{r^2 \cdot \cos^2(\frac{\phi_2}{2}) \cdot \{\cos^2(\theta_1) \cdot \cos^2(\frac{\phi_2}{2}) \cdot [ctg(\theta_1) - tg(\frac{\phi_2}{2})]^2 - tg^2(\frac{\phi_2}{2})\} \cdot \cos^2(\theta_1 + \frac{\phi_2}{2})}{\cos^2(\theta_1 + \frac{\phi_2}{2})}$$
$$= 0$$

Because M is in the second quadrant, the abscissa is negative, and the positive value is discarded:

$$x_0 = -r \cdot \cos\frac{\phi_2}{2}[\frac{\cos\theta_1 \cdot ctg\alpha \cdot \cos^2\frac{\phi_2}{2} \cdot (ctg\theta_1 - tg\frac{\phi_2}{2})^2}{\cos^3(\theta_1 + \frac{\phi_2}{2}) + \cos(\theta_1 + \frac{\phi_2}{2}) \cdot ctg^2\alpha \cdot \cos^2\frac{\phi_2}{2} \cdot (ctg\theta_1 - tg\frac{\phi_2}{2})^2}$$
$$+ \frac{\sqrt{\begin{array}{c} -\cos^2\frac{\phi_2}{2} \cdot \cos^2\theta_1 \cdot (ctg\theta_1 - tg\frac{\phi_2}{2})^2 \cdot \cos^2(\theta_1 + \frac{\phi_2}{2}) + tg^2\frac{\phi_1}{2}\cos^4(\theta_1 + \frac{\phi_2}{2}) \\ + tg^2\frac{\phi_1}{2} \cdot \cos^2\frac{\phi_2}{2} \cdot \cos^2(\theta_1 + \frac{\phi_2}{2}) \cdot ctg^2\alpha \cdot (ctg\theta_1 - tg\frac{\phi_2}{2})^2 \end{array}}}{\cos^3(\theta_1 + \frac{\phi_2}{2}) + \cos(\theta_1 + \frac{\phi_2}{2}) \cdot ctg^2\alpha \cdot \cos^2\frac{\phi_2}{2} \cdot (ctg\theta_1 - tg\frac{\phi_2}{2})^2}]$$

$$y_0 = -ctg\alpha \cdot x_0 - \frac{r \cdot \cos\theta_1 \cdot \cos\frac{\phi_2}{2}}{\cos(\theta_1 + \frac{\phi_2}{2})} \tag{2.98}$$

As the transmitter elevation angle increases, the forward-scattering arc gradually dips toward the transmitter, whereas the backward-scattering arc gradually expands

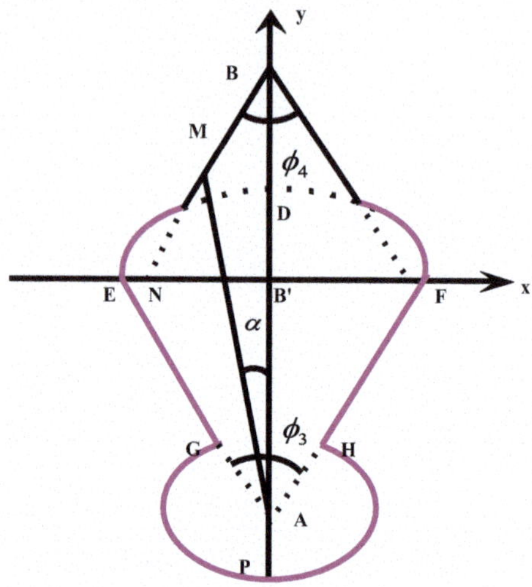

Fig. 2.50 Plan view of NLOS(C) coverage

as the elevation angle increases to 90°. For NLOS(A) communication, the coverage area is circular.

(3) NLOS(C) communication

In this mode, the elevation angles of the transmitter and receiver are both less than 90°. Figure 2.50 shows the projected coverage of UV communication on the ground. The communication effect is better than those of NLOS(A) and NLOS(B) communication. The need for NLOS communication gradually decreases in NLOS(B); that is, it becomes NLOS(C). Because of the coverage of NLOS(B), a triangular coverage area is added. The backscattering is very small and can be ignored.

$$|AM| = \frac{|Ax_0 + By_0 + C|}{\sqrt{A^2 + B^2}} = |\sin\frac{\phi_4}{2}| \cdot |\frac{r \cdot \cos\theta_1 \cdot \cos\frac{\phi_2}{2}}{\cos(\theta_1 + \frac{\phi_2}{2})} + \frac{r \cdot \sin\theta_1 \cdot \cos\theta_2}{\sin\theta_s}| \tag{2.99}$$

2.5.3.2 Coverage in LOS Mode

When the transmitter and receiver are essentially aligned, the LOS communication model is used for analysis, and the coverage area is diamond-shaped, as shown in Fig. 2.51. The effective coverage of LOS communication can also be analyzed as that of NLOS(C) communication. The LOS mode has higher communication capacity and a longer communication distance, but the omnidirectionality of UV communication is not available. The range is the distance from point A to the straight line BE, and the calculation formula is similar to that in (2.99).

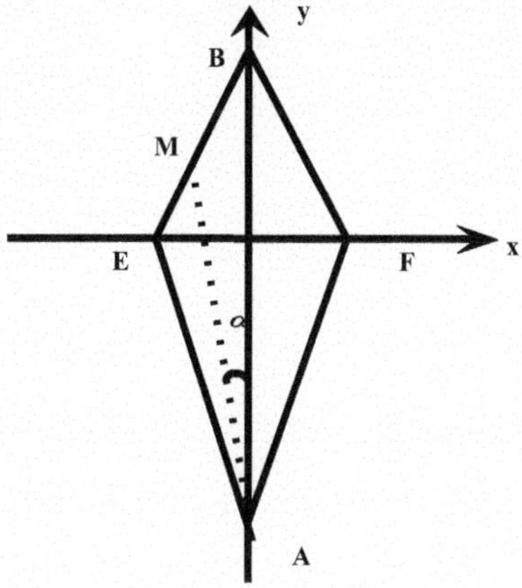

Fig. 2.51 Plan view of LOS coverage

In the communication process, many complex situations must be resolved. For example, the optical axes of the receiver and transmitter may not intersect, and the receiver and transmitter are oriented in the same direction.

2.5.4 Model and Simulation of UV NLOS Single-Scattering Link Considering Height

As shown in Fig. 2.52, they become θ_1' and θ_2' respectively. Therefore, the calculated received power does not correspond to R, but to a point on a circle with T as the center, *r* as the radius, and the same height as T, which is indicated as *R′* in Fig. 2.53. In this case, the received power is denoted as P_r'.

The parameters affecting the received power of receiver R should take the plane containing the line between T and R as the reference plane, and other parameters such as the transmit divergence angle and receiver field of view receive field angle remain unchanged.

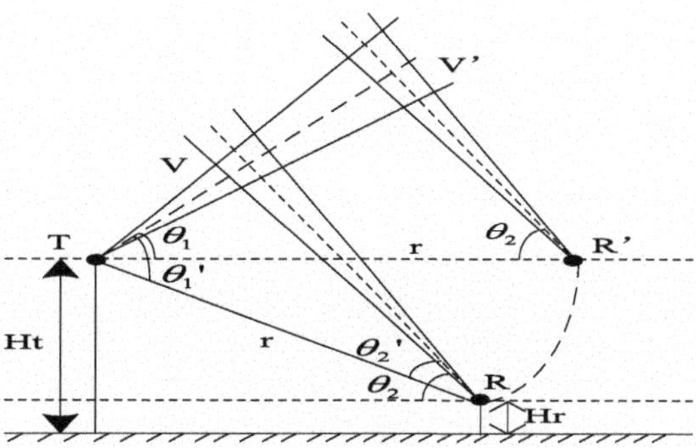

Fig. 2.52 UV light single scattering link model with height

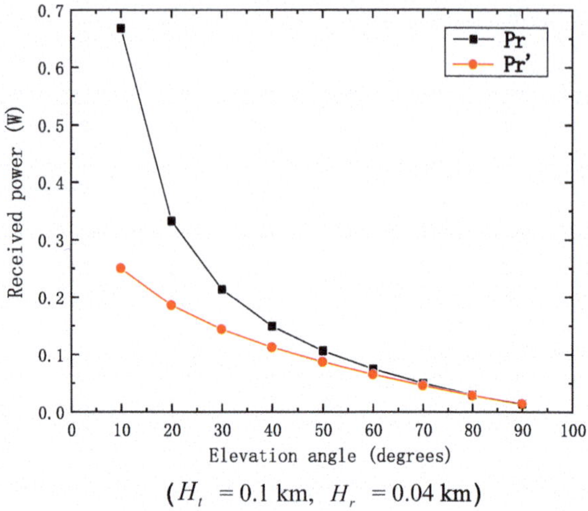

Fig. 2.53 Fixed received power comparison chart

$$\Delta\theta_1 = \arcsin(\frac{H_t - H_r}{r}) \tag{2.100}$$

$$\Delta\theta_2 = \Delta\theta_1 \tag{2.101}$$

$$\theta_1' = \theta_1 + \Delta\theta_1 = \theta_1 + \arcsin(\frac{H_t - H_r}{r}) \tag{2.102}$$

$$\theta_2' = \theta_2 - \Delta\theta_2 = \theta_2 - \arcsin(\frac{H_t - H_r}{r}) \tag{2.103}$$

From (2.102), (2.103), and (2.25), the received power of the UV NLOS single-scattering link depending on height can be obtained according to (2.63):

$$P_r' = \frac{P_t A_r K_s P_s \varphi_2 \varphi_1^2 \sin(\theta_1' + \theta_2')}{32\pi^3 r \sin\theta_1'(1 - \cos\frac{\varphi_1}{2})} e^{-\frac{K_e r(\sin\theta_1' + \sin\theta_2')}{\sin(\theta_1' + \theta_2')}} \tag{2.104}$$

For solar-blind UV light with a wavelength of 250 nm, the corrected received power P_r' at heights of $H_T = 0.1$ km, $H_R = 0.04$ km and $H_T = 0.2$ km, $H_R = 0.05$ km was obtained. Equation (2.25) was used to obtain P_r values for comparison; the results are shown in Figs. 2.53, 2.54, 2.55 and 2.56 (Table 2.4).

The simulation results for these height configurations reveal the following:

(1) The uncorrected and corrected algorithms [(2.25) and (2.63), respectively] clearly give different results. When the receiver elevation angle is fixed at a relatively large value, the transmitter elevation angle changes from small to large, and the difference between P_r and P_r' decreases gradually, because when both angles are large, the total received power does not change greatly.

Fig. 2.54 Fixed received power comparison chart

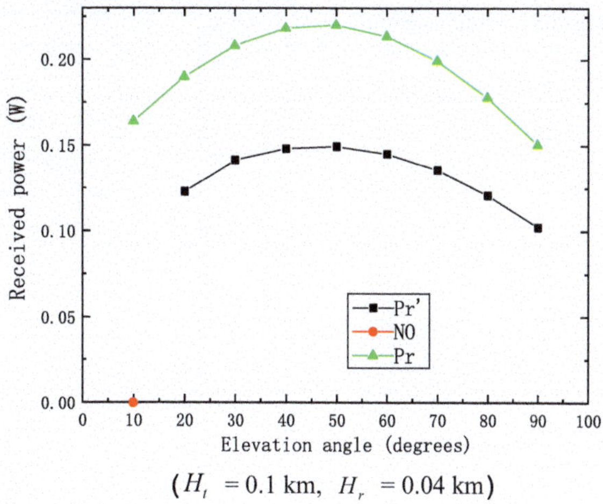

Fig. 2.55 Fixed received power comparison chart

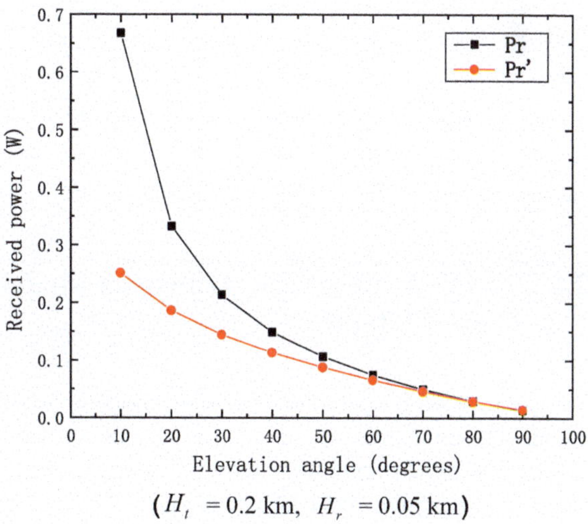

(2) Figures 2.53, 2.54, 2.55 and 2.56 show that overall, P_r is greater than P_r'. Therefore, when UV communication equipment is networked, each node should try to find a node with a similar height for forwarding during routing. If no such node can be found, the node should try to reduce the transmitter elevation angle during communication and increase it appropriately. The received power increases with increasing receiver elevation angle.

(3) When $H_t > H_r$ and $\theta_2 - \Delta\theta_2 < 0$, the two ends of the transceiver cannot form a common scatterer well in air, and thus they cannot communicate, as shown in Figs. 2.55 and 2.56. When $H_t < H_r$, the value of θ_1 becomes constrained. In practical applications, we should try to avoid these elevation angles on the basis of the height of the transceiver so that communication becomes possible.

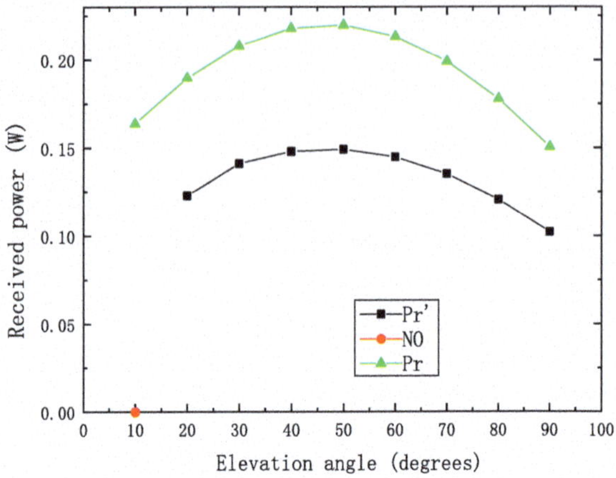

Fig. 2.56 Fixed received power comparison chart

(H_t = 0.2 km, H_r = 0.05 km)

Table 2.4 Scattering, absorption, and extinction coefficients of molecules and aerosols

λ (nm)	$K_{s,m}$	$K_{a,m}$	K_m	$K_{s,a}$	$K_{a,a}$	K_a	K_S	K_A	K_E
200	0.95	7.2	8.12	1.6	0.49	2.1	2.6	7.7	10.2
250	0.34	0.79	1.12	1.5	0.24	1.7	1.8	1.0	2.8
300	0.15	0.02	0.17	1.4	0.10	1.5	1.6	0.12	1.7

2.6 NLOS Atmospheric Transport Model Based on Monte Carlo Method

The atmospheric radiative transfer models LOWTRAN, MODTRAN, and FASCODE can numerically simulate and analyze the transmission characteristics of UV LOS scattering communication in various complex atmospheric environments. However, they cannot be used to simulate UV NLOS scattering communication [48]. The main methods used in the theoretical simulation of NLOS UV atmospheric transport are the single-scattering approximation, Monte Carlo, and discrete coordinate methods. Single-scattering simulation has the advantages of fast calculation speed and ease of use. However, for a complex atmosphere and long-distance transmission, because of the obvious effects of multiple scattering, the single-scattering simulation results have a large error. In the single-scattering simulation proposed by Van de Hulst, the approximate simulation condition of subscattering is that the product of the attenuation coefficient and transmission distance is less than 0.1 [49]; however, when the transmission distance exceeds 100 m, this condition can rarely be satisfied. The theory of the discrete coordinate method for scattering transmission is not mature; however, Monte Carlo methods can be used study light transmission in arbitrary environments and under multiple scattering conditions [61].

2.6.1 Monte Carlo Method

The Monte Carlo method is a computational method that differs greatly from general numerical calculation methods. It is based on probability and statistical theory, also known as random sampling or the statistical test method. For more than 50 years, with the development of science and technology and the invention of electronic computers, this method has been used as an independent method; it was first applied in the testing and development of nuclear weapons. Because the Monte Carlo method can describe the characteristics of materials and the physical experimental process more realistically and solve some problems that are difficult to solve by numerical methods, its application field is expanding [50].

2.6.2 Basic Concept

When the probability of a random event or the expected value of a random variable must be determined experimentally, the probability of the random event is estimated from the frequency of the event, or numerical characteristics of the variable are obtained to solve the problem. This is the basic concept of the Monte Carlo method.

When a random variable can be only 1 or 0, its mathematical expectation is the probability of an event [51]. Therefore, it can be said that the Monte Carlo method calculates the integral by means of a random experiment; that is, the integral to be calculated is regarded as the mathematical expectation of the random variable $g(r)$ obeying a certain distribution density function $f(r)$.

$$E(g(r)) = \int_{0}^{\infty} g(r)f(r)dr \tag{2.105}$$

The results of some types of experiments can be used as follows as an estimate of the integral (approximate value) [52]:

$$\overline{g}_N = \frac{1}{N}\sum_{i=1}^{N} g(r_i) \tag{2.106}$$

2.6.3 Convergence and Error of Monte Carlo Method

The convergence and error of the Monte Carlo method are important issues of general concern.

(1) Convergence

The Monte Carlo method calculates the arithmetic mean of a simple subsample $X_1, X_2, \ldots, X_N$ of the random variable X:

$$\overline{X}_N = \frac{1}{N}\sum_{i=1}^{N} X_i \tag{2.107}$$

as an approximation to the solution:

$$P\left(\lim_{N\to\infty} \overline{X}_N = E(X)\right) = 1 \tag{2.108}$$

(2) Error

Regarding the error between the approximate value obtained by the Monte Carlo method and the true value, the central limit theorem states that if the random variable sequence $X_1, X_2, \ldots, X_N$ is independent and identically distributed and has a finite nonzero variance, then

$$0 \neq \sigma^2 = \int (x - E(X))^2 f(x)dx < \infty \tag{2.109}$$

However,

$$\lim_{N\to\infty} P\left(\frac{\sqrt{N}}{\sigma}\left|\overline{X}_N - E(X)\right| < x\right) = \frac{1}{\sqrt{2\pi}}\int_{-x}^{x} e^{-t^2/2}dt \tag{2.110}$$

When N is sufficiently large, the following approximations can be used:

$$P\left(\left|\overline{X}_N - E(X)\right| < \frac{\lambda_\alpha \sigma}{\sqrt{N}}\right) \approx \frac{2}{\sqrt{2\pi}}\int_{0}^{\lambda_\alpha} e^{-t^2/2}dt = 1 - \alpha \tag{2.111}$$

In general, the error ε of the Monte Carlo method is defined as

$$\varepsilon = \frac{\lambda_\alpha \sigma}{\sqrt{N}} \tag{2.112}$$

After determining the confidence level according to the requirements of the problem, check the standard normal distribution table to determine λα [63, 64]. Several commonly used α and λ_α values are given in Table 2.5.

Table 2.5 Commonly used α and λ_α values [50]

α	0.5	0.05	0.003
λ_a	0.6745	1.96	3

2.6.4 Features of Monte Carlo Method

The Monte Carlo method has the following advantages:

(1) It can realistically describe the characteristics of random objects and a physical experimental process. Consequently, the Monte Carlo method can partially replace physical experiments and can even be used to obtain results that are difficult to obtain experimentally. The use of the Monte Carlo method to solve practical problems can start directly from the problem itself, not from equations or mathematical expressions. It supports the use of intuition and images.

(2) It is limited by the geometric conditions.
The integral over any region D_s in s-dimensional space is calculated:

$$g = \int \underset{D_s}{\cdots} \int g(x_1, x_2, \cdots, x_s)dx_1dx_2 \cdots dx_s \tag{2.113}$$

The integral is approximated as

$$\overline{g}_N = \frac{V_s}{N}\sum_{i=1}^{N} g(x_1^{(i)}, x_2^{(i)}, \cdots, x_s^{(i)}) \tag{2.114}$$

In addition, in problems that involve random properties, for example, when the shape of the system under consideration is very complex, it is difficult to solve the problem by general numerical methods. However, in principle there is no difficulty when the Monte Carlo method is used.

(3) The convergence rate is independent of the dimension of the problem.
According to the central limit theorem, for a given confidence level, the convergence order of the Monte Carlo method is independent of the dimension of the problem [53]. The dimension affects only the sampling time and the calculation time of the estimator; it does not affect the error. That is, when the Monte Carlo method is used, the total number of samples N is independent of the dimension s. Consequently, the Monte Carlo method is suitable for solving multidimensional problems. In general numerical methods, such as the calculation of definite integrals, the calculation time increases with the power of the dimension, and because the number of points is proportional to the power of the dimension, more computer memory is needed. This is an insurmountable problem in the computation of high-dimensional integrals by general numerical methods.

(4) It can calculate multiple schemes and multiple unknowns simultaneously.
For problems where multiple schemes must be calculated, the Monte Carlo method sometimes does not need to make calculations sequentially as in conventional methods but can calculate all the schemes simultaneously, and the total number of calculations is almost equal to that required for one scheme. In addition, by using the Monte Carlo method, several required quantities can be obtained simultaneously. For example, in a simulation of particles, the flux, energy spectrum, angular distribution, etc. in different regions can be obtained simultaneously, in contrast to conventional methods, where they must be calculated sequentially.

(5) The error is easy to determine.
For the Monte Carlo method, in contrast to the general calculation method, the difference between the calculation result and true value is easily determined. According to the error formula of the Monte Carlo method, the error can be calculated at the same time as the desired quantity. It is also easy to determine the computational problems of the Monte Carlo method, which are very complicated.
Common calculation methods often have the problem of loss of significant digits, and it is sometimes difficult to solve this problem, but Monte Carlo methods do not have this problem.

(6) The program structure is simple and easy to implement.
When the Monte Carlo method is implemented on a computer, the program structure is simple, the block is strong, and it is easy to implement.

The disadvantages of the Monte Carlo method are as follows:

(1) The convergence speed is slow.

As mentioned above, the convergence rate of the Monte Carlo method is low, and it is generally not easy to obtain approximate results with high accuracy. For problems with few (less than three) dimensions, it is not as good as other methods.

(2) The error is probabilistic.

Because the error of the Monte Carlo method is estimated at a certain confidence level, its error is probabilistic rather than absolute.

(3) In particle transport problems, the calculation result is related to the size of the system.

Experience shows that the results calculated by the Monte Carlo method are satisfactory only when the size of the system is comparable to the mean free path of particles (usually approximately 10 mean free paths). However, for the calculation of large systems or low-probability events, the calculation results are often lower than the true values. For large systems, numerical methods are preferable [52, 53].

2.6.5 Monte Carlo Simulation of NLOS UV Transmission

For UV communication in complex geometries and inhomogeneous media, the Monte Carlo method is generally used to study the problem of scattering transmission. The main parameters considered in the NLOS UV transport model based on the Monte Carlo method include the photon beam emitted by the light source, NLOS transport of photons in the atmosphere, and statistical results.

2.6.5.1 Photon Initialization

Photons are initialized mainly to determine the emission point and direction of photons emitted by the light source assuming the light source is a point light source and to limit the emission of photons by the light source within a certain solid angle of the field of view. A plan view of NLOS UV transmission is shown in Fig. 2.57. Figure 2.58 shows a perspective view of NLOS UV transmission, where the central axis of the point light source is in the y–z plane.

The azimuth [54] is given by

$$\psi = 2\pi\xi^{(\psi)} \tag{2.115}$$

and can be directly sampled in the simulation.

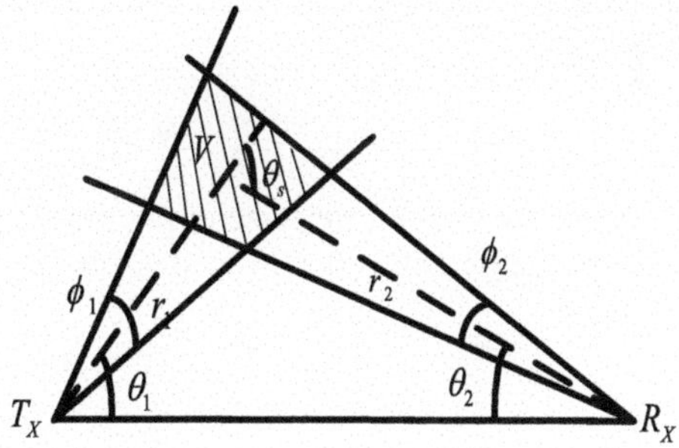

Fig. 2.57 NLOS scattered light link analysis [31]

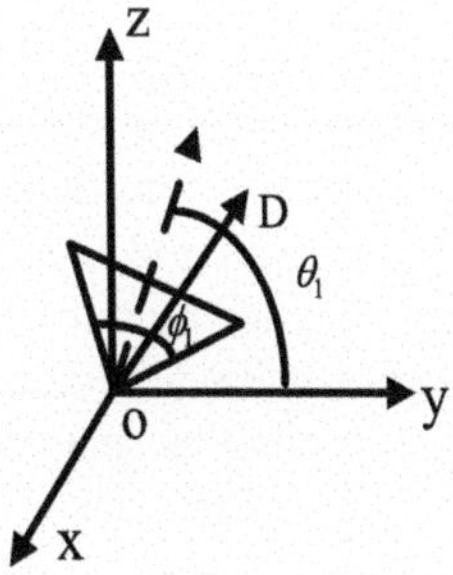

Fig. 2.58 Emission solid angle of light source

$$\cos(\theta) = 1 - \xi^{(\theta)}(1 - \cos(\phi_1/2)) \tag{2.116}$$

Then the direction cosine is expressed as [54]

$$\begin{aligned} u_x &= \sin(\theta)\cos(\psi) \\ u_y &= \sin(\theta)\sin(\psi) \\ u_z &= \cos(\theta) \end{aligned} \tag{2.117}$$

2.6.5.2 NLOS Transport of Photons in the Atmosphere

The Monte Carlo method was used to simulate the NLOS transport of photons in the atmosphere, with a focus on the interaction of photons with particles and obstacles in the atmosphere. Atmospheric molecules, aerosols, and suspended particles in the atmosphere absorb and scatter photons, mainly by determining their next collision point, transmission direction after collision, and weight statistics.

We determine the initial value of the emitted photon, and we know the cosine of the photon's emission direction. Next, we determine the random step size of photon transmission and coordinate position at the next moment.

By using the Bouguer–Lambert law and Monte Carlo method, the random step size is sampled to obtain

$$\Delta s = -\ln \xi^{(s)}/k_s \tag{2.118}$$

When the random step size of photon transmission is determined, the initial state value of the photon is determined according to the light source:

$$\begin{aligned} x' &= x + u_x \Delta s \\ y' &= y + u_y \Delta s \\ z' &= z + u_z \Delta s \end{aligned} \tag{2.119}$$

The next step in the NLOS transmission process is to determine the scattering process. Generally, only Rayleigh scattering by molecules and Mie scattering by aerosols are considered in the simulation. Photons are scattered by particles in the atmosphere, and the direction of scattering is determined by the scattering phase function and random numbers. The scattering of photons by atmospheric molecules occurs by Rayleigh scattering, and the cosine function of the deflection angle obtained by sampling its scattering phase function is [55]

$$\cos(\theta) = \sqrt[3]{(4\xi - 2) - \sqrt{(2 - 4\xi)^2 + 1}} + \sqrt[3]{(4\xi^{(r)} - 2) + \sqrt{(2 - 4\xi)^2 + 1}} \tag{2.120}$$

The scattering of UV photons by aerosols in the atmosphere is a typical instance of Mie scattering, and the cosine function of the deflection angle obtained by sampling its scattering phase function is [56]

$$\cos(\theta) = [1 + g^2 - (1 - g^2)^2/(1 - g + 2g\xi^{(m)})^2]/2g \tag{2.121}$$

After scattering, the photon transmission direction is different, and the new direction cosine is expressed as [55]

$$u'_x = \sin(\theta)(u_x u_y \cos(\psi) - u_y \sin(\psi))/\sqrt{1 - u_z^2} + u_x \cos(\theta) \tag{2.122}$$

$$u'_y = \sin(\theta)(u_y u_z \cos(\psi) - u_x \sin(\psi))/\sqrt{1 - u_z^2} + u_y \cos(\theta)$$

$$u'_z = -\sin(\theta)\cos(\psi)\sqrt{1 - u_z^2} + u_z \cos(\theta)$$

where u_x,u_y, u_z is the direction cosine of the exit point; $\sin(\theta)$, $\cos(\theta)$ is the sine–cosine function of the deflection angle $\sin(\psi)$; and $\cos(\psi)$ is the sine–cosine function of the azimuthal angle.

2.6.5.3 Termination of Photon Transmission

The above process is a single-scattering process; when it is repeated, multiple scattering occurs. We must also take into account the probability of attenuation due to absorption and scattering throughout the simulation.

Next, we describe the probability of detection when the photon is scattered. The angle of the detected photon must be determined. Figures 2.59 and 2.60 show simulated single and double scattering, respectively, in the same coordinate system.

The probability that the photon is scattered toward the detection surface after the nth scattering is

$$p_{1n} = \frac{1}{4\pi} \int [s_r p_r(\theta) + s_m p_m(\theta)] d\Omega \tag{2.123}$$

After the nth scattering, the probability that the photon reaches the detection surface without extinction is

$$p_{2n} = e^{-k_e |r_n - r'|} \tag{2.124}$$

Fig. 2.59 Single scattering

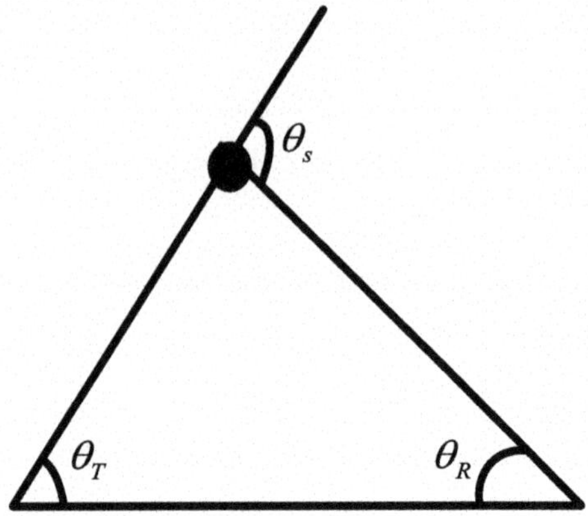

Fig. 2.60 Double scattering

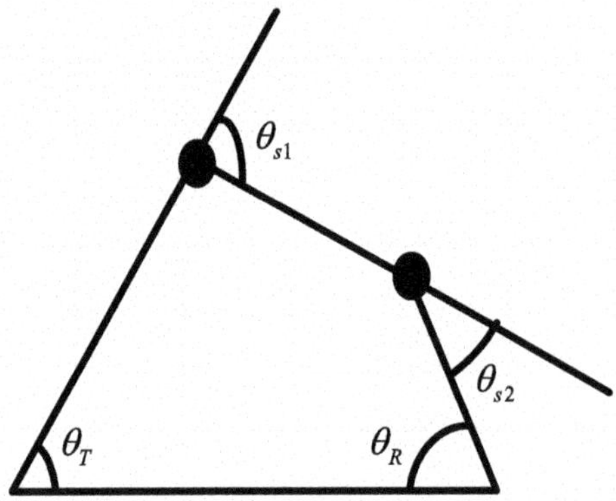

The weight of the photon after the *n*th scattering is

$$w_n = (1 - p_{1n})e^{-k_a|r_n - r_{n-1}|}w_{n-1} \tag{2.125}$$

In summary, the probability that the detector receives the photon is [55, 56]

$$P = \sum_{n=1}^{\infty} w_n p_{1n} p_{2n} \tag{2.126}$$

The Monte Carlo method is used to simulate the outgoing direction of photons, random step size of photon transmission, scattering type and scattering phase angle of photons, and probability of a photon entering the detector. After a certain number of photons are simulated, the probability of detection in different time regions is accumulated.

2.6.6 Statistical Results and Analysis

We used MATLAB to write the program for the Monte Carlo simulation. A flow chart of the main functions is shown in Fig. 2.61, and the key program code is given in Appendix A.

The Monte Carlo method is used to simulate and analyze the path loss of single-scattering and multiple-scattering UV optical links with transmission distances ranging from 20 to 100 m.

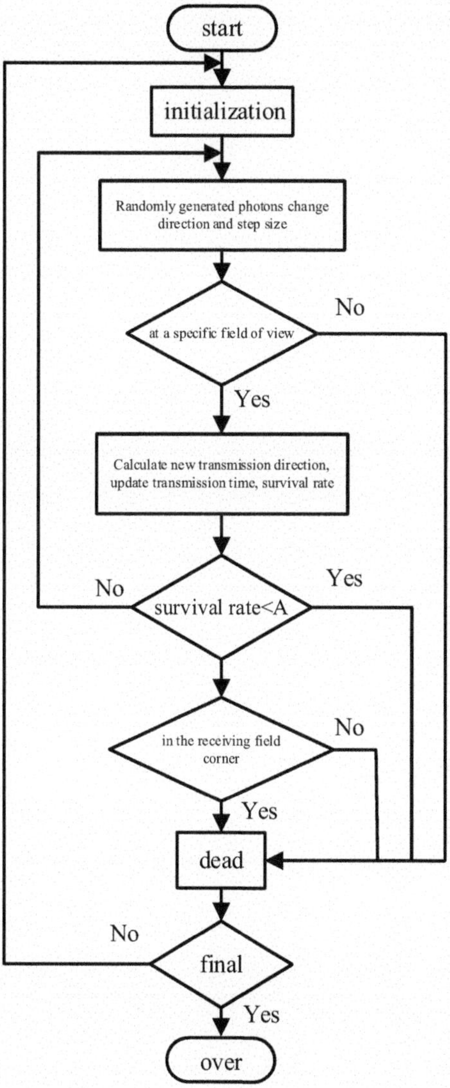

Fig. 2.61 Flowchart of main functions in Monte Carlo simulation

In Figs. 2.62 and 2.63, we compare the single and multiple transmission losses using two sets of transmitter and receiver elevation angles. It has been proposed, on the basis of experimental data, that the analysis of UV transmission using multiple scattering is more accurate than that using single scattering and can better provide a theoretical basis [57]. Our analysis shows that there is a discrepancy between the path losses for single and multiple scattering, and the data can provide a reference for system design.

Figure 2.64 shows the path loss of the system versus transmission distance for transmitter elevation angles of 20°, 45°, and 90°. The emission field of view of the transmitter is 17°, the elevation angle of the receiver is 20°, and the field of view of the receiver is 30°. Under these conditions, increasing the transmitter elevation angle significantly increases the path loss. The reason is that increasing the transmitter elevation angle increases the scattering angle of light scattered from the effective scatterer. According to scattering theory, the forward scattering is strongest, and the scattering energy decreases as the deviation of the angle from the transmission direction increases.

Figure 2.65 shows the path loss of the system versus transmission distance for receiver elevation angles of 20°, 45°, and 90°. The field of view of the transmitter is 17°, the elevation angle is 20°, and the field of view of the receiver is 30°. Under these

Fig. 2.62 Path loss for $(\theta_1, \theta_2) = (20°, 20°)$

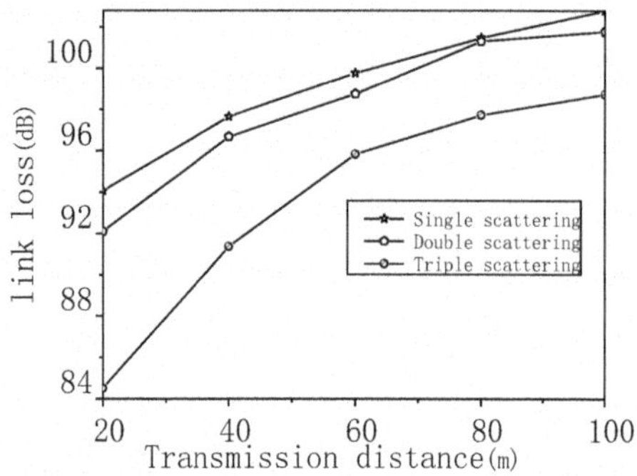

Fig. 2.63 Path loss for $(\theta_1, \theta_2) = (40°, 15°)$

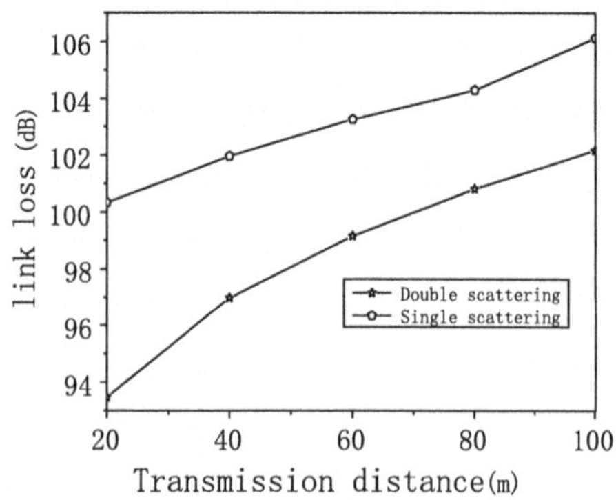

Fig. 2.64 Path loss at different transmitter elevation angles

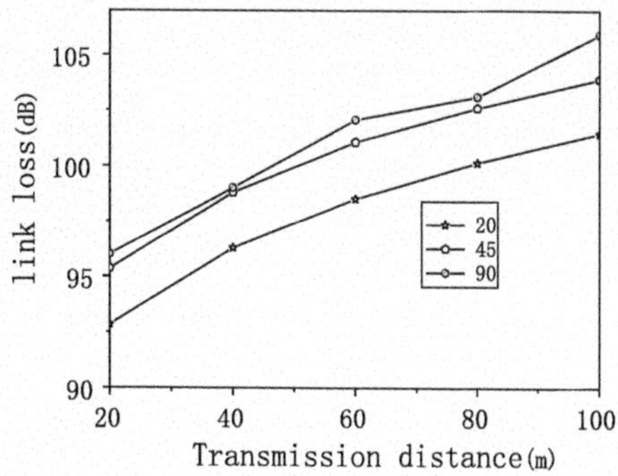

conditions, increasing the receiver elevation angle increases the scattering angle of light scattered from the scatterer, thus increasing the path loss of the system.

Figure 2.66 shows the path loss of the system versus transmission distance for transmitter fields of view of 10°, 15°, and 30°. The transmitter elevation angle is 20°, the receiver elevation angle is 20°, and the field of view of the receiver is 30°. With increasing transmission distance, the path transmission loss increases sharply. The reason is that as the transmission distance increases, the transmission loss of light in the atmosphere increases. In addition, for certain values of the other parameters, increasing the divergence angle of the transmitter has little effect on the path loss. The reason is that increasing the transmitter divergence angle increases the common area between the transmitter and receiver but also reduces the unit power density of the scatterer.

Figure 2.67 shows the path loss of the system versus transmission distance when the field of view of the receiver is 10°, 30°, and 50°. The field of view and elevation angle of the transmitter are 17° and 20°, respectively, and the elevation angle of the receiver is 20°. Under these conditions, increasing the field of view of the receiver clearly reduces the path loss of the system. However, in practical systems, the field of view of the receiver is limited by the detector and optical antenna.

The link loss of the channel is reportedly very large and increases with increasing transmitter and receiver elevation angles [54, 62]; the transmission pulse is broadened

Fig. 2.65 Path loss at different receiver elevation angles

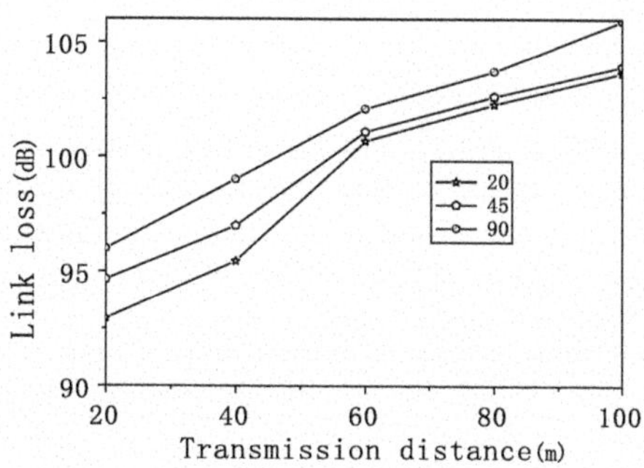

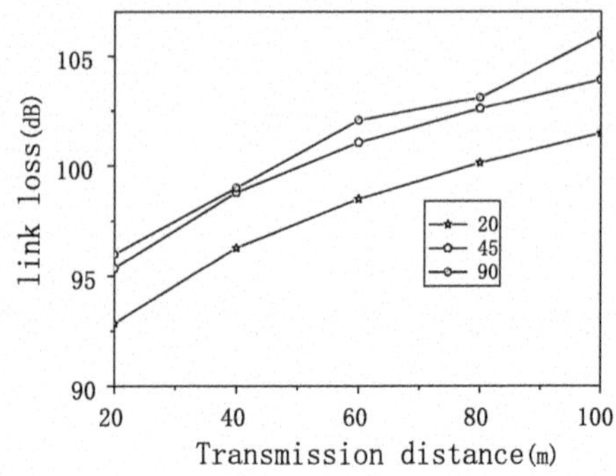

Fig. 2.66 Path loss for various transmitter fields of view

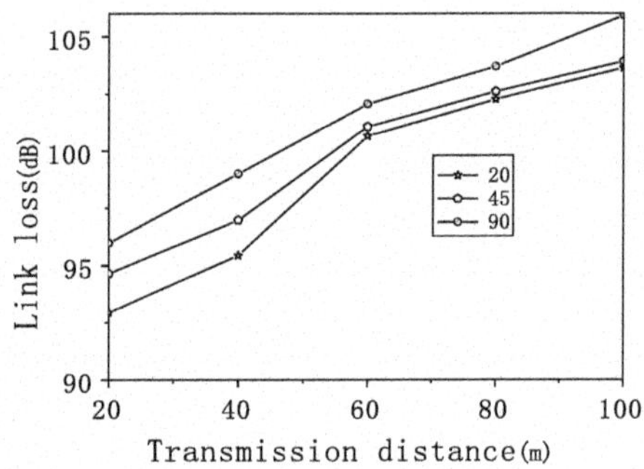

Fig. 2.67 Path loss for various receiver fields of view

after transmission through the NLOS scattering channel, which limits the channel bandwidth of the communication system and may cause intersymbol interference [54]. A Monte Carlo simulation was used to calculate the probability of reception at the receiving end. The statistical results show that the coverage of photons emitted according to the scattering model is consistent with our geometric analysis in the previous section, as shown in Figs. 2.68 and 2.69 for NLOS(B) and NLOS(C) communication, respectively.

To obtain the results shown in Figs. 2.68 and 2.69, we used 1015 emitted photons to calculate the reception probability of NLOS(B) and NLOS(C) communication, respectively, at transmission distances of 20–1000 m. Because the ground coverage, which is equivalent to the projection of the x–y surface, is incorporated into the geometric model, we count the probability of reception from 0° to 180° with respect to the x axis. As shown in the figures, for transmission distances of 60 and 80 m, the reception probability is relatively unstable; at other transmission distances, the reception probability is essentially stable. The simulation results further validate the coverage model and ensure that the model provides a reliable network for the upper layer.

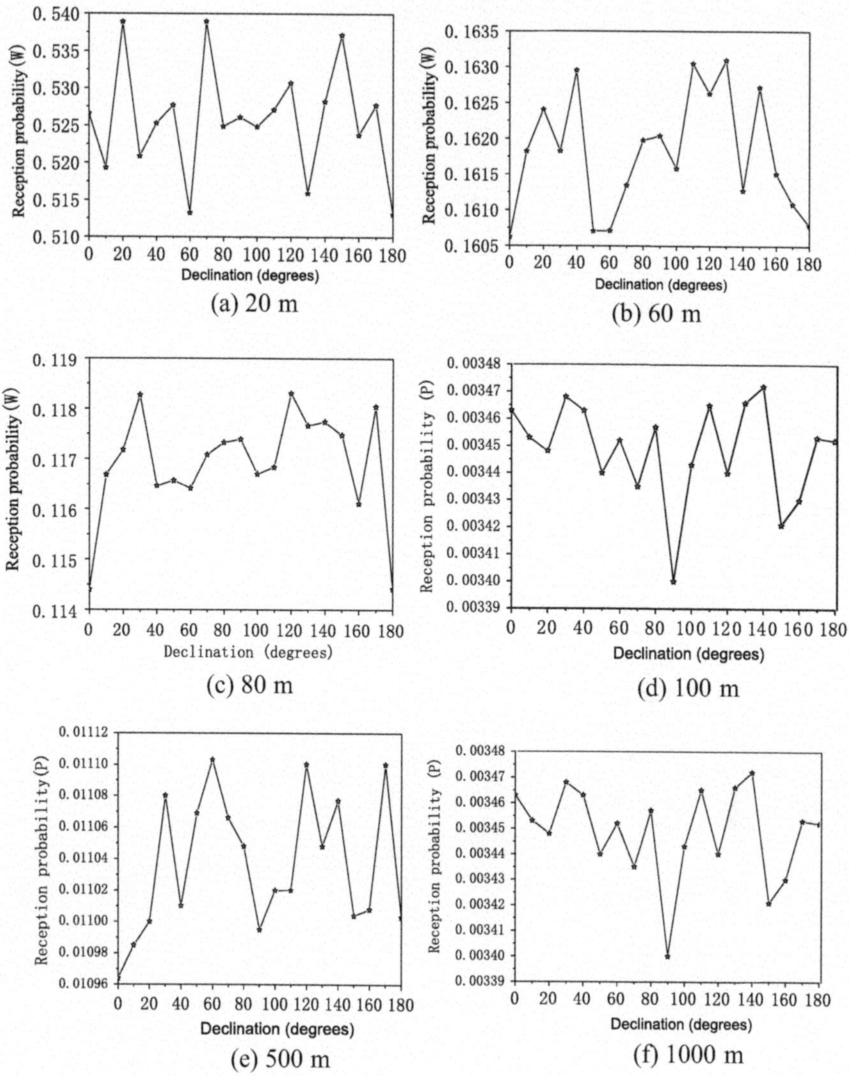

Fig. 2.68 Reception probability for NLOS(B) communication at different transmission distances

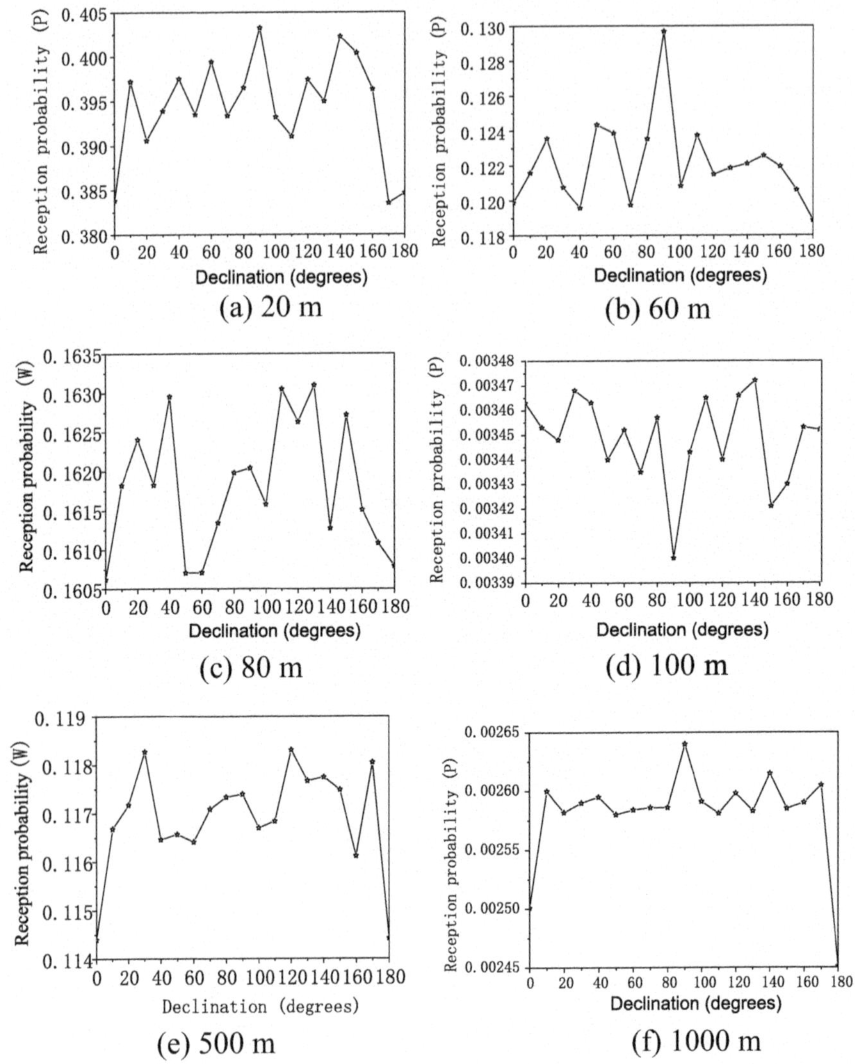

Fig. 2.69 Reception probability of NLOS(C) communication at different transmission distances

References

1. Kahn JM, Barry JR (1997) Wireless infrared communication. Proc IEEE 2(17):265–298
2. Gfeller FR, Bapst UH (1979) Wireless in-house data communication via diffuse infrared radiation. Proc IEEE 12(67):1474–1486
3. Kotzin MD (1981) Short-range communications using diffusely scattered infrared radiation. North western University 6(20):1–10
4. Chang Sl, Yang JK, Yang JC et al (2004) The experimental research of UV communication. Proc SPIE 5284:344–348

5. Huffman RE (1992) Atmospheric ultraviolet remote sensing. Academic Press, Boston
6. David MR, Daniel TM, John AM (2004) Unique properties of solar blind ultraviolet communication systems for unattended ground sensor networks. Proc SPIE 5611:244–254
7. Liu X, Ju M (2002) A preliminary study on ultraviolet light communication and countermeasures. Appl Optoelectron Technol 5(20):8–9
8. Wu J, Yang C, Liu J (2006) Optical Transport theory in atmosphere. Beijing University of Posts and Telecommunications Press, Beijing
9. http://baike.baidu.com/view/287541.htm?fr=ala0_1_1
10. Xu Z, Shen L, Wang J et al (2009) Ultraviolet scattering communication experimental system and its performance analysis. Southeast University 6(39):1087–1092
11. Linxing C, Xi Z, Yi C (2006) Mobile ad hoc network—ad hoc wireless network technology. Electronic Industry Press, Beijing
12. Lavigne C, Durand G, Roblin A (2006) Ultraviolet light propagation under low visibility atmospheric conditions and its application to aircraft landing aid. Appl Opt 9140–9150
13. Junhong C, Xiaoli Y (2008) Research of the atmospheric factors of solar blind ultraviolet communication. Laser Mag 4:38–39
14. Gary AS, Melissa N, Mrinal I et al (2000) NLOS UV Communication for distributed sensor systems. Proc SPIE 4126:83–96
15. Patterson EM, Gillespie JB (1989) Simplified ultraviolet and visible wavelength atmospheric propagation model. Appl Opt 28(3):425–429
16. Li X (2007) Principle and technology of satellite-to-ground laser communication link. National Defense Industry Press, Beijing
17. Zhu M, Zhang H, Jia H et al (2007) Research on the phase function of ultraviolet light scattering based on Mie scattering theory. J Light Scatt 3:225–228
18. Yubao C, Xiwei Y (2005) Calculation method of Mie scattering factor and its application. Optoelectron Technol Appl 20(5):12–14
19. Ye Y, Zhenxi Z, Dazong J (1997) Numerical calculation of Mie scattering physical quantity. Appl Opt 18(4):17–19
20. Yuan Y, Ren D, Hu X (2006) Mie theory recursive formula to calculate scattering phase function. J Light Scatt 17(4):266–371
21. Wiscombe W J. Mie Scattering Calculations:advances in Technique and Fast, Vector 2speed Computer codes//National center for Atmospheric Research, Boulder, 1979: 177–180
22. Lin H (2007) Mie scattering and Brillouin scattering properties of ocean suspended particles. Doctoral dissertation of Huazhong University of Science and Technology, Wuhan
23. Li H, Qin S, Hu X et al (2008) Analysis of Mie scattering in 1.06um laser simulation test. J Nat Univ Defense Technol 30(3):5–10
24. Shiying Z (2008) Study on light scattering properties of graphite microsphere particles based on Mie's scattering theory. Infrared Laser. 29(11):8–11
25. Xu Z (2007) Research on modulation and coding of free space ultraviolet speech communication. Chongqing University Master's Thesis, Chongqing
26. Wu L, Ying J, Geng B (2008) Influence of atmospheric turbulence on laser transmission. Infrared Laser 3(10):578–779
27. Qian X, Zhu W et al (2008) Simulation research on atmospheric transmission characteristics of ultraviolet communication. Acta Photonica Sinica 4(10):789–790
28. Lingyu W, Lijuan J (2002) Influence of light intensity scintillation in turbulent atmosphere on optical communication link. Opt Commun Technol 2(6):17–20
29. Li J, Koni Q (2005) Application of ultraviolet optical communication in military communication system. Opt Photoelectr Technol 3(4):19–21
30. Neer ME, Schlupf JM, Fishburne ES et al (1979) The development and testing of an UV voice communication system. Nav Electron Syst Command 12(7):125–126
31. Zachor AS (1978) Aureole radiance field about a source in a scattering-absorbing medium. Appl Opt 12(17):1911–1922
32. Xu Z (2007) Approximate performance analysis of wireless ultraviolet links. Proc IEEE 4(15):15–20

33. Shaw GA, Siegel AM, Mode J (2005) Recent progress in short-range ultraviolet communication. Proc IEEE 6(2):214–225
34. Lan L, Ni G (2003) Simulation of atmospheric transmission characteristics for ultraviolet communication. J Beijing Inst Technol 4(23):19–23
35. Tang Y, Ni G, Zhang L et al (2007) Research on single scattering transmission model of non-direct-viewing ultraviolet light communication. Opt Technol 5(33):759–760
36. Mark RL, Jeffrey HS, David MR (1991) Non-line-of-sight single-scatter propagation model. J Opt Soc Am 8(12):1964–1972
37. Ding H, Chen G, Majumdar AK, Xu Z (2008) A parametric single scattering channel model for non-line-of-sight ultraviolet communications. Proc SPIE 19(8):1–6
38. Shute RD (1995) Electrodeless ultraviolet communications system. Proc IEEE 10(11):1–4
39. Xu S (1996) Application of Monte Carlo method in experimental nuclear physics. Atomic Energy Press, Beijing
40. Wang J, Luo T, Dai M, Tian Y (2009) UV nlos communications atmospheric channel model and its performance analysis. Proc IEEE 9(31):85–88
41. Jia H, Chang S, Yang J et al (2007) Monte Carlo simulation of atmospheric transmission characteristics of non-line-of-sight ultraviolet communication. Acta Photonica Sinica 5(36):955–957
42. Lei J, Zhang Z, Liu B et al (2005) Application of Monte Carlo method in polymer science. Polym Bull 6(21):2–124
43. Feng Y (2011) Research on atmospheric channels of ultraviolet optical communication. Master's Thesis of Xi'an University of Technology, Xi'an
44. Zhao T, Feng Y, Ke X et al (2010) Research on node coverage in "solar blind" ultraviolet optical communication network. Optics 30(8):2229–2235
45. Guidelines on limits of exposure to ultraviolet radiation of wavelengths between 180 nm and 400 nm (Incoherent Optical Radiation). Health Physics 2:171–186 (2004)
46. Zhengzheng S, Liquan Z, Shengli C (2006) Research status of non-line-of-sight optical transmission model. Infrared Laser Eng 36(Supplement):496–497
47. Tang Y, Ni G, Lan L et al (2007) Simulation calculation of transmission distance of solar-blind ultraviolet communication system. Opt Technol 33(1):27–30
48. Light HV, Particles SBS (1957) The first edition. Wiley, New York, pp 107–185
49. Wang P, Zhou J, Yang Y et al (2004) Multi-particle Monte Carlo study on high-field transport properties of 6H-SiC. Acta Photonica Sinica 3(15):322–325
50. Chen G, Abou-Galala F, Xu Z, Sadler BM (2008) Experimental evaluation of LED-based solar blind NLOS communication links. Opt Express 16(19):15059–15068
51. Kahn H (1960) On thermonuclear war. Free Press, New York
52. Xu Z (1985) Monte Carlo method. Shanghai Science and Technology Press, Shanghai
53. http://wenku.baidu.com/view/9fb6d9c789eb172ded63b799.html
54. Xu Z, Chen G, Abou-Galala F, Leonardi M (2007) Experimental performance evaluation of non-line-of-sight ultraviolet communication system. Proc SPIE 16(19):1–12
55. Tumei S (1984) (trans by Wang M, Wang G et al) Atmospheric aerosols. Science Press, Beijing
56. Shao Z, Chang S, Lan Y et al (2006). Numerical simulation of the transmission time characteristics of non-line-of-sight ultraviolet signals. Opt Optoelectron Technol 6(4):18–19
57. Feng T, Xiong F, Ye Q et al (2007) Non-line-of-sight optical scattering communication based on solar-blind ultraviolet light. Proc SPIE 10(12):1–5
58. Lindner M, Elstein S, Panitkova E, et al. UV devices for solar blinol & background limited detection. Proceedings of The International Society for Optical Engineering, 10(2):527–534
59. Yin Zhi-guo, HE Jun, HUANG Chang-chun, DENG Dai-zu, et al (2005) Experiment Analysis of Influence of Atmosphere Stochastic Channel on Laser Transmission. J Appl Opt, 26(4):33–35
60. Tang Y, Ni GQ, et al (2007) Study of channel character of solar blind UV communication. Proceedings of SPIE, 10(12):1–8
61. Elkamchouchi HM, El-Shimy (2006) Monte Carlo simulation of light propagation through the troposphere for free space optical communication. Radio Science Conference National

62. Ding H, Chen G, Majumdar AK, Sadler BM Xu Z (2009) Modeling of non-line-of-sight ultraviolet scattering channels for communication. Proc IEEE 9(27):1535–1538
63. Pearson ES, Hartley HO (1966) Biometrika tables for statistians. London
64. Yunru X (1994) Probability and statistics calculation methods. Nankai University Press, Tianjin

Chapter 3
Node Location Algorithm for Ultraviolet Communication Network

Using ultraviolet (UV) communication not only avoids the expenditure of cost and time to lay cable for wired communication, but also makes wireless communication less insecure and greatly reduces the time required to set up and dismantle communication equipment and lines. The design of UV non-line-of-sight (NLOS) communication systems must consider the physical characteristics of the communication channel as well as transmitter and receiver, digital communication, and network technology [1]. This chapter discusses the node location algorithm used in UV communication networks.

3.1 UV Communication Networks

The use of solar-blind UV light as the transmission medium in a communication network can avoid the effects of obstacles and increase the transmission distance.

3.1.1 Overview of Wireless Communication Networks

Solar-blind UV light is suitable only for short-distance communication because of strong absorption by molecules and particles in the near-Earth atmosphere and the characteristics of the transmission power and scattering channel. To expand the coverage of UV communication and make it suitable for more extensive application, we can form networks with different topological structure depending on user needs. Three topological structures are typically used for UV wireless optical networks: ring, star, and mesh networks, as shown in Fig. 3.1. The choice depends on user needs and specific applications.

X. Ke, *UV Light Self Organizing Network Theory*, Optical Wireless Communication Theory and Technology, https://doi.org/10.1007/978-981-97-7854-6_3

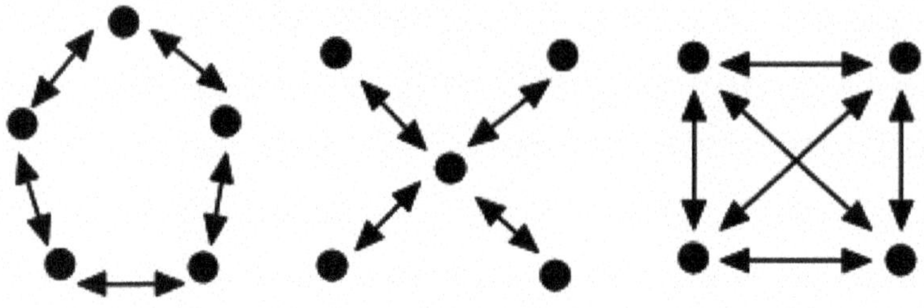

Fig. 3.1 UV optical network topologies

The transmission medium in the ring network extends from one end user to another until all end users are connected in a ring, as shown in Fig. 3.1a. Because the structure is circular, each end user is connected with two adjacent end users. The one-way point-to-point links in the wireless optical system can therefore be used to provide two-way transmission of information. This structure clearly eliminates reliance on a central system for end-user communication. However, any message transmitted on the ring must pass through all endpoints, and if a point on the ring breaks, communication between all endpoints on the ring stops.

Communication between end users in a star network must pass through a central station for centralized control, as shown in Fig. 3.1b. When terminal user equipment is shut down because of failure, communication between other terminal users will not be affected, but the central system must have high reliability. The advantage of this structure is that a business can be centralized to a point (hub or central node) and then access the core network, which is more efficient and economical. The disadvantage is that the link provides less bandwidth and has no redundancy protection, resulting in poor reliability.

Wireless mesh networks (WMNs) are multihop, self-organizing broadband wireless networks that are generally composed of a mesh router (MR) and mesh client (MC) nodes. In the mesh network structure, each communication node is connected to multiple other communication nodes, as shown in Fig. 3.1c. Communication nodes on the network can provide multiple detour routes almost in real time, which not only provide redundancy protection for network services, but also converge services to certain points for more effective network access.

3.1.2 Wireless Mesh Networks

Mesh topology, which can also be said to be a multipoint-to-multipoint network topology, is adopted in WMNs [2]. In this mesh network structure, each network node is connected to the others through adjacent network nodes in wireless multihop mode. A WMN generally consists of an MR, a mesh terminal, and a mesh gateway. However, depending on the network configuration, a WMN may not contain all of these types of nodes. An MR is a mesh node with routing functionality. It is responsible for terminal access and data forwarding. MRs are interconnected to form a wireless backbone network. Mesh terminals, which access the Internet through the MR, are used directly by users [3]. Some mesh terminals also provide routing functions. In special cases, they can provide routing and forwarding for other terminal users who cannot directly access the WMN. Mesh terminal devices can be common desktop and laptop computers, personal digital assistants, IP telephones, and so on. A mesh gateway connects a WMN to a wired network and provides routing and gateway functions. A WMN can have multiple gateways, and the most suitable gateway can be selected for data transmission to communicate with the wired network. According to node function, WMNs can be divided into three network structures: dual-equality, hierarchical, and hybrid network structure [4].

3.1.2.1 Pair-Equality Wireless Mesh Network

An equalization wireless mesh, which is also called a terminal device mesh, is the simplest WMN structure. As shown in Fig. 3.2, all nodes are peers with identical features, and each uses the same MAC, routing, management, and security protocols. In this network structure, packets sent by any node can be forwarded by multiple nodes to the destination node. Although nodes do not need to have gateway functionality, routing and self-organization capability are necessary. Clients in this network architecture typically use only one wireless interface technology. This structure is suitable for applications where nodes have little data and do not need to be connected to the core network. In this network architecture, nodes are enhanced end-user devices that function as MRs. The MC itself is equipped with a radio-frequency transceiver device to form a point-to-point network connected by wireless channels. Two MCs that cannot communicate directly can communicate through the forwarding function of other MCs. This type of mesh topology is arbitrary; that is, nodes in the network can move arbitrarily, which changes the network topology. Terminal device mesh networks are essentially mobile ad hoc networks; they provide a communication support environment when existing network infrastructure is unavailable or inconvenient to use.

3.1.2.2 Hierarchical Wireless Mesh Network

Hierarchical wireless mesh structure, also known as infrastructure/backbone mesh network structure, can be divided into upper and lower layers, as shown in Fig. 3.3. MRs constitute an infrastructure to which MCs can connect to form a self-configurable and self-healing network. The MR connects to the Internet through its gateway function. This method provides the backbone for common users. Through the gateway/bridge function of the MR, the WMN can be integrated with an existing wireless network. Conventional clients with Ethernet interfaces can connect to the

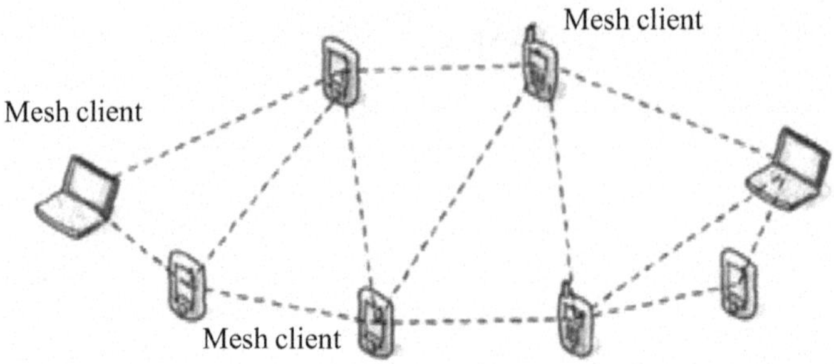

Fig. 3.2 Structure of pair equality WMN [5]

MR through Ethernet links. If an MC uses the same wireless technology as the MR, it can communicate directly with the MR. If different wireless technologies are used, the client must communicate with the base station, which has an Ethernet connection to the MR. Backbone mesh network architecture is the most commonly used type. The mobile MC node is connected to the backbone network of the upper mesh structure through the routing selection and relay functions of the wireless MR, forming a wireless loop between the mesh gateway and mesh end user and realizing communication between network nodes [6].

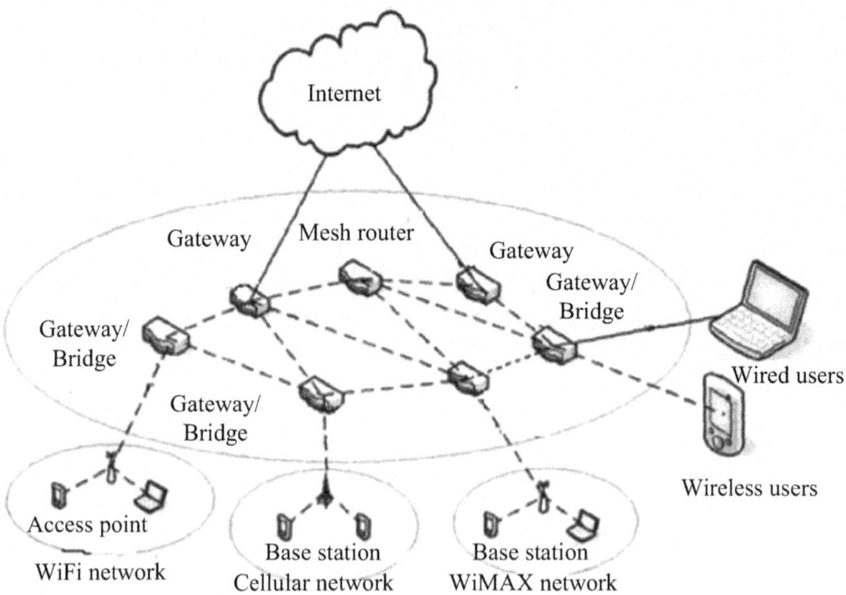

Fig. 3.3 Hierarchical WMN structure. WiMAX: Worldwide Interoperability for Microwave Access

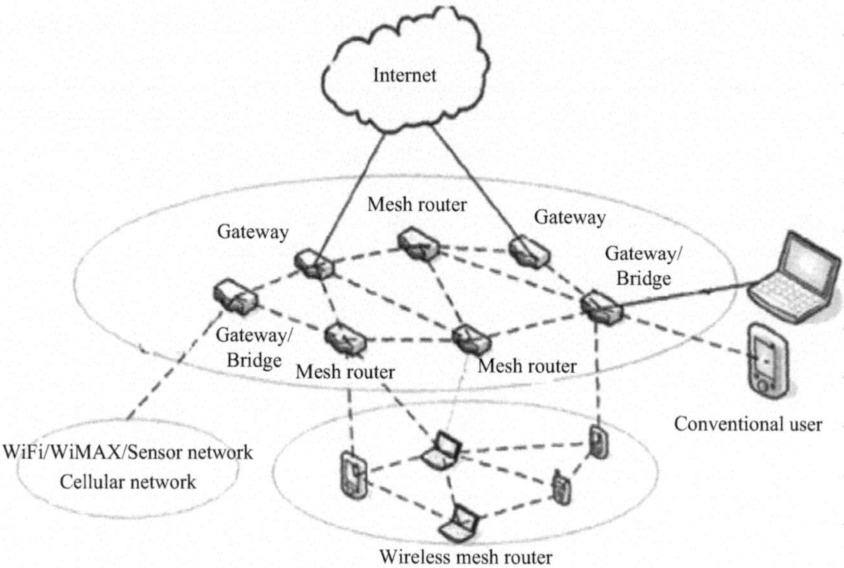

Fig. 3.4 Structure of hybrid WMN. WiMAX: Worldwide Interoperability for Microwave Access

3.1.2.3 Hybrid Wireless Mesh Network

Figure 3.4 shows the hybrid network structure [7], which is a mixture of infrastructure/backbone mesh and terminal standby mesh network structure. In this architecture, terminal nodes not only support ordinary wireless local area network (WLAN) devices, but are also mesh devices with forwarding and routing functions. Devices can be connected in ad hoc mode. Terminal device nodes can generally access both the upper-layer mesh agent and the peer node of the local layer. Because the first two network modes are complementary, the hybrid mesh network supporting the two modes will provide multihop wireless communication in a wide area. The terminal device not only connects with other networks to achieve wireless broadband access, but also communicates with other users; it can be used as a router to forward data from other nodes and send it to the destination node. The hybrid WMN has a wider range of applications and better adaptability. For example, in emergency rescue operations, rescuers can communicate with each other using a temporary MC network they carry with them and send important data on rescue operations in a timely manner to headquarters through the Internet.

3.1.2.4 Characteristics of Wireless Mesh Networks

Compared with conventional wireless networks, WMNs have the following characteristics [8–10]:

(1) Rapid deployment and easy installation. Mesh node installation is as simple as removing the device from the box and connecting it to power. Because installation is greatly simplified, users can easily add new nodes to expand wireless network coverage and network capacity. In a WMN, not every mesh node needs to be connected by cable, which is the largest difference between a WMN and a wired access point (AP). WMNs are designed to minimize the number of wired devices and APs, which greatly reduces the total construction cost and installation time. The cost savings from this measure alone are significant. The configuration and other network management functions of a WMN are the same as those of a conventional WLAN. Users can easily apply their WLAN experience to a mesh network.
(2) NLOS transmission. An NLOS configuration can be easily implemented using WMN technology, which thus has wide application prospects in outdoor and public places. A user with a direct line of sight to the transmitter receives the wireless signal and forwards it to a user with an indirect line of sight. In this way, the best path is automatically chosen so that the signal continuously passes from one user to another, eventually reaching the target user without a direct line of sight. Users with a direct line of sight thus provide wireless broadband access to nearby users without one. WMNs greatly expand the application field and coverage of wireless broadband because of their NLOS transmission capability.
(3) Strong robustness. A common approach to network robustness is to use multiple routers to transfer data. If one router fails, the information is sent by other routers over alternative paths. Mesh networks are more robust than single-hop networks because they do not depend on the performance of a single node. In a single-hop network, if one node fails, the entire network collapses. In the mesh network structure, each node has one or several paths available for data transmission. If the nearest node fails or is disturbed, data packets are automatically routed to an alternative path for continuous transmission, and the operation of the entire network is not affected.
(4) Flexible structure. In a single-hop network, devices must share APs. If several devices want to access the network at the same time, it may cause congestion and slow down the system. In a multihop network, devices can connect to the network through different nodes at the same time; consequently, the system performance will not be degraded. Mesh networks also provide greater redundancy and communication load-balancing. In a WMN, multiple transmission paths are available to each device, and the network can dynamically allocate communication routes according to the communication load of each node, thus effectively avoiding the congestion of nodes. Single-hop networks cannot handle communication interference and AP overload dynamically.
(5) High bandwidth. The physical characteristics of wireless communication determine that it is easier to obtain high bandwidth at shorter transmission distances. The reason is that with increasing wireless transmission distance, all types of interference and other factors leading to data loss increase. Therefore, data transmission through multiple short hops is an effective method of obtaining higher network bandwidth, which is an advantage of mesh networks. In a mesh

network, a node can not only transmit and receive information, but also act as a router to forward information to nearby nodes. As more nodes are connected to each other and the number of possible paths increases, the total bandwidth increases greatly.

3.1.3 Classification of Wireless Communication Networks

Mesh network structure and UV communication each have strong advantages. A UV communication network combined with a WMN using solar-blind UV light for communication between all nodes exhibits good performance. In this network, UV communication expands the range of information transmission owing to the multihop transmission performance of the mesh network, which addresses the limited transmission distance of UV communication. The network structure is shown in Fig. 3.5. The typical wireless UV mesh network has a hierarchical network structure in which MRs are interconnected to form a multihop wireless backbone network, which is responsible for data relay. Backbone networks are generally connected to other networks through gateway nodes, whereas MC nodes are connected to WMNs through MRs.

Wireless UV mesh networks can provide smooth communication in complex terrains, can meet the requirements of modern communication with strong security and flexible mobility, and can be quickly deployed and easily installed when short-range communication is needed. Therefore, wireless UV mesh networks have good potential for applications such as broadband access in residential areas and wireless networking between buildings. They also meet the requirements of short-range communication between underwater equipment and the take-off and landing of helicopters from aircraft carriers.

Because of the short wavelengths of solar-blind UV light, it is strongly scattered by molecules and aerosols in the atmosphere during transmission. According to the relationship between the wavelength of transmitted UV light and the diameter of scattering particles, scattering can be divided into three types. Rayleigh scattering

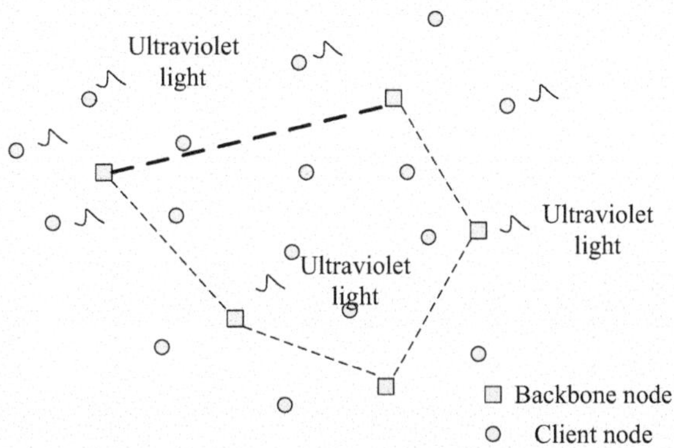

Fig. 3.5 Wireless UV mesh network structure diagram [11]

usually occurs when the wavelength of light is longer than the size of the scattering particles; Mie scattering typically occurs when the wavelength is comparable to the size of the scatterer. Finally, nonlinear scattering occurs when the wavelength is longer than the size of the scattering particles. The scattering properties of UV light enable it to provide NLOS communication.

NLOS communication differs from line-of-sight (LOS) communication. LOS communication (for example, infrared communication) requires strict alignment between sender and receiver; however, precise alignment is difficult under poor environmental conditions. In addition, even if the environmental conditions are favorable, two-way alignment is not easy. Communication equipment that is easily set up is often used under harsh conditions, in severe disasters, or on complex terrain, and it is difficult to have no shielding between the transmitter and receiver. In these cases, NLOS communication is particularly important. NLOS capability is also an important advantage of UV communication.

Figure 2.1 illustrates three UV NLOS communication models with different elevation angles of the transmitter and receiver. When the elevation angles of both the transmitter and receiver are 90°, the best omnidirectional performance is obtained, but the transmission distance is the shortest. When both elevation angles are less than 90°, the worst omnidirectional performance is obtained, but the transmission distance is the longest. When one elevation angle is less than 90° and the other is equal to 90°, intermediate omnidirectionality and transmission distance are obtained. Figure 2.22 and Table 3.1 describe the point-to-point UV communication mode and performance.

UV communication networks consist of multiple point-to-point UV links. Therefore, the communication model for each node in the network is based on the performance of its point-to-point link model. When the nodes in the network are in the group sending or searching state, they can operate in NLOS(A) mode, which gives them the best omnidirectionality and enables them to interact with many nodes in the surrounding coverage area. When network nodes need to conduct multinode communication in a certain direction, they can operate in NLOS(B) mode because they require a certain level of omnidirectionality to ensure a given bandwidth. They can operate in NLOS(C) mode when it is necessary for network nodes to communicate by point-to-point NLOS broadband in a certain direction, because the NLOS communication distance between nodes is the longest and the bandwidth is the widest.

Table 3.1 Performance comparison of three communication configurations [12]

Mode	Transmitter elevation angle (°)	Receiver elevation angle (°)	Omnidirectionality	Trans. distance (km)	Common scatterer	Bandwidth
NLOS(A)	90	90	Best	1	Infinite	Minimum
NLOS(B)	< 90	90	Good	1.5 ~ 2	Finite	Smaller
NLOS(B)	< 90	< 90	Worse	2 ~ 2.5	Finite	Biggest

3.2 Wireless Communication Network Location Algorithm

In UV mesh communication networks, the effectiveness and suitability of the functionality of the access and routing layers are the key to successful communication from transmitter to receiver. The node localization algorithm is important for high-quality, highly efficient wireless communication. To this end, it is necessary to give the access and routing layers additional node location information so that each node can estimate a communication path to the target node, which may require forwarding through other nodes.

3.2.1 Conventional Location Algorithms

Node location algorithms can be divided into two types according to the information they use: ranging and non-ranging algorithms. Nodes whose coordinates are known in the network are called beacon or anchor nodes. Ranging algorithms use point-to-point distance or angle information to calculate the coordinates of unknown nodes. This type of algorithm usually requires three or four beacon nodes for a two-dimensional or three-dimensional node network, respectively. Non-ranging algorithms calculate the geometric coordinates of unknown nodes using information such as connections with neighboring beacons [13].

3.2.1.1 Non-ranging Algorithms

(1) APIT algorithm

The approximate point-in-triangulation test (APIT) algorithm is a node localization algorithm based on regional partition [14]. As shown in Fig. 3.6, a network area is divided into many triangular area, which generally intersect, and the location of an unknown node is determined by identifying whether it is located within each triangular area. Possible locations are gradually eliminated, and the location of the unknown node is determined. The point-in-triangulation test (PIT) is used to determine whether a node is within a triangular region. During the test, an unknown node randomly selects three of the detectable beacon nodes to form a triangle and decides whether it is located within the triangle. The APIT algorithm repeats the PIT until a triangular combination of any three beacons is obtained or accurate positioning is achieved. In this way, all triangles satisfying the condition that their points are located in a given triangular region can be found, and the intersection of these triangular regions, which is a polygon, can obtained. The center-of-gravity coordinate of the polygon can be calculated using the center of gravity, which is the estimated position of the unknown node. However, because the APIT uses information on the received signal strength to determine whether the node is inside the triangle, noise

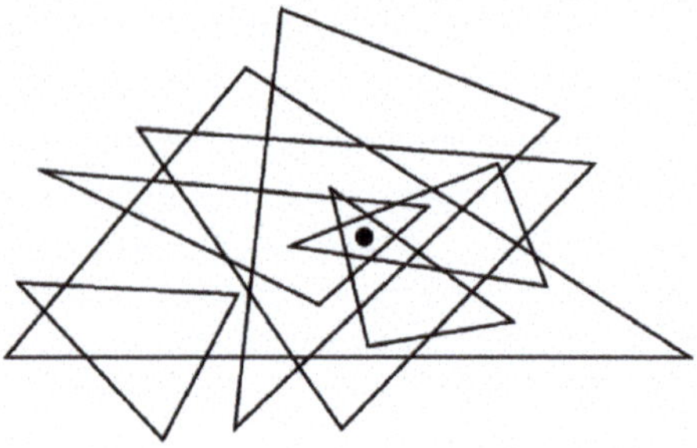

Fig. 3.6 Schematic diagram of APIT algorithm

in the received signal strongly affects the performance of the APIT algorithm. With increasing noise, the positioning accuracy of the algorithm decreases rapidly.

(2) DV-Hop [15]

Distance Vector-Hop (DV-Hop) is a distributed location algorithm based on the communication hop count. First, the beacon node broadcasts its own coordinate information to its surroundings. Each node in the network maintains a minimum hop-count table for these beacon nodes. Then each beacon node calculates its own hop distance and broadcasts it to its surroundings, and the unknown node chooses the hop distance of the nearest beacon node as its own hop distance. Then, according to the minimum hop table recorded previously, it can calculate its distance to all the beacon nodes, and its coordinates. The DV-Hop algorithm requires flood communication and therefore has a high communication overhead.

As shown in Fig. 3.7, the distance and hop number between anchor nodes L1, L2, and L3 are known, and the correction value (i.e., the average hop distance) can be calculated by L2 as $(50 + 75)/(2 + 5) = 17.85$. In this example, if node A obtains the correction value from L2, its distance from the three beacon nodes is L-3 * 17.85, L2-2 * 17.85, and L3-3 * 17.85, respectively, and thus the position of node A is determined by three-point measurement. The location accuracy of the DV-Hop algorithm is approximately 33% in isotropic networks with an average connectivity of 10 and anchor node ratio of 10%. Its disadvantage is that the average hop density can be estimated reasonably well from the correction values only in dense isotropic networks [16].

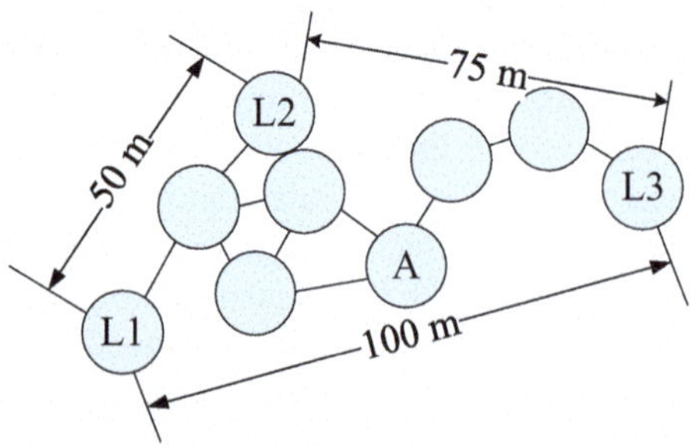

Fig. 3.7 Schematic diagram of DV-Hop algorithm

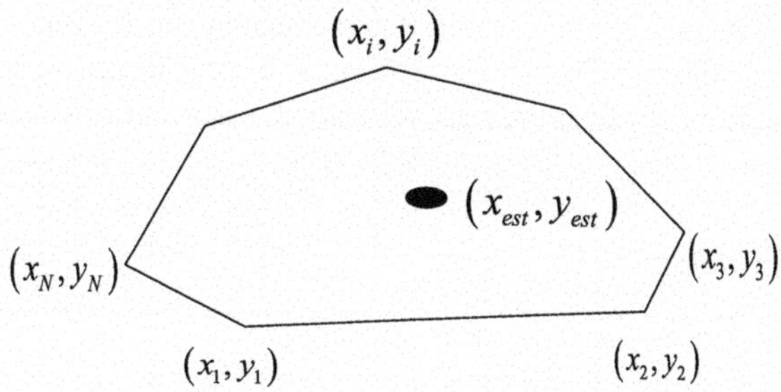

Fig. 3.8 Schematic diagram of Centroid algorithm

(3) Centroid algorithm

Centroid (Fig. 3.8) is a non-ranging coarse-grained algorithm proposed by Bulusu [17]. This algorithm estimates the coordinates of unknown nodes using only the coordinates (x_i, y_i) of beacon nodes. Beacon nodes periodically broadcast their positions to their surroundings, and unknown nodes receiving beacon node packets can estimate their position by calculating the overlapping regions of these unknown nodes. After the unknown node receives the beacon node information, the coordinates of the unknown node are calculated as follows:

$$(x_{est}, y_{est}) = \left(\sum x_i/N, \sum y_i/N\right) \tag{3.1}$$

This algorithm is simple and easy to implement because it does not require coordination between beacon nodes and unknown nodes. However, it assumes that all nodes follow an ideal spherical wireless signal propagation model, which is not the case. In addition, the accuracy of the algorithm depends strongly on the density and distribution of beacon nodes, where the positioning accuracy is higher for denser and more uniformly distributed networks. Moreover, the location error of unknown nodes at the edge of the network is very large.

3.2.1.2 Ranging Algorithms [16, 18]

Although non-ranging algorithms have the advantages of economy and wide use, for some applications their accuracy is low. In these cases, a ranging algorithm can be used [16, 18]. Ranging algorithms are divided into three subclasses: time-of-arrival (TOA), angle-of-arrival (AOA), and received signal strength index (RSSI) algorithms.

TOA algorithms determine the distance by measuring the signal travel time. The positioning system used for this purpose is GPS, which requires expensive, power-intensive electronics for the precise synchronization of satellite clocks. Therefore, GPS and other TOA technologies are nearly impossible to implement in wireless UV communication networks.

AOA technology is used to estimate the direction to neighboring nodes for signal transmission. It can be realized using an antenna array or multiple receivers. In addition to positioning, AOA algorithms can also provide the direction information of nodes. However, AOA technology is affected by the external environment, for example, by noise, and NLOS transmission and other problems will affect the measurement results. In addition, AOA algorithms requires additional hardware, which may not be suitable for UV communication network nodes in terms of hardware size and power consumption.

RSSI technology measures the received power at the receiving node to calculate the propagation loss by comparing it with the known transmitted power and converts the propagation loss into distance using theoretical or empirical signal propagation models. Its main error source is the complexity of signal propagation models resulting from environmental effects; reflection, multipath propagation, NLOS transmission, antenna gain, and other problems will produce significantly different propagation losses for the same distance.

3.2.2 *Two Common Node Location Algorithms*

(1) Three-point measurement [19]

The theoretical basis of this method is that the coordinates of the unknown node can be determined when the coordinates of three anchor nodes and the distances between an unknown node and the anchor nodes are determined. As shown in Fig. 3.9, in two-dimensional space, the coordinates of the three anchor nodes are written as $A(x_1, y_1)$, $B(x_2, y_2)$, and $C(x_3, y_3)$; the distances between the unknown node and the anchor nodes are d_1, d_2, and d_3, respectively, and the coordinates of the unknown node are denoted as $D(x, y)$. According to the formula for calculating the distance between two points, the equations can be written as

$$\begin{cases} d_1 = \sqrt{(x - x_1)^2 + (y - y_1)^2} \\ d_2 = \sqrt{(x - x_2)^2 + (y - y_2)^2} \\ d_3 = \sqrt{(x - x_3)^2 + (y - y_3)^2} \end{cases} \tag{3.2}$$

From (3.2):

$$\begin{bmatrix} x \\ y \end{bmatrix} = \begin{bmatrix} 2(x_2 - x_1) & 2(y_2 - y_1) \\ 2(x_2 - x_3) & 2(y_2 - y_3) \end{bmatrix}^{-1} \begin{bmatrix} x_2^2 - x_1^2 + y_2^2 - y_1^2 + d_1^2 - d_2^2 \\ x_2^2 - x_3^2 + y_2^2 - y_3^2 + d_3^2 - d_2^2 \end{bmatrix} \tag{3.3}$$

(2) Triangulation [20]

The principle of triangulation is illustrated in Fig. 3.10. Because the coordinates of nodes A, B, and C are (x_1, y_1), (x_2, y_2), and (x_3, y_3), respectively, the angles of node D

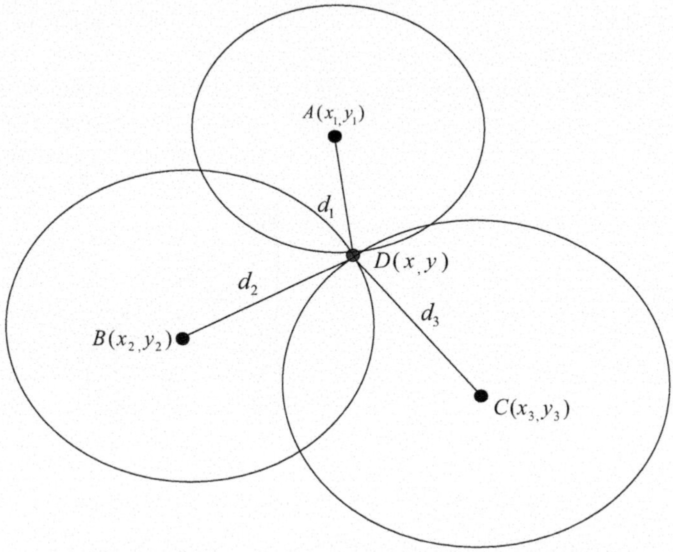

Fig. 3.9 Schematic diagram of three-point measurement and positioning algorithm

relative to nodes A, B, and C are $\angle ADB$, $\angle ADC$, and $\angle BDC$, where the coordinates of node D are (x, y). For nodes A and C and angle $\angle ADC$, if arc AC is within ΔABC, a circle can be uniquely determined. Let the center and radius of the circle be $O_1(x_{o1}, y_{o1})$ and r_1, respectively; then $\alpha = \angle AO_1C = (2\pi - 2\angle ADC)$, and

$$\begin{cases} \sqrt{(x_{O1} - x_1) + (y_{O1} - y_1)} = r_1 \\ \sqrt{(x_{O1} - x_3) + (y_{O1} - y_3)} = r_1 \\ (x_1 - x_2)^2 + (y_1 - y_3)^2 = 2r_1^2 - 2r_1^2 \cos\alpha \end{cases} \tag{3.4}$$

The coordinates of the center O_1 and radius r_1 can be calculated using (3.4). Similarly, for nodes A, B and $\angle ADB$ and nodes B, C and $\angle BDC$, the corresponding coordinates O_2, r_2 and O_3, r_3 are obtained by making the appropriate changes in (3.4).

Finally, the coordinates of node D, (x, y); O_1, (x_{O1}, y_{O1}); O_2, (x_{O2}, y_{O2}); and O_3, (x_{O3}, y_{O3}) are determined by three-point measurement.

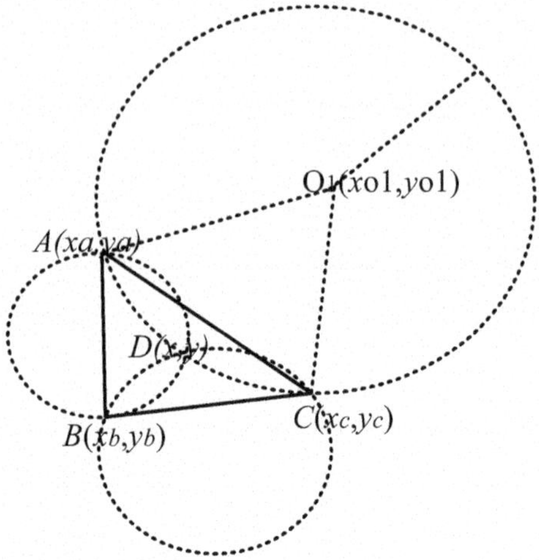

Fig. 3.10 Schematic illustration of triangulation method

3.3 Application of Three-Point Measurement Method in UV Mesh Communication Network

In NLOS communication, two terminals in communication can still communicate even in the presence of obstacles because of the scattering effect of the atmosphere. Figure 3.11 shows the channel model of UV transmission [1] assuming that the energy arriving at the receiver results mainly from the single scattering of photons and the effects of multiple scattering can be ignored.

Here θ_1 is the elevation angle of the transmitter, θ_2 is the elevation angle of the receiver, ϕ_1 is the aperture angle of the transmitted beam, ϕ_2 is the receiving field of view, V is the effective scatterer volume of the overlap between the transmitter and receiver elevation angles, and r is the distance from the transmitter Tx to the receiver Rx. In addition, r_1 and r_2 are the distances from the transmitter and receiver, respectively, to V. The scattering angle θ_s is the angle between θ_1 and θ_2 ($\theta_s = \theta_1 + \theta_2$). In this model, the received power of the NLOS UV single-scattering link is defined as follows [1]:

$$P_r = \frac{P_t A_r K_s P_s \varphi_2 \varphi_1^2 \sin(\theta_1 + \theta_2)}{32\pi^3 r \sin\theta_1 (1 - \cos\frac{\varphi_1}{2})} e^{-\frac{K_e r(\sin\theta_1 + \sin\theta_2)}{\sin(\theta_1 + \theta_2)}} \tag{3.5}$$

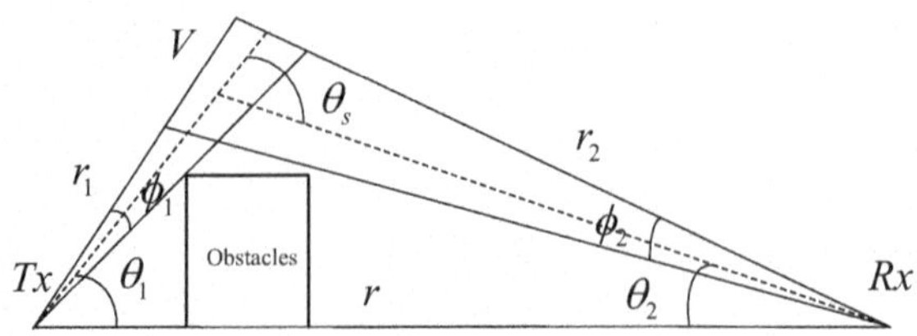

Fig. 3.11 Schematic illustration of single scattering in NLOS UV communication

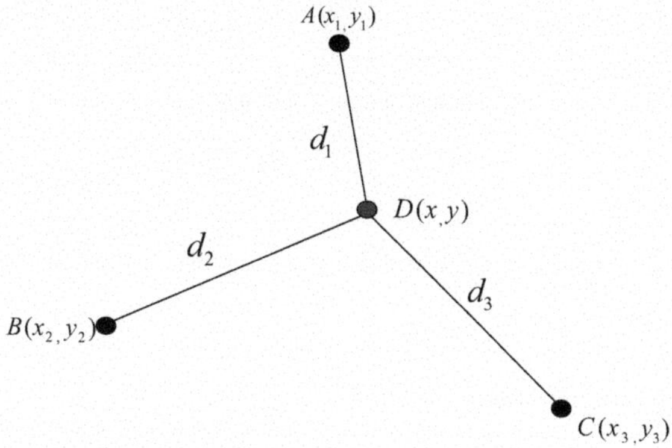

Fig. 3.12 Schematic diagram of three-point measurement and positioning algorithm

where P_r is the received power, P_t is the transmitted power, P_s is the scattering phase function, A_r is the area of the receiver aperture, and K_e is the atmospheric attenuation coefficient, which is the sum of the Rayleigh scattering coefficient k_{sR}, Mie scattering coefficient k_{sM}, and atmospheric absorption coefficient K_a ($K_e = k_{sR} + k_{sM} + K_a$).

In location algorithms based on distance measurement, trilateral measurement is the most basic method of determining the coordinates of unknown nodes. The theoretical basis of this method is that the coordinates of the unknown node can be determined when the coordinates of the three beacon nodes and the distance between the unknown node and the beacon nodes are determined. As shown in Fig. 3.12, in two-dimensional space, the coordinates of the beacon nodes are denoted as $A(x_1, y_1)$, $B(x_2, y_2)$, and $C(x_3, y_3)$, and their distances from the unknown node are d_1, d_2, and d_3, respectively.

The coordinates of the unknown node D are denoted as (x, y). According to the formula for calculating the distance between two points, the following system of equations can be obtained:

$$\begin{cases} d_1 = \sqrt{(x - x_1)^2 + (y - y_1)^2} \\ d_2 = \sqrt{(x - x_2)^2 + (y - y_2)^2} \\ d_3 = \sqrt{(x - x_3)^2 + (y - y_3)^2} \end{cases} \tag{3.6}$$

From (3.6):

$$\begin{bmatrix} x \\ y \end{bmatrix} = \begin{bmatrix} 2(x_2 - x_1) & 2(y_2 - y_1) \\ 2(x_2 - x_3) & 2(y_2 - y_3) \end{bmatrix}^{-1} \begin{bmatrix} x_2^2 - x_1^2 + y_2^2 - y_1^2 + d_1^2 - d_2^2 \\ x_2^2 - x_3^2 + y_2^2 - y_3^2 + d_3^2 - d_2^2 \end{bmatrix} \tag{3.7}$$

The coordinates (x, y) of the unknown node are obtained. The problem then becomes how to obtain the distance between the unknown node and beacon nodes. According to (3.5), the distance between the transmitter and detector can written as

$$r = \frac{\sin(\theta_2+\theta_2)}{k_e(\sin\theta_1+\sin\theta_2)} \times \log_e\left[\frac{1}{P_r}\frac{P_t A_r K_s P_s \varphi_2 \varphi_1^2 \sin(\theta_1+\theta_2)}{32\pi^3 \sin\theta_1 (1-\cos\frac{\varphi_1}{2})}\right] \times \frac{-K_e(\sin\theta_1+\sin\theta_2)}{\sin(\theta_1+\theta_2)} \tag{3.8}$$

Thus, r can be calculated as long as the received power is obtained, if the transmitted power is known and the other transceiver state parameters are set well.

3.4 Simulation Test

The atmospheric channel simulation software MODTRAN was used to simulate solar-blind UV communication with different elevation angles under light haze conditions. A transmission wavelength of 266 nm, transmitted power of 5 W, observation altitude of 0.5 km, and visibility of 5 km were used. The transmission distance and received power under different conditions were obtained, as shown in Figs. 3.13 and 3.14.

When the observation height and the elevation angles of the transmitter and receiver are determined, the transmission distance can be obtained, as shown in Figs. 3.13a and 3.14a. The transmission distance exhibits a variation trend opposite to that of the received power, which is consistent with the basic rule that the attenuation increases with increasing transmission distance.

According to the simulation results in Figs. 3.14b and 3.15b, we used (3.8) to calculate r in the simulation. The following parameters were used: $A_r = 1.4\ \text{cm}^2$, $P_s = 1$, $\phi_1 = 10°$, $\phi_2 = 60°$, $K_a|_{266\ \text{nm}} = 0.75\ \text{km}^{-1}$, $K_s = K_{sM} = (3.91/R_v) \times (266/\lambda_0)^q$,$q = 0.585{R_v}^{1/3}$ is the correction factor [21]. The visibility R_v is 5 km, and $\lambda_0 = 550$ nm). $K_e = K_a + K_{sM}$ (because $R_v = 5$ km corresponds to light haze, the molecular concentration at low altitude is very low; therefore, Rayleigh scattering

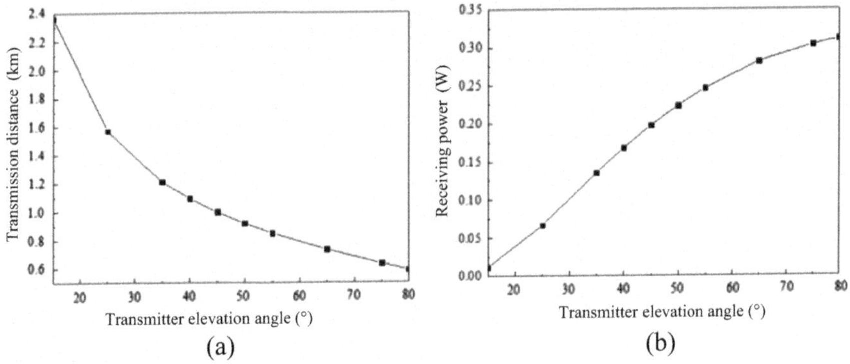

Fig. 3.13 Simulation results for fixed receiver elevation angle ($\theta_2 = 45°$)

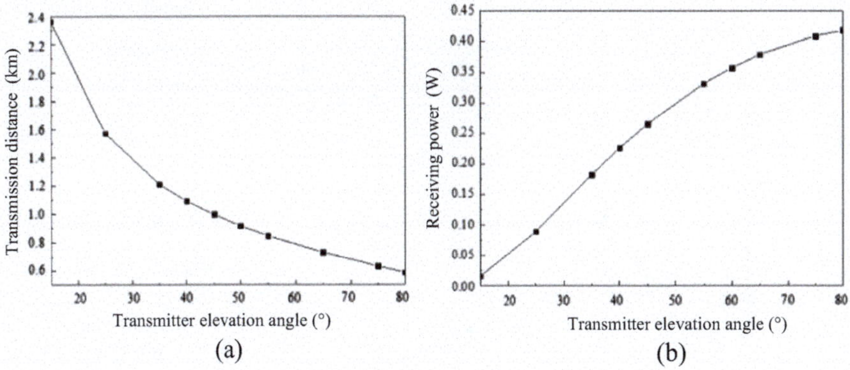

Fig. 3.14 Simulation results for fixed transmitter elevation angle ($\theta_1 = 60°$)

is ignored here [21]. The results are compared with the transmission distances in Figs. 3.13a and 3.14a, as shown in Figs. 3.15 and 3.16, respectively.

If the elevation angle of either the receiver or the transmitter is too low, the transmission distance error is large. The reason is that in this case, the communication process is almost equivalent to LOS communication; therefore, the use of (3.8), which was derived assuming an NLOS communication link to calculate r, will result in a large error.

According to (3.6), the final error in the coordinates of the unknown node is d^2, that is, r^2. As shown in Figs. 3.15 and 3.16, the error of r can be as low as 0.01 km, and the error of r^2 is even lower if the transmitter and receiver elevation angles are

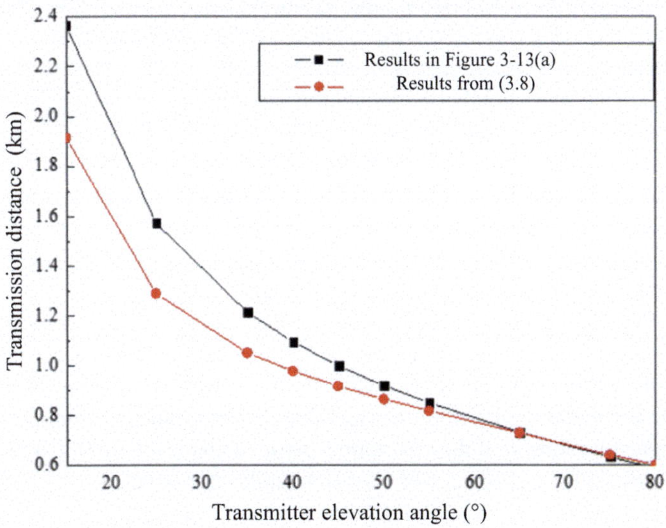

Fig. 3.15 Comparison of transmission distance r at $\theta_2 = 45°$

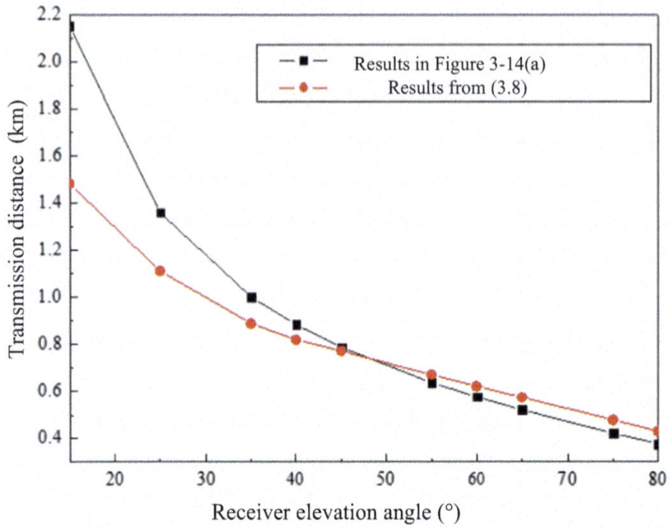

Fig. 3.16 Comparison of transmission distance r at $\theta_1 = 60°$

not too low. Because of the scattering characteristics of UV light, communication can be conducted as long as effective scatterers can be formed at the transmitting and receiving ends, but not as accurate as required by line-distance communication, even if there is a certain error. Therefore, it is feasible to apply a localization algorithm using three-point measurement to solar-blind UV communication networks.

UV communication has the advantages of strong anti-interference ability, good confidentiality, NLOS transmission, and so on, which are difficult to obtain using light in other frequency bands. However, it also has the disadvantage of limited transmission distance. To increase the UV communication range so that it is more widely applicable, UV point-to-point communication is combined with a wireless network. Here, we chose a WMN, which has many advantages. Specifically, it is a multihop, self-organizing, self-healing, multipoint-to-multipoint network that does not depend on a central node, and the failure of one node does not affect the communication performance of the entire network. Combining UV communication and a WMN fully exploits their respective advantages and avoids the disadvantage of limited transmission distance, offering an ideal combination.

For WMNs using solar-blind UV communication, it is necessary to give the access and routing layers additional information on the coordinates of nodes. To obtain high-quality highly efficient wireless communication, we used three-point measurement in the node localization algorithm to calculate the coordinates of unknown nodes. Simulation results show that the positioning error can be less than 0.01 km when the appropriate transmitter and receiver elevation angles are selected.

References

1. Xu ZY, Sadler MB (2008) Ultraviolet communications: potential and state-of-the-art. IEEE Commun Mag 3
2. Kang XY (2007) Research on node cooperation and distributed scheduling in 802.16Mesh Networks. Master's thesis, Huazhong University of Science and Technology, Wuhan
3. Wang LH (2005) Research on Routing Protocol of Wireless Mesh Networks. Xi 'an: master's thesis, Chang 'an University
4. Shan BL (2009) Research on internet gateway layout technology of wireless mesh backbone network. Master's thesis of Harbin Engineering University, Harbin
5. Tang T (2007) Multi-channel MAC protocol and channel allocation mechanism in wireless mesh networks. Master thesis of Xi'dian University, Xi 'an
6. Zhao JL (2009) Research on hybrid routing protocol for wireless mesh networks. Master thesis of Xi'dian University, Xi 'an
7. Xue DQ (2009) Research on routing technology of wireless mesh network. Master's thesis of Chongqing University, Chongqing
8. Su J (2008) Research on routing protocol of wireless mesh network based on cross-layer design. Master's thesis, Xiamen University, Xiamen
9. Shi HY (2008) Research on routing algorithm and load balancing in multi-interface wireless mesh networks. Master's thesis of Tianjin University, Tianjin
10. Wang DP (2008) Research on efficient and fair communication protocol in wireless mesh networks. Master's thesis, University of Science and Technology of China, Hefei
11. He H, Ke XZ, Zhao TF et al (2010) Research on location in wireless "sun blind" UV grid. Laser Technol 34(5):607–610
12. Tang Y, Ni GQ, Zhang LJ et al (2007) Single scattering transmission model for non-direct ultraviolet communication. Opt Technol 33(5):759–765
13. Wang WD (2008) Research on mobile sensor Network identification technology. Doctoral dissertation of University of Electronic Science and Technology of China, Chengdu
14. Chen X (2007) Research on wireless sensor network communication protocol and localization algorithm. Doctoral dissertation of Fudan University, Shanghai
15. Niculescu D, Nath B (2003) DV-based positioning in ad hoc networks. Telecommun Syst 22(1):267–280
16. Wang Y (2007) Research on wireless Sensor network localization technology. Doctoral dissertation of University of Science and Technology of China, Hefei
17. Bulusu N, Heidemann J, Estrin D (2000) GPS-less low cost out door localization of very small devices. IEEE Wirel Commun 7(5):28–34
18. Yu N (2008) Localization optimization method for wireless sensor networks. Doctoral dissertation of Beijing University of Posts and Telecommunications, Beijing
19. Wang JC (2009) Study on node determination in wireless sensor networks. Doctoral dissertation of University of Science and Technology of China, Hefei
20. Ji WW (2008) Research on node location and coverage technology of wireless sensor networks. Doctoral dissertation, Nanjing University of Science and Technology, Nanjing
21. Chen JH, Yang XL (2008) Atmospheric factors for non-line-of-sight "sun-blind" ULTRAVIOLET communication. Laser Mag 29(4):38–39

Chapter 4
Multiuser Detection Technology in Ultraviolet Ad Hoc Networks

Ultraviolet (UV) communication is greatly affected by the channel environment. The distributed operation of ad hoc networks, time variation of channels, random movement of nodes, and other factors make it difficult to avoid collision and interference between data packets, which limits the network capacity. This chapter studies the causes of multiple-access interference (MAI) and describes the use of multiuser detection to address these problems and improve the performance of UV ad hoc networks.

4.1 Spread Spectrum and Multiple-Access Interference

4.1.1 Spread Spectrum Sequence

Spread spectrum technology is used to transmit information in a bandwidth much wider than the signal bandwidth. In spread spectrum communication [1], the information data to be transmitted are encoded and modulated by a spreading sequence [also known as a pseudo-noise (PN) code] and then transmitted after spectral expansion. At the receiving end, the same code is used for demodulation and correlation processing to recover the original data. The basic principle of spread spectrum communication is derived from the well-known Shannon formula:

$$C = W \log_2(1 + S/N) \tag{4.1}$$

where C is the channel capacity, W is the channel bandwidth, and S/N is the power signal-to-noise ratio (ratio of the channel output signal power to the output noise power).

The Shannon formula shows that the maximum transmission rate C of a communication channel is limited by the channel bandwidth and signal-to-noise ratio.

X. Ke, *UV Light Self Organizing Network Theory*, Optical Wireless Communication Theory and Technology, https://doi.org/10.1007/978-981-97-7854-6_4

Increasing the channel bandwidth or signal-to-noise ratio can increase channel capacity. When the channel capacity C remains unchanged, the signal-to-noise ratio can be reduced by increasing the channel bandwidth, that is, reducing the power of the transmitted signal. Note that increasing the channel bandwidth does not increase the channel capacity indefinitely. Code division multiple-access (CDMA) is a typical spread spectrum broadband communication method [2].

An important parameter reflecting the characteristics of spread spectrum communication is the spread gain G. G is defined as the ratio of the signal bandwidth W_2 after spectral expansion to the signal bandwidth W_1 before expansion:

$$G = W_2/W_1 \tag{4.2}$$

In spread spectrum communication, after the receiver performs spread spectrum demodulation, only the signal components of the bandwidth W_1 after pseudo-random code correlation processing are extracted, and the MAI components extending to W_2 are excluded. The spread spectrum gain G accurately reflects the capacity for spread spectrum communication. The principle of spread spectrum communication is illustrated in Fig. 4.1 [3].

A simplified model of the spread spectrum communication system is shown in Fig. 4.2 [3]. The transmitter is simplified so that it performs only modulation and spectral spreading, and the receiver is simplified so that it performs only despreading and demodulation. There are identical pseudo-random code generators on the receiving and transmitting sides. The PN codes of the receiver and transmitter must be accurately synchronized, and other links in the system are added to the channel. Communication systems using spread spectrum technology have six main functions: inputting the original information, source coding and decoding, channel coding and decoding (error control), carrier modulation and demodulation, spread spectrum modulation and despreading, and transmission via the channel. The purpose of source coding is to remove redundant information, compress the digital rate of the source, and improve the transmission efficiency of the channel. Error control is used to increase the redundancy of information in channel transmission, provide the ability to detect or correct errors, and improve the quality of channel transmission. The modulation component enables symbols encoded by the channel to be transmitted in appropriate frequency bands, such as the microwave or short-wave frequency band. Spread spectrum modulation and despreading is a spectral broadening and restoration technology used for various purposes. In contrast to that in conventional communication systems, the signal transmitted in the channel is a broadband signal with low spectral density.

In optical CDMA (OCDMA), a direct sequence spread spectrum (DSSS) system is typically used. In a DSSS system, the information to be transmitted is spread into a wide frequency band with a pseudo-random sequence (or PN code). At the receiving end, the original information is recovered by correlating the received spread spectrum signal with the same pseudo-random sequence used at the transmitting end. Because the interference signal is not related to the pseudo-random sequence, it is expanded at the receiving end. Consequently, the power of the interference signal falling into the

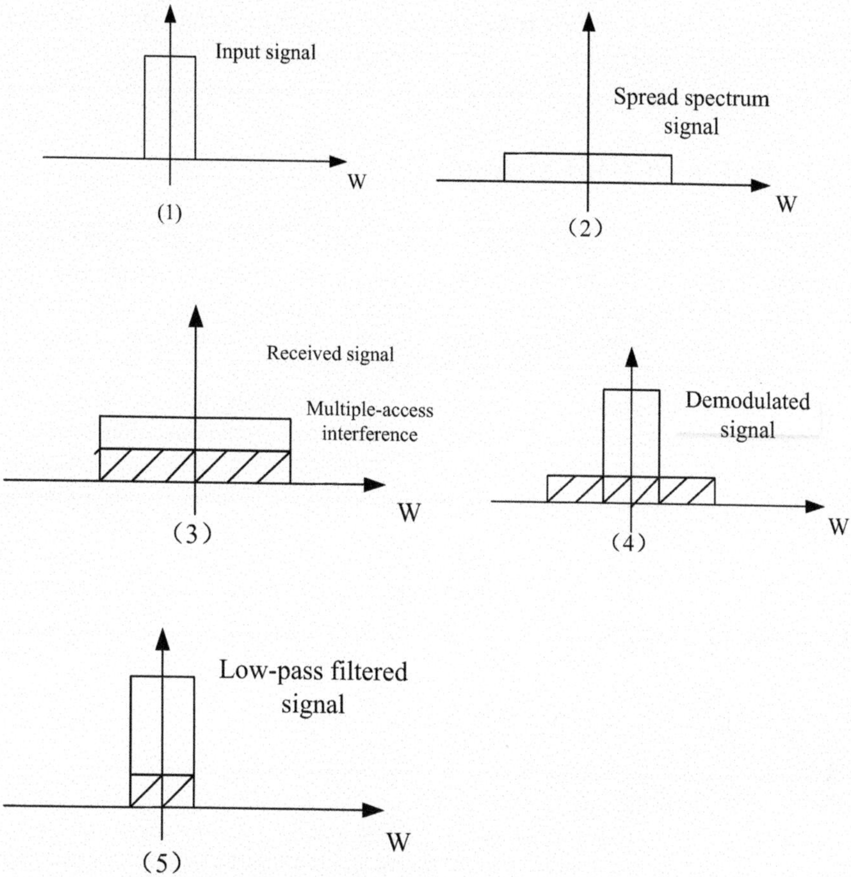

Fig. 4.1 Principle of spectral spreading [3]

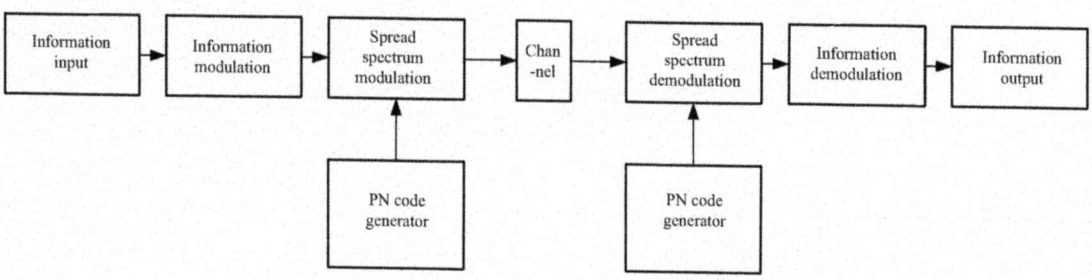

Fig. 4.2 Simplified model of spread spectrum communication [3]

signal frequency band is greatly reduced, which improves the output signal-to-noise ratio of the system and realizes anti-interference capability.

4.1.2 MAI in OCDMA Systems

Each user in a CDMA system is assigned a pseudo-random sequence. At the transmitter, the appropriate pseudo-random sequence is used to spread the spectrum to the user. At the receiving end, the same pseudo-random sequence is used for correlation to correctly decode the information. These pseudo-random sequences are called the user address codes.

MAI refers to interference between different user signals in the same system. It is one of the main types of interference in CDMA communication systems. Because the spread spectrum sequences used in CDMA systems are generally not completely orthogonal, nonzero cross-correlation coefficients result in mutual interference among users.

In OCDMA systems, the main cause of MAI is the incomplete orthogonality of address codes. For N users in a chip-synchronous OCDMA system, the received signal can be expressed as

$$r(t) = \sum_{k=1}^{N} \lambda_k b_k s_k(t) + \lambda_d \tag{4.3}$$

where $b_k \in \{0, 1\}$ is the information bit value of the kth user; λ_k is the transmitted optical power; $s_k(t)$ is the user address code, which normalizes the spread spectrum signal for the kth user (assuming that all user address codes have unit energy); and λ_d is the background light noise power. The cross-correlation between user address codes ρ_{ik} is given by

$$\rho_{ik} = \frac{1}{T_b} \int_0^{T_b} s_i(t) s_k(t) dt \tag{4.4}$$

In OCDMA systems, conventional signal receiving technology is used to directly correlate the received signal with the user address code; then photoelectric detection is performed, and the data are recovered after evaluation, as shown in Fig. 4.3.

The output r_k of the correlator is given by [4]

$$r_k = \frac{1}{T_b} \int_0^{T_b} r(t) s_k(t) dt$$

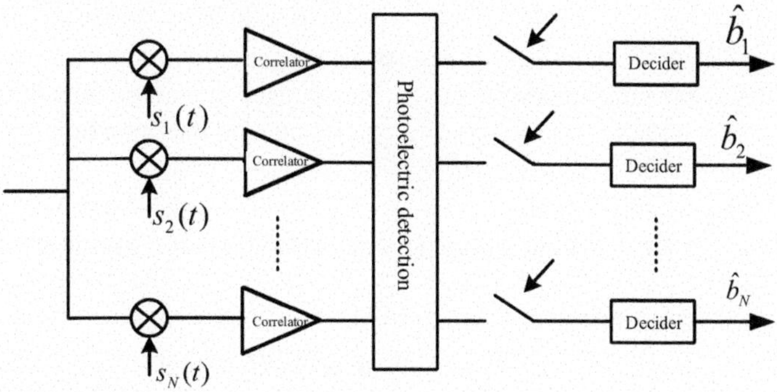

Fig. 4.3 Structure of conventional signal receiver

$$= \lambda_k b_k + \sum_{\substack{i=1 \\ i \neq k}}^{K} \lambda_i b_i \rho_{ik} + n_k \tag{4.5}$$

where $\lambda_k b_k$ is the expected user signal; $\sum_{\substack{i=1 \\ i \neq k}}^{K} \lambda_i b_i \rho_{ik}$ is an MAI term; n_k is the background light noise term, which is given by $n_k = \int_0^{T_b} \lambda_d s_k(t) dt = \lambda_d \sqrt{W}$; and W is the code weight of the user address code.

The output r_k of the correlator is described in matrix form as [4]

$$\underset{N}{\vec{r}} = \vec{R}\underset{N}{\vec{b}} + \underset{N}{\vec{n}} = \underset{N}{\vec{b}} + \vec{Q}\underset{N}{\vec{b}} + \underset{N}{\vec{n}} \tag{4.6}$$

where $\underset{N}{\vec{r}} = [r_1, r_2, \ldots, r_N]^T$, $\underset{N}{\vec{b}} = [\lambda_1 b_1, \lambda_2 b_2, \ldots, \lambda_N b_N]^T$, and $\underset{N}{\vec{n}} = [n_1, n_2, \ldots, n_N]^T$.

$$\vec{R} = \begin{bmatrix} 1 & \rho_{12} & \ldots & \rho_{1N} \\ \rho_{12} & 1 & \ldots & \rho_{2N} \\ & & \ddots & \\ \rho_{1N} & \rho_{2N} & \ldots & 1 \end{bmatrix} \tag{4.7}$$

$$\vec{I} = \begin{bmatrix} 1\,0 \ldots 0 \\ 0\,1 \ldots 0 \\ \ddots \\ 0\,0 \ldots 1 \end{bmatrix} \tag{4.8}$$

$$\vec{Q} = \begin{bmatrix} 0 & \rho_{12} & \dots & \rho_{1N} \\ \rho_{12} & 0 & \dots & \rho_{2N} \\ & & \ddots & \\ \rho_{1N} & \rho_{2N} & \dots & 0 \end{bmatrix} \tag{4.9}$$

where $\underset{N}{\vec{b}}$ indicates the data item required for demodulation after amplitude weighting of the received signal, and $\vec{Q}\underset{N}{\vec{b}}$ refers to MAI.

If the user address codes are orthogonal to each other, that is, if $\rho_{ik} = 0\,(i \neq k)$, the MAI term in (4.5) is zero, and the conventional detection method is the best choice. If the user address codes are not completely orthogonal, the MAI, and thus the detection performance of the system, increase sharply with increasing number of users.

In direct-detection OCDMA systems, because the address codes adopted by different users cannot be completely orthogonal, the nonzero cross-correlation coefficient results in mutual interference among users. In addition, because the wavelength or time slot is not isolated, the MAI is severe. It greatly affects the link performance of the system and is worse in environments with an asynchronous transmission channel and multipath propagation. OCDMA communication systems are thus self-interfering. Because MAI affects both the capacity and performance of the system, it is an important issue for OCDMA systems [5].

The following methods are used to eliminate MAI: (1) select address codes with good orthogonality; (2) use power control; (3) use an error correction mechanism; (4) place an optical hard limiter (OHL) at the receiving end; and (5) use multiuser detection technology, which is a more active method than methods 1–4. First, the selection of the OCDMA address code is analyzed.

4.2 Analysis and Construction of Address Code

In CDMA communication systems, all users share the same frequency band and time in the same channel. The signals used by different users to transmit information are distinguished by their coding sequences. At the transmitter, each information bit of a user is encoded into a series of pulses according to the user pseudo-random sequence. The user at the receiving end uses the same pseudo-random sequence for correlation to recover the transmitted information. These pseudo-random sequences are called address codes, and each coding pulse is called a chip. Figure 4.4 illustrates the concept of the address code and coding process [6].

In CDMA systems, the selection and physical implementation of address codes is an important core issue and is directly related to the system capacity, bit error rate (BER), system complexity, system synchronization, and many other system

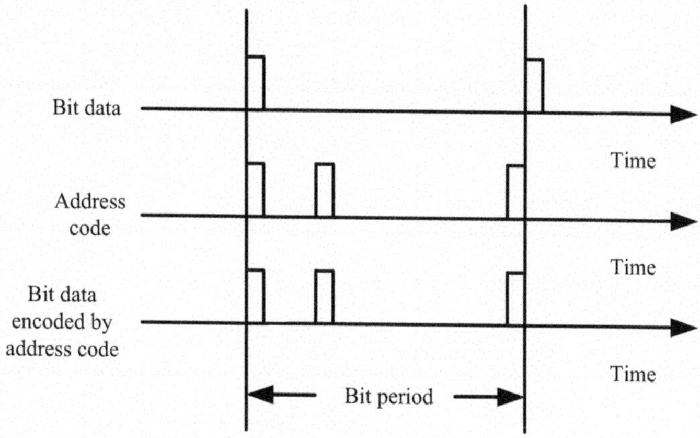

Fig. 4.4 Concept of address code and coding process [6]

characteristics. To minimize interference from other users and obtain a high signal-to-noise ratio, these code sequences should have good autocorrelation and cross-correlation characteristics; that is, the main peak value of the autocorrelation function should be large, and the cross-correlation should be low. In addition, the side peak of the autocorrelation function should be as small as possible to make it easy for both the transmitter and receiver to obtain synchronization.

Similarly, in OCDMA systems, each user is assigned a unique optical address code with good correlation characteristics to identify and distinguish the user from other users and thus realize shared-channel random communication. Therefore, the overall performance of OCDMA systems depends largely on the optical address codes adopted by the system. Therefore, it is crucial to identify optical address codes with high capacity and good correlation characteristics, which are suitable for OCDMA systems.

Direct-detection OCDMA systems adopt the incoherent optical communication mode of power intensity detection. Therefore, by sending an optical pulse to represent 1 and not sending a pulse to represent 0 in one bit period, the receiving end decodes the information by determining whether the received pulse optical field appears in each bit period. The optical signal power is nonnegative; therefore, it can be regarded as a sequence with a value in the (0,1) binary range. However, in conventional CDMA systems, bipolar positive and negative levels are used, which can be regarded as sequences in the $(-1, +1)$ binary domain. Therefore, bipolar codes with good correlation characteristics in the electrical domain (such as the M and Gold sequences) cannot be used in OCDMA systems, and new address codes need to be designed. It is particularly important to select a coding scheme suitable for OCDMA [7].

OCDMA address codes can be classified into unipolar and bipolar codes according to polarity. Unipolar codes are used mainly in incoherent optical communication, and bipolar codes are used mainly in coherent optical communication. In the study of OCDMA systems, many unipolar codes with quasi-orthogonal characteristics (i.e., nonzero cross-correlation) have been proposed. Prime codes (PCs), optical orthogonal codes (OOCs), frequency-hopping codes, and so on are commonly used.

4.2.1 Prime Sequence Codes (PCs)

Prime numbers are constructed as follows [8]:

(1) Let p be a prime number greater than 1, and construct a set of residual classes in the finite field $GF(p)$:

$H = \{0, 1, 2, \ldots, (p-1)\}$.

(2) In $GF(p)$, a family of prime sequences $S_i = \{S_{i0}, S_{i1}, \ldots S_{ij}, \ldots S_{i(p-1)}\}$ can be constructed, where

$$S_{ij} = \{i * j\}(\mathrm{mod}\, p), \quad 0 \le i, j \le p - 1 \tag{4.10}$$

(3) Prime sequence codes can be constructed from the prime sequence S_i:

$C_i = \{C_{i0}, C_{i1}, \ldots C_{ik}, \ldots C_{i(p^2-1)}\}$, where

$$C_{ik} = \begin{cases} 1 \ \forall k = S_{ij} + j * p \\ 0 \ \forall k \neq S_{ij} + j * p \end{cases} \quad 0 \le i, j \le p - 1, 0 \le k \le p^2 - 1 \tag{4.11}$$

Table 4.1 shows the prime sequence and prime sequence codes for $p = 5$ [8].

The PC length, code weight, and address code capacity for this method are $L = p^2$, $w = p$, and $C = p$, respectively. The peak value of the autocorrelation function is p, the autocorrelation limit is $\lambda_\alpha = p - 1$, and the cross-correlation limit is $\lambda_c = 2$.

The prime sequence code has the advantage of a simple structure; its main disadvantage is that the autocorrelation limit is very close to the correlation peak, and thus it is difficult to synchronize the code sequence at the receiver. The sidelobe of the autocorrelation function of the prime sequence is reportedly too large [9]. In practical applications, light interference causes the system to misjudge the detection of the

Table 4.1 Prime sequence and codes for $p = 5$ [8]

Codeword	Prime sequence	0,1 codeword sequence
C_0	$S_0 = \{0, 0, 0, 0, 0\}$	10,000 10,000 10,000 10,000 10,000
C_1	$S_1 = \{0, 1, 2, 3, 4\}$	10,000 01,000 00,100 00,010 00,001
C_2	$S_2 = \{0, 2, 4, 1, 3\}$	10,000 00,100 00,001 01,000 00,010
C_3	$S_3 = \{0, 3, 1, 4, 2\}$	10,000 00,010 01,000 00,001 00,100
C_4	$S_4 = \{0, 4, 3, 2, 1\}$	10,000 00,001 00,010 00,100 01,000

autocorrelation peak and degrade the BER of the system. Therefore, it is not suitable for use as the spread spectrum sequence of OCDMA systems. In fact, the premise of systems using prime sequence codes is generally strict synchronization, which is difficult to achieve in UV scattering communication. In addition, as the number of users communicating simultaneously in the network increases, the intersymbol interference becomes severe, and the system performance is sharply degraded [7]. Next, we analyze the possibility of 1-code collision between prime address codewords [8].

For two prime digital words, C_a and C_b, because the cross-correlation limit of prime digital words is 2, the number of 1-code collisions between C_a and C_b can be only 0, 1, or 2. The code length $L = p^2$ and code weight $w = p$ apply to PCs; in most communication systems, it can be reasonably assumed that a user signal has equal probability of being 1 or 0. Therefore, the collision probability between a specific 1 code in codeword C_a and a 1 code in codeword C_b in the channel is $w/2L = 1/2p$. The total collision probability is given by [8]

$$\overline{\lambda} = \frac{1}{2} \tag{4.12}$$

Assuming that the number of collisions with C_b after the codeword C_a moves $s(s = xp + y)$ chip cycles is 2, the codeword must satisfy the following equations [8]:

$$\begin{cases} [i \otimes (k - x)(\mathrm{mod}p)] + y = j \otimes k(\mathrm{mod}p) \\ [i \otimes (k - x - 1)(\mathrm{mod}p)] + y - p = j \otimes k(\mathrm{mod}p) \end{cases} \tag{4.13}$$

In the cyclic shifting of codewords, the probability that the above equations are satisfied at the same time is [8]

$$\overline{q}_2 = \frac{\sum_{i=1}^{p-1}\left(\sum_{y=0}^{p-1} y(p-y) - i(p-i)\right)}{2p^3(p-1)} = \frac{(p+1)(p-2)}{12p^2} \tag{4.14}$$

From (4.12) and (4.13), it can be obtained that the number of 1-code collisions between steps (1) and steps (2) is one and the probability of no collision is [8]

$$\overline{q}_1 = \overline{\lambda} - 2 \times \overline{q}_2 = \frac{1}{2} - \frac{(p+1)(p-2)}{12p^2} = \frac{2p^2 + p + 2}{6p^2} \tag{4.15}$$

$$\overline{q}_0 = 1 - \overline{q}_1 - \overline{q}_2 = \frac{7p^2 - p - 2}{12p^2} \tag{4.16}$$

The variance of PC MAI can be obtained [8]:

$$\overline{\sigma^2} = \frac{5p^2 - 2p - 4}{12p^2} \tag{4.17}$$

According to the central limit theorem, when the number of users n is large, it can be estimated by a Gaussian distribution, and its expectation and variance are $\lambda_0 = (N-1)\overline{\lambda}$ and $\sigma_0^2 = (N-1)\overline{\sigma^2}$, respectively. When n is large, the codeword capacity P should also be larger, and the variance is $\overline{\sigma^2} \approx \frac{5}{12}$. Therefore, when the user capacity is large, the BER formula can be approximated as [8]

$$PE_{PC} = \phi\left(\frac{-p}{\sqrt{4(N-1)\sigma_0^2}}\right) = \phi\left(\frac{-p}{\sqrt{1.67(N-1)}}\right) \tag{4.18}$$

where $\phi(x) = \frac{1}{2\pi}\int_{-\infty}^{x} e^{-\frac{t^2}{2}}dt, \quad -\infty < x < +\infty$

4.2.2 Modified Prime Codes (MPCs)

Because the capacity of the PC is the same as the code weight, when the system needs to accommodate more users, the code weight P increases. In fact, the increase in code weight results in a decrease in the BER. When the user capacity is large, the BER caused by MAI is very low. If the PC weight P is still used, the power efficiency of the system decreases. In this case, extracting a part of the 1 code from the PC with high code weight can be considered to reduce the code weight and number of optical pulses so as to improve the power efficiency and thus ensure a low BER. This PC with reduced code weight is called the modified PC (MPC). When the number of users n is large, a Gaussian distribution can also be used to estimate the BER [8]:

$$PE_{EPC} = \phi\left(\frac{-p}{\sqrt{4(N-1)\sigma_0^2}}\right) = \phi\left(\frac{-p}{\sqrt{0.75(N-1)}}\right) \tag{4.19}$$

The MPC address code has the same code length, codeword capacity, and cross-correlation characteristics as the original PC sequence, whereas the code weight $w < p$ and autocorrelation characteristics are improved compared with those of the original PC sequence.

4.2.3 Optical Orthogonal Codes (OOCs)

An $(L, w, \lambda_\alpha, \lambda_c)$ OOC C is a set of (0, 1) sequences with length L and code weight W. The cycle of each codeword $(x_0, x_1, \ldots x_{n-1})$ is still a codeword. The cyclic autocorrelation function of each codeword and the cyclic cross-correlation function between any two different codewords and sums satisfy [10]

$$\sum_{i=0}^{n=1} x_i \cdot x_{i\oplus\tau} = \begin{cases} w & \tau = 0 \\ \leq \lambda_\alpha & \tau = 1, 2, \ldots, L-1 \end{cases} \tag{4.20}$$

$$\sum_{i=0}^{n=1} x_i \cdot y_{i\oplus\tau} \leq \lambda_c \qquad \tau = 0, 1, \ldots, L-1 \tag{4.21}$$

respectively, where the symbol $\oplus$ represents the addition of module L. λ_α and λ_c are autocorrelation and cross-correlation constraints, respectively; they can be any positive integer. $(y_0, y_1, \ldots y_{n-1})$ is a different codeword of $(x_0, x_1, \ldots x_{n-1})$.

The autocorrelation coefficient λ_α (the maximum side peak of the autocorrelation of any codeword) affects mainly the difficulty of synchronization between sender and receiver. The correlation number λ_c (the maximum cross-correlation between any two different codewords) affects not only the synchronization difficulty, but also the number of simultaneous users. $\lambda_\alpha = \lambda_c = 1$ is the minimum value of the autocorrelation and cross-correlation limits; that is, the interference between users reaches the minimum.

In this OOC, the number of codewords $|C|$ is called the capacity of the code. $|C|_{\max} = \varphi(L, w, \lambda_\alpha, \lambda_c)$ is the maximum possible capacity. Let $\lambda = \max\{\lambda_\alpha, \lambda_c\}$; then [10]

$$\varphi(L, w, \lambda_\alpha, \lambda_c) \leq \frac{(L-1)(L-2)\ldots(L-\lambda)}{w(w-1)\ldots(w-\lambda)} \tag{4.22}$$

In this case, the capacity $|C|$ of the OOC meets the Johnson limit; that is, $|C| = \frac{(L-1)(L-2)\ldots(L-\lambda)}{w(w-1)\ldots(w-\lambda)}$ [11].

If the codeword capacity of an OOC is $|C| = \phi(L, w, \lambda_\alpha, \lambda_c)$, the OOC is called the optimal OOC. For OOCs with good characteristics, the autocorrelation function of each codeword sequence has a narrow main lobe and a sufficiently small sidelobe, and the cross-correlation value of two different code sequences remains very small. There is a large difference between the numbers of zeros and ones in codewords with good correlation (the number of zeros is greater than or equal to the number of ones); thus, the OOC is a type of sparse quasi-orthogonal code. Compared with the PC, the OOC has better autocorrelation and cross-correlation. It is a promising coding scheme in OCDMA applications. In this section, the performance of OOCs is analyzed and simulated.

4.2.3.1 Types of OOCs

OOCs are divided into equal-weight symmetric, equal-weight asymmetric, variable-weight symmetric, and variable-weight asymmetric OOCs according to their code weight and the relationship between autocorrelation and cross-correlation. The classification principles are as follows.

① Equal-weight OOC

An OOC $(L, w, \lambda_\alpha, \lambda_c)$ is called an equal-weight OOC when w is constant, that is, when all codewords have the same weight. When $\lambda_\alpha = \lambda_c$, the autocorrelation limit of the code is the same as the cross-correlation limit, and the code is a symmetric OOC.

② Equal-weight asymmetric OOCs

When w is constant (that is, all codewords have the same weight), but $\lambda_\alpha \neq \lambda_c$, the OOC $(L, w, \lambda_a, \lambda_c)$ is an equal-weight asymmetric OOC.

③ Variable-weight symmetric OOCs

The OOC $(L, w, \lambda_\alpha, \lambda_c)$ is called a variable-weight symmetric OOC when the code weight w; it is represented by (L, w, F, λ_c, Q). W, F, and Q represent the sets $\{w_0, w_1, \ldots w_p\}$, $\{\lambda_\alpha^0, \lambda_\alpha^1, \ldots \lambda_\alpha^p\}$, and $\{q_0, q_1, \ldots q_p\}$, respectively, where q_i represents the percentage of the number of codewords with weight w_i in the total capacity, and λ_α^i represents the autocorrelation limit of codewords with code weight w_i. The OOC $(L, w, F, \lambda_c, Q)(\lambda_\alpha^i = \lambda_c, i = 0, 1, \ldots p)$ is a variable-weight symmetric OOC with the same autocorrelation and cross-correlation values.

④ Variable-weight asymmetric OOCs

The OOC $(L, w, F, \lambda_c, Q)(\lambda_\alpha^i = \lambda_c, i = 0, 1, \ldots p)$ is a variable-weight asymmetric OOC, and its autocorrelation value differs from its cross-correlation value.

Compared with PCs, equal-weight OOCs with a code length L, code weight w, and autocorrelation sidelobe and cross-correlation peak values of 1 are more suitable for use as address codes in OCDMA communication.

4.2.3.2 BER of OOCs

The possibility of 1-code collision between OOC address codewords is analyzed assuming that the OOC cross-correlation limit is 1. It is assumed that the number of 1-code collisions between two OOC codewords, C_a and C_b, can be only 0 or 1.

Because the length of the OOC is l and the weight is w, the probability of 1-code collisions between C_a and C_b is [8]

$$\overline{\lambda} = \frac{w^2}{2L} \tag{4.23}$$

That is, the probability of collision between C_a and C_b is $\overline{\lambda}$, whereas the probability of no collision is $1-\overline{\lambda}$. The BER formula [8] for n simultaneous users can be obtained from the classical probability formula:

$$PE_{OOC} = \frac{1}{2}\int_{Th}^{\infty} P_{I_1}(I_1)dI_1 = \frac{1}{2}\sum_{i=Th}^{N-1}\binom{N-1}{i}\left(\frac{w^2}{2L}\right)^i\left(1-\frac{w^2}{2L}\right)^{N-1-i} \tag{4.24}$$

where *Th* is the decision threshold of the receiver.

4.2.3.3 Optimal OOC for λ = 1

The optimal OOC is generally the OOC with the best correlation characteristics (that is, the correlation parameter $\lambda = 1$) and contains $\frac{(L-1)}{w(w-1)}$ codewords. Next, we study $(L, w, 1)$ codewords.

(1) Code capacity analysis of optimal OOCs

Figure 4.5 shows the capacity curve of the optimal OOC for $\lambda = 1$ given by (4.22).

When the code weight w is constant, the code capacity increases with increasing code length. When the code length is fixed, the code capacity decreases with increasing code weight. To improve the capacity of OOCs, it is necessary to increase the code length L or reduce the code weight w. However, increasing the code length requires that the pulse width be reduced; consequently, the requirements for the light sources become very stringent. Reducing the code weight decreases the peak value of the autocorrelation and increases the probability of bit errors during detection, degrading the system performance. Therefore, orthogonal codewords with low code weight are typically used without increasing the complexity of the encoder.

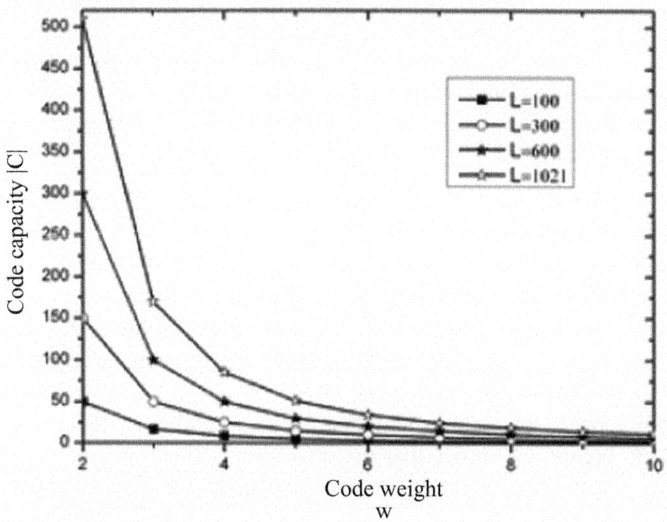

Fig. 4.5 Capacity versus code weight for $\lambda = 1$

There are many algorithms for constructing OOCs. Next, we analyze several construction methods and the relevant characteristics of the generated codewords.

(2) Construction of OOCs using difference matrices

Consider codewords $X = (x_0, x_1, \ldots, x_{L-1})$ and $Y = (y_0, y_1, \ldots, y_{L-1})$, where $i_x = (i_{x,0}, i_{x,1}, \ldots, i_{x,w-1})$ and $i_y = (i_{y,0}, i_{y,1}, \ldots, i_{y,w-1})$ represent the codeword blocks of codewords X and Y, respectively. For code length L and code weight w, the difference matrix of codeword X is [12]

$$S_{xx} = (i_{x,k} - i_{x,j})(\mathrm{mod}\, n) \quad 0 \le k \le w-1 \quad 0 \le j \le w-1 \tag{4.25}$$

The matrix is a $w \times w$ square matrix. The elements in the square matrix indicate that the value of the autocorrelation function at the tangent position indicated by the element is 1. If there are m identical elements in the matrix, the value of the autocorrelation function at the tangent position indicated by the element is m. Thus, the autocorrelation function can be obtained by only w^2 subtractions. The process is simplified, and the number of calculations is reduced. Similarly, the difference matrix of codewords X and Y is [12]

$$S_{xy} = (i_{y,k} - i_{x,j})(\mathrm{mod}\, n) \quad 0 \le k \le w-1 \quad 0 \le j \le w-1 \tag{4.26}$$

The algorithm is constructed as follows, and the process is illustrated in Fig. 4.6 [12].

The capacity of codeword $(L, w, 1)$ is calculated according to the Johnson limit:

(a) Codewords with an autocorrelation limit of 1 are obtained from the difference matrix.
(b) Codewords with a cross-correlation limit of 1 are selected from the codewords with an autocorrelation limit of 1.
(c) The Johnson limit is used to verify whether the constructed codeword exceeds its maximum capacity.

Table 4.2 shows the code block constructed by this algorithm assuming a code length of 30 and code weight of 3.

Codewords (30, 3, 1) are the best equal-weight OOCs with a code length of 30 and code weight of 3. We obtained the autocorrelation and cross-correlation curves of each codeword using a computer simulation, as shown in Fig. 4.7.

Figure 4.8 shows the simulated codeword cross-correlation curve.

Figure 4.7 shows that for the optimal OOC, (30, 3, 1), the autocorrelation peak is 3, which is equal to the code weight, and the autocorrelation sidelobe is less than or equal to 1. Figure 4.8 shows that the cross-correlation value between the two codewords is less than or equal to 1, and its correlation characteristics are consistent with those of the OOC $C(L, w, 1)$.

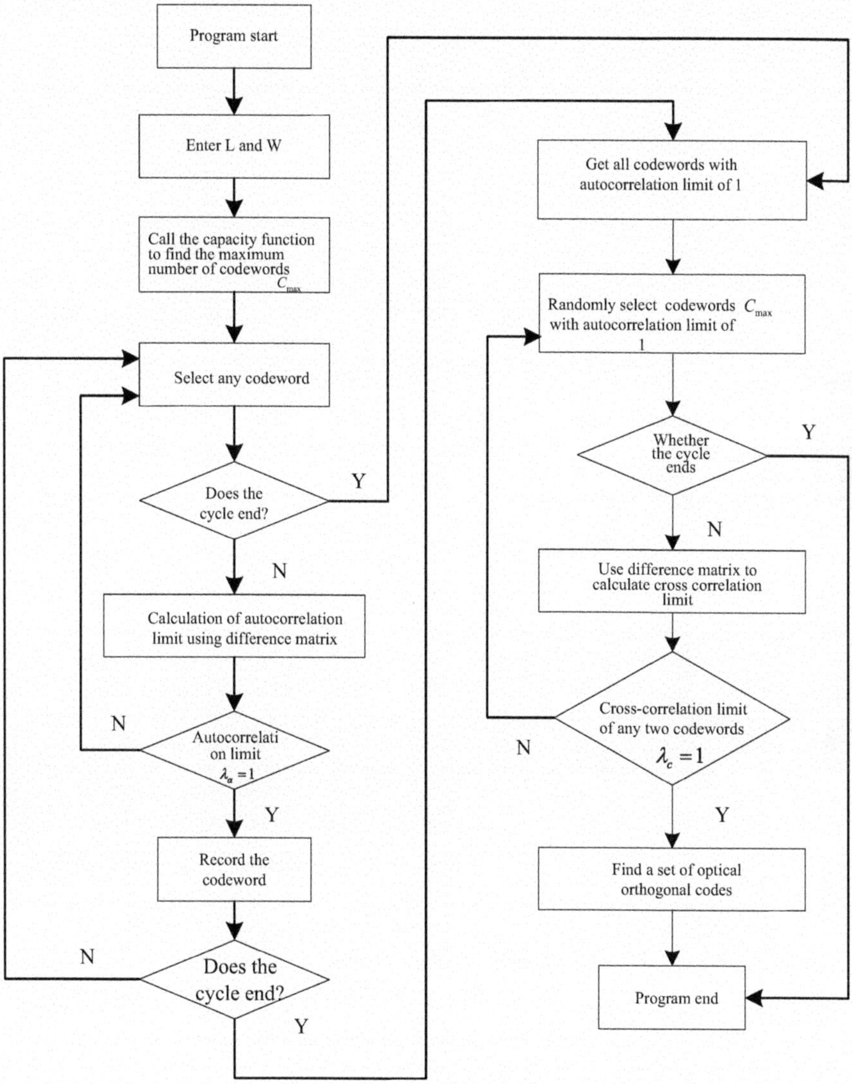

Fig. 4.6 Flow chart for obtaining optimal (L, w, 1) structure [12]

Table 4.2 Codewords (30, 3, 1)

Number of users	Codeword	Number of users	Codeword
1	0, 1, 3	2	0, 4, 9
3	0, 6, 13	4	0, 8, 18

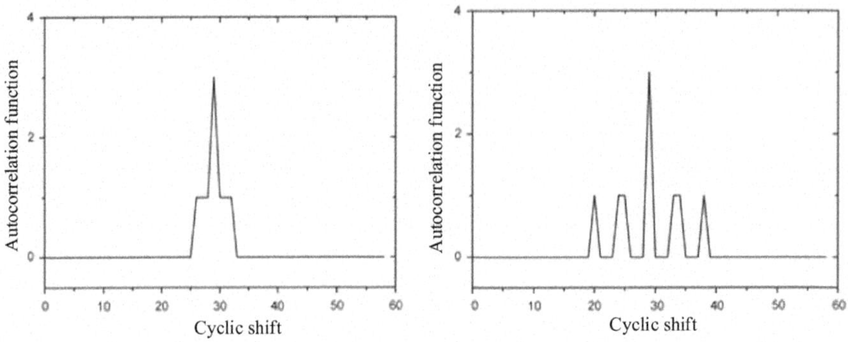

Fig. 4.7 Autocorrelation curves of **a** codeword (30, 3, 1) and **b** codeword (0, 4, 9)

Fig. 4.8 Simulated cross-correlation curve of codewords (0,4,9) and (0,6,13)

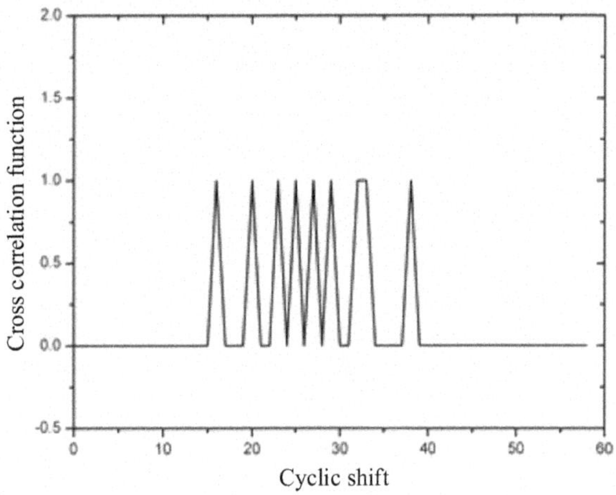

(3) Generation of OOCs by genetic algorithm (GA)

Reference [13] proposed a method of generating OOCs with arbitrary code length and code weight using a GA. For N codewords and an algorithm for searching for $(L, w, 1)$, the method (GA-OOC) is as follows [13].

① Construction of new codeword matrix

Each row of the matrix is designed as an independent code group composed of L codewords. Let the population size (the number of code groups in the new algorithm) be m. The codeword matrix is shown in Fig. 4.9. The length of each codeword ($1 \leq i \leq L$, $1 \leq j \leq m$) is w, and the value is an integer between 0 and $N - 1$ that does not repeat and represents the position of 1 in the codeword. Because the objective function value is calculated only between the codewords of each independent code group, the amount of computation is proportional to $M \cdot L^2$. Because m is no longer required to

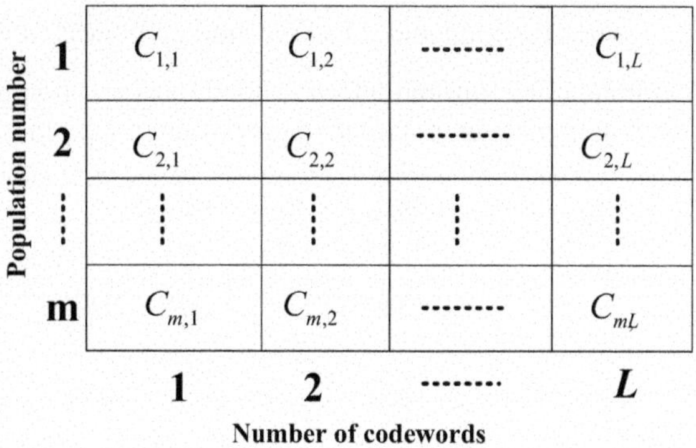

Fig. 4.9 Codeword matrix [13]

be much larger than L, the new algorithm does not significantly increase the amount of computation when alternative codewords are added, which is convenient for the convergence of the algorithm.

② Design of objective function

According to the correlation characteristics between codewords, the objective function $ObjV(x)$ of each codeword x is defined as [13]

$$ObjV(x) = \begin{cases} 0 & if\ A_{x,x} = 0 \\ \frac{1}{N}\sum_{1}^{N} A_{x,y} & if\ A_{x,x} = 1 \end{cases} \tag{4.27}$$

Here y refers to other codewords in the same code group. In addition [13],

$$A_{x,x} = \begin{cases} 1\ if & \sum_{t=0}^{n-1} x_t x_{t+\tau} \leq 1 \quad 0 < \tau < n \\ 0\ otherwish & \end{cases} \tag{4.28}$$

$$\underset{x \neq y}{A_{x,y}} = \begin{cases} 1\ if & \sum_{t=0}^{n-1} x_t y_{t+\tau} \leq 1 \quad 0 < \tau < n \\ 0\ otherwish & \end{cases} \tag{4.29}$$

The average objective function value of each code group (the objective function value hereafter) $\overline{ObjV}$ is defined as the arithmetic average of the objective function values of N codewords. During calculation, the autocorrelation criterion of the codeword is first detected according to (4.28). If the codeword does not meet the autocorrelation criterion, the cross-correlation operation related to it will not be performed, which improves the convergence speed of the algorithm.

③ Fitness function and selection strategy Fitness function

The fitness function and selection strategy fitness function are used to map the objective function value to a relative fitness measure value, which determines the probability that an individual will be copied to the next generation. In this algorithm, the following mapping function [13] is selected:

$$FitnV(P) = 2 - SP + 2(SP - 1)\frac{P - 1}{M - 1} \tag{4.30}$$

The value of pressure SP is a real number between (0, 2], and P is the number of individuals in the sorted population, $P = 1, 2, \cdots, M$.

In this algorithm, the random universal sampling method is used to select the number of individuals that will be copied to the next generation according to the fitness value.

④ Crossover and mutation strategies

The two-point crossing scheme is selected, and a crossing point at an integral multiple of w (the code weight) is selected to avoid invalid codewords caused by breaking in the middle of codewords. Random numbers are used to control the chips in the codeword matrix to mutate according to the mutation probability P_m. After mutation, the presence of duplicate numbers in the codeword is detected; that is, the code weight is w.

⑤ Introduction of new immigrants

This algorithm introduces a migration mechanism; that is, in each generation, the codewords that do not meet the autocorrelation conditions given in (4.28) are replaced with probability P_i, and new codewords that meet the autocorrelation characteristics are randomly generated.

⑥ Iteration termination criteria

When the number of codewords is less than or equal to the Johnson limit, the maximum value of the objective function is 1. The algorithm terminates when the value of the objective function approaches 1 or when the predetermined number of iterations is completed.

The codeword (100, 3, 1) was constructed by the GA-OOC as described above (Table 4.3).

Figure 4.10 shows the autocorrelation and cross-correlation curves of some codewords obtained by a computer simulation using this group of codewords.

The optimal OOC (100, 3, 1) has an autocorrelation peak of 3, which is equal to the code weight, and the autocorrelation sidelobe is less than or equal to 1. The cross-correlation value of two codewords is less than or equal to 1, and its correlation characteristics meet the requirements.

Table 4.3 Codewords of (100, 3, 1) [13]

Number of users	Codeword	Number of users	Codeword
1	0, 3, 46	2	0, 4, 70
3	0, 5, 6	4	0, 7, 31
5	0, 9, 37	6	0, 12, 14
7	0, 13, 14	8	0, 18, 44
9	0, 21, 29	10	0, 25, 35
11	0, 33, 55	12	0, 41, 61
13	0, 42, 53	14	0, 48, 84
15	0, 49, 81	16	0, 62, 85

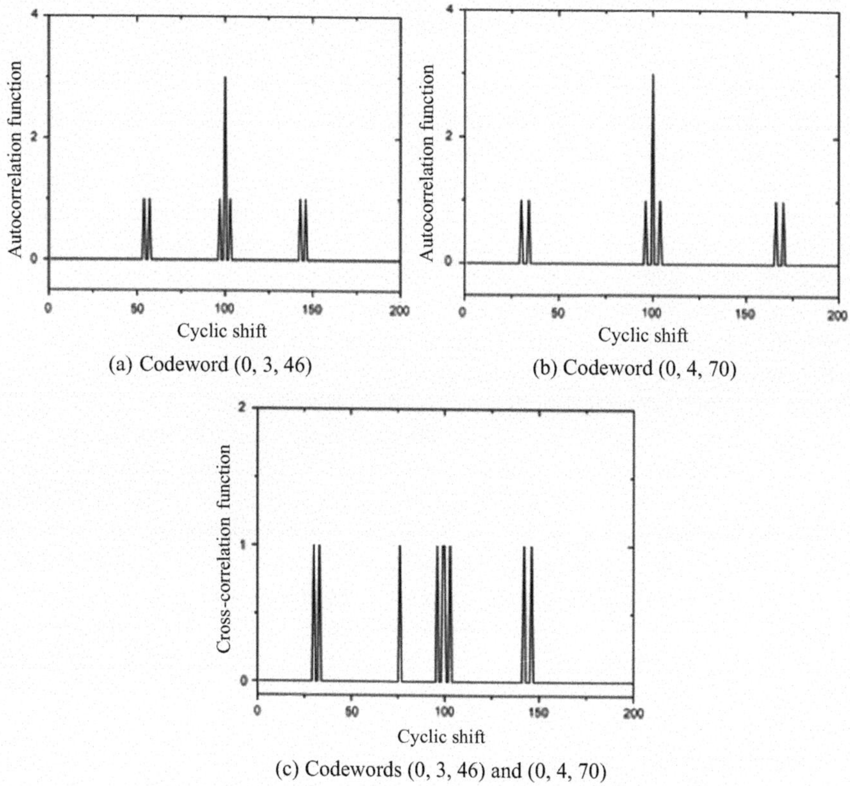

Fig. 4.10 Simulated **a**, **b** autocorrelation curves and **c** cross-correlation curve of codewords of (100, 3, 1)

4.2.3.4 Optimal OOC for $\lambda = 2$

The optimal OOC capacity and the use of codeword selection have been extensively studied, but most studies have considered the BER from the mathematical or physical point of view and ignored the bandwidth utilization and equipment cost in actual network applications. In address code design, it is necessary to balance effectiveness and reliability.

(1) Codeword capacity analysis

For optimal OOCs, a correlation limit of $\lambda = 1$ is generally set; that is, the autocorrelation and cross-correlation limits are both 1. This limit indicates a large autocorrelation peak and the smallest maximum cross-correlation peak. It directly reflects the facts that the correlation between codewords is optimal, the decision is the most accurate, decision error is least likely to occur, and the BER is low. In terms of the effectiveness and reliability of communication, the performance is optimal. However, effectiveness and reliability have a trade-off relationship in communication systems; improving the reliability decreases the effectiveness, and vice versa. Therefore, although both effectiveness and reliability should be considered, more attention should be given to one of them than the other, as appropriate; the results can often have unexpected effects on the system [14]. Figure 4.11 shows the codeword capacity versus code weight for OOCs with different code lengths under different correlation limits.

The code capacity of the optimal OOC is closely related to the code weight and length. In addition, if we consider increasing the correlation limit appropriately, that is, reducing the reliability of the system appropriately, the code capacity of the optimal OOC increases rapidly, and its effectiveness increases considerably. In Fig. 4.11c, where the code length L is 300, the code capacity increases by two orders of magnitude with the appropriate reduction in reliability. With increasing code length, when $L = 1021$ [Fig. 4.11a], the code capacity increases by three orders of magnitude or even more. When $\lambda = 1$, the number of codewords of the OOC is relatively small. However, in an actual network system, the number of codewords can be greatly increased by appropriately increasing the size according to the system's robustness to various types of interference (including MAI). For example, for an OOC with a code length L of 1021, a weight w of 4, and $\lambda = 1$, the number of codewords is $\lambda = 2$, and the number of codewords is 43,307. The notable increase greatly increased the number of users, which may be useful for addressing sudden or urgent congestion.

(2) Correlation of OOCs of (L, 4, 2).

Reference [15] reported the construction method and performance of (L, 4, 2) OOCs. Next, we simulate and analyze the correlation and capacity of (L, 4, 2) codes. Tables 4.4 and 4.5 show two (L, 4, 2) OOC groups [15].

Figures 4.12 and 4.13 show the autocorrelation and cross-correlation curves of codewords obtained by simulations of the codeword groups shown in Tables 4.4 and 4.5, respectively. The capacity of the (L, 4, 1) and (L, 4, 2) codewords is analyzed, and the code capacity versus code length is shown in Fig. 4.14.

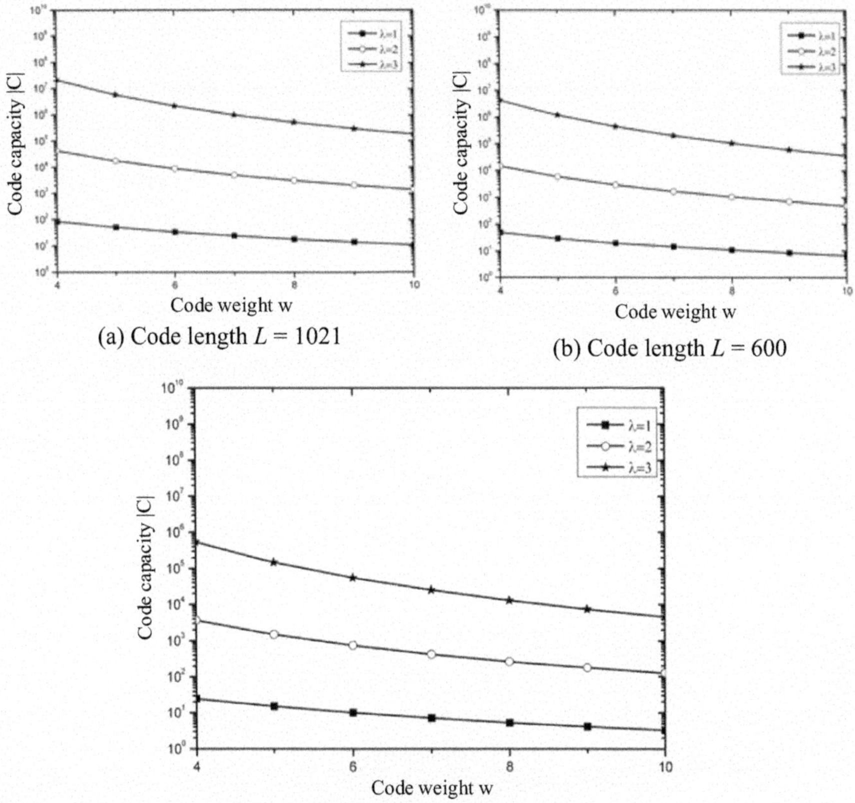

(a) Code length $L = 1021$

(b) Code length $L = 600$

(c) Code length $L = 1021$

Fig. 4.11 Relationship between OOC capacity and correlation limit

Table 4.4 Codewords of (10, 4, 2) [15]

Number of users	Codeword
1	0, 2, 4, 7
2	0, 1, 3, 4
3	0, 1, 2, 6

Table 4.5 Codewords of (20, 4, 2) [15]

Number of users	Codeword	Number of users	Codeword
1	0, 1, 2, 11	2	0, 1, 3, 18
3	0, 1, 4, 17	4	0, 1, 5, 6
5	0, 1, 7, 8	6	0, 1, 9, 12
7	0, 2, 4, 12	8	0, 2, 6, 8
9	0, 2, 7, 9	10	0, 3, 6, 13
11	0, 3, 8, 15	12	0, 3, 9, 14
13	0, 4, 8, 14	14	0, 4, 9, 13

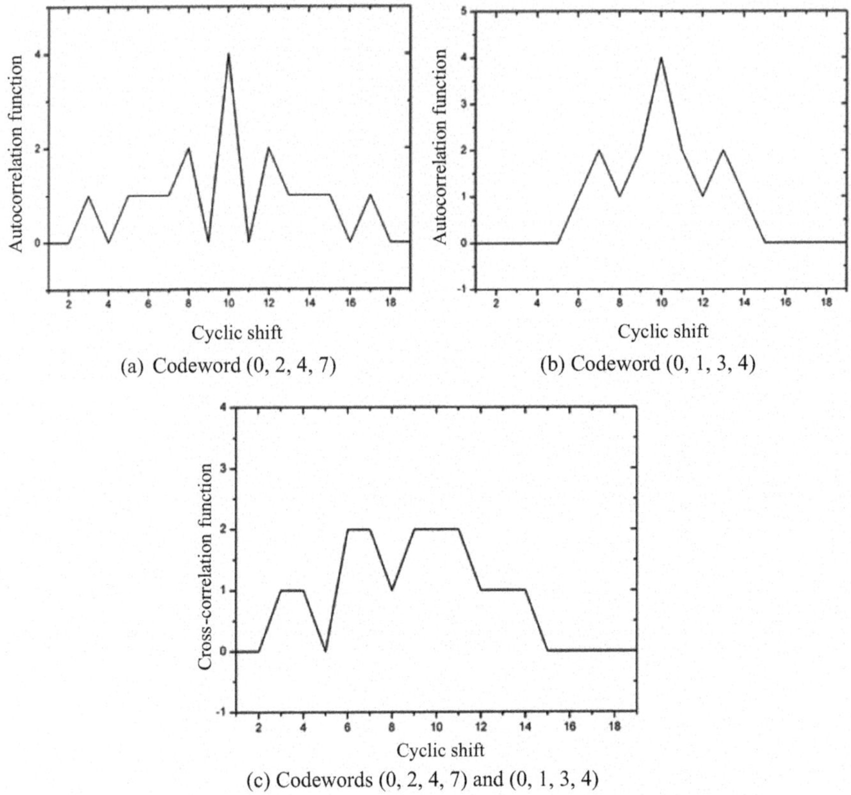

Fig. 4.12 Simulated **a**, **b** autocorrelation and **c** cross-correlation curves of codewords of (10, 4, 2)

As shown in Figs. 4.12 and 4.13, the autocorrelation limit of the codeword is 4, which is equal to the code weight; the autocorrelation sidelobe value is less than or equal to 2, and the cross-correlation limit of the two codewords is less than or equal to 2. In addition, the correlation of the codeword with $\lambda = 2$ is lower than that of the codeword with $\lambda = 1$. Figure 4.14 shows that the capacity of the (L, 4, 2) codes increases rapidly with increasing code length. The capacity of codewords with the same code length is significantly higher than that of the (L, 4, 1) codes, and the (L, 4, 2) codes can greatly improve the capacity by increasing the correlation limit of codewords. In actual networks, the appropriate OOC can be selected according to the demand.

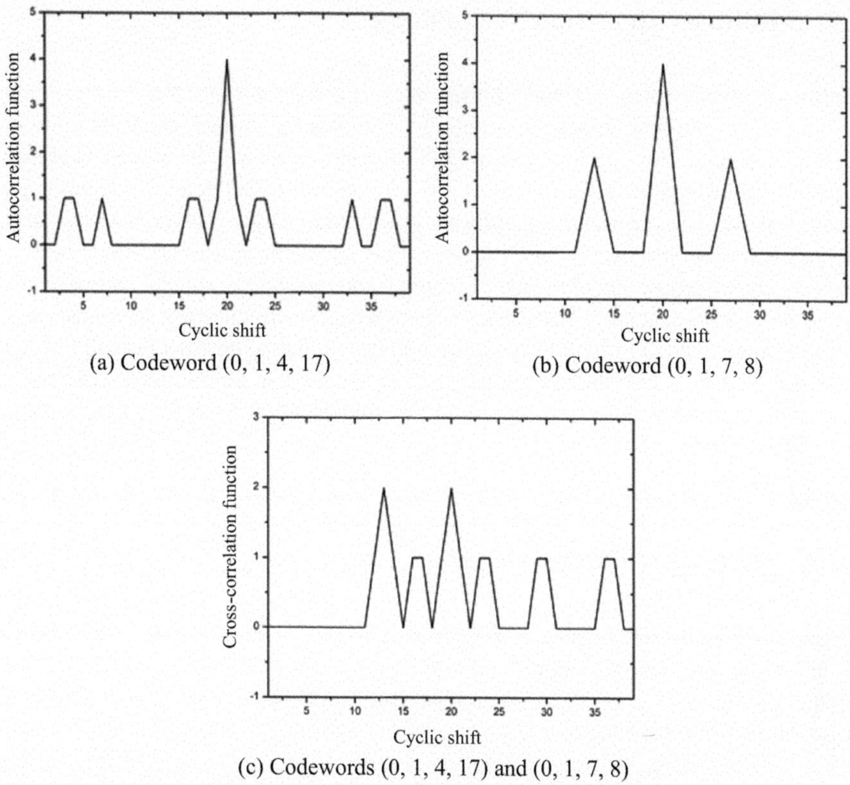

(a) Codeword (0, 1, 4, 17)

(b) Codeword (0, 1, 7, 8)

(c) Codewords (0, 1, 4, 17) and (0, 1, 7, 8)

Fig. 4.13 Simulated autocorrelation and cross-correlation curves of codewords of (20, 4, 2)

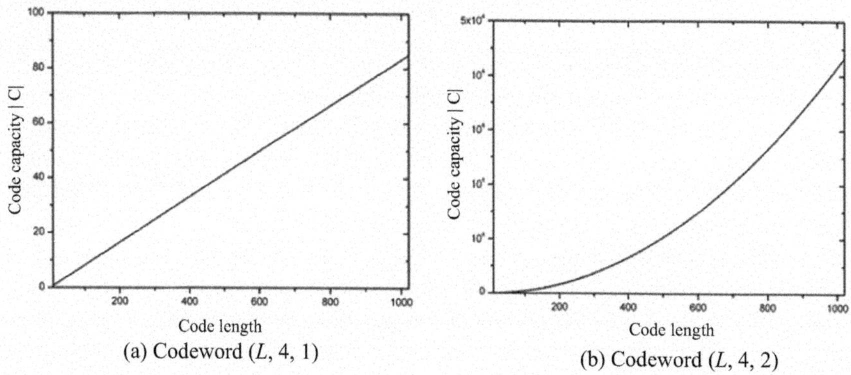

(a) Codeword (L, 4, 1)

(b) Codeword (L, 4, 2)

Fig. 4.14 Code capacity versus code length L for codewords (L, 4, 1) and (L, 4, 2)

4.3 Multiuser Detection Technology

Multiuser detection is the key anti-interference technology in CDMA communication systems. Conventional detection technology matches the spread spectrum code of each user signal according to classical DSSS theory; therefore, the anti-MAI ability is poor. Multiuser detection technology based on conventional detection technology makes full use of all the user signal information, including that causing MAI, to detect the signal of a single user. Consequently, it has excellent anti-interference performance, solves the problem of the near/far effect, and reduces the system requirements for power control accuracy. Therefore, it can make more effective use of uplink spectrum resources and significantly improve the system capacity. Theoretically, if the MAI received by users can be eliminated, the capacity can be improved. The basic concept of multiuser detection is to treat the signals of all users as useful signals rather than interference signals.

4.3.1 *Optimal Multiuser Detector*

The optimal multiuser detector was proposed by Verdu in 1986 [16]. It is based on the maximum Bayes posterior probability. Essentially, it seeks the maximum input sequence $b = [b_1, b_2, \ldots, b_N]$ according to maximum likelihood sequence estimation to minimize the output noise [16]:

$$\left[\hat{b}_1 \ldots \hat{b}_N\right] = \arg \max_{b_1 b_2 \ldots b_N} \exp\left\{-\frac{1}{2\sigma^2}\int_0^T \left[r(t) - \sum_{k=1}^{N} \lambda_k b_k s_k(t)\right]^2 dt\right\} \tag{4.31}$$

The optimal multiuser detector can achieve the optimal performance for a single user under ideal conditions, it can theoretically obtain the minimum BER and provide the best detection performance. However, it is not suitable for practical application. The main reason is that it needs to know almost all the information of the expected and interfering users, including the characteristic waveform, timing information, and relative amplitudes. This information cannot be obtained a priori and must be estimated, which inevitably causes errors and affects the detection performance. The computational complexity of the optimal multiuser detector is exponentially related to the number of users, and the amount of computation is too large for practical application. Although the optimal multiuser detection algorithm is difficult to implement physically, some suboptimal multiuser detectors have been proposed. The performance of a suboptimal detector is very similar to that of the optimal detector, but suboptimal detector algorithms are much less complex and require less information from users. Suboptimal detectors can be realized by engineering.

4.3.2 Linear Multiuser Detector

Linear multiuser detection is a suboptimal method of multiuser detection. It is based on certain decision criteria; a linear operator is added to the back of the matched filter, and then the output sequence is evaluated to obtain the required signal. The complexity of this type of multiuser detector is linear with the number of users. The specific principle is as follows. Let *t* represent the linear transformation matrix. When the output signal vector of the matched filter is $\underset{N}{\vec{r}}$, the output signal of the linear multiuser detector is $T\underset{N}{\vec{r}}$. Figure 4.15 shows its structure block diagram, $\hat{b} = (\hat{b}_1, \hat{b}_2, \ldots, \hat{b}_N)^T$, which shows the output signal $\underset{N}{\vec{r}}$ of the receiver matched filter bank through the linear transformation matrix.

The purpose of linear multiuser detection is to find a linear transformation matrix to map sufficient statistics to the set of multiuser symbol sequences. The output matrix of the conventional correlator is transformed linearly, and the transformed output sequence is evaluated. When the linear transformation matrix takes different forms, it can be divided into decorrelation detection and minimum mean square error (MMSE) detection.

4.3.2.1 Decorrelation Detector

The decorrelation multiuser detector has been used in a wireless OCDMA system to eliminate MAI [17].

As shown in Fig. 4.16, the input signal $r(t)$ is multiplied by the user address code $s_k(t)$, and the multiplied signal is input into the correlator. That is, the signal is integrated to obtain n correlation values, and the n values are linearly transformed (that is, multiplied by the inverse of the correlation matrix) to obtain n linearly transformed values, which are then evaluated. The correlator output is

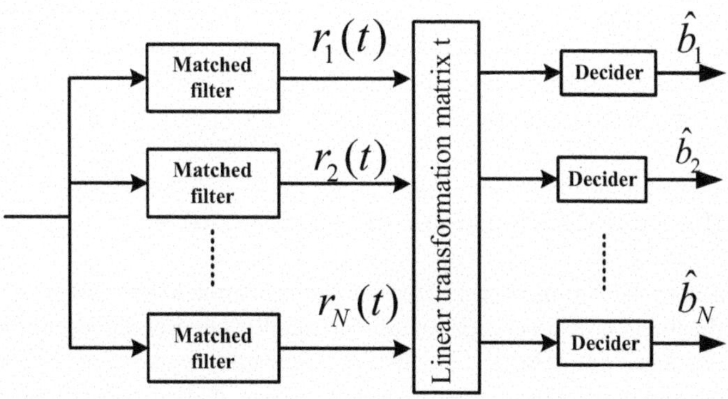

Fig. 4.15 Block diagram of linear multiuser detector

Fig. 4.16 Structure of decorrelation detector [17]

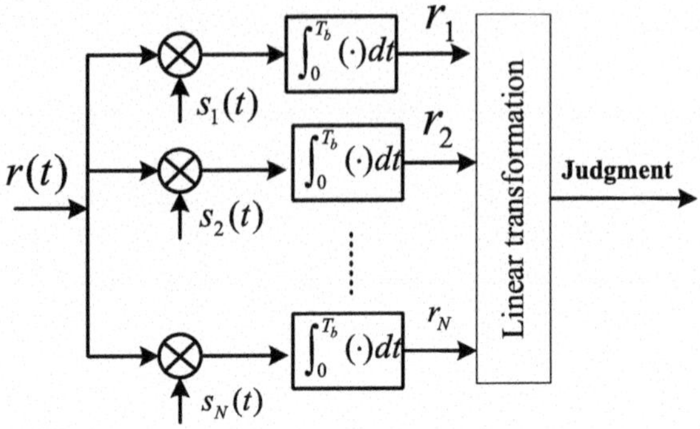

$$\vec{r}_N = \vec{R}\,\vec{b}_N + \vec{n}_N \tag{4.32}$$

where $\vec{b}_N = [\lambda_1 b_1, \lambda_2 b_2, \ldots, \lambda_N b_N]^T$, and $\vec{n}_N = [n_1, n_2, \ldots, n_N]^T$. The correlation matrix $\vec{R}$ of the address code is

$$\vec{R} = \begin{bmatrix} 1 & \rho_{12} & \cdots & \rho_{1N} \\ \rho_{12} & 1 & \cdots & \rho_{2N} \\ & & \ddots & \\ \rho_{1N} & \rho_{2N} & \cdots & 1 \end{bmatrix} \tag{4.33}$$

Suppose the matrix $\vec{R}$ is reversible, and the linear transformation matrix $T_{dec} = \vec{R}^{-1}$. Then, the output of the decorrelation detector is [18]

$$\vec{R}^{-1}\vec{r}_N = \vec{R}^{-1}(\vec{R}\vec{b}_N) = \vec{b}_N + \vec{R}^{-1}\vec{n}_N \tag{4.34}$$

Equation (4.34) shows that other interference is set to 0, the only interference source is background noise, and decorrelation detection completely eliminates MAI. However, the cost of noise power improvement is that the computational complexity is linear with the number of users N.

4.3.2.2 Minimum Mean Square Error (MMSE) Detector

Linear multiuser detection is regarded as a linear estimation problem, which is to transform the linear transformation required to find the kth user into the minimization of the mean square error (MSE). The MMSE algorithm regards the received signal vector as the sum of each user signal vector and noise; it attempts to obtain the coefficient value of each function vector using the correlation matrix and received signal amplitude. The mean square value of the error between the kth user transmitted signal b_k and its estimated value after detection is minimized. The linear transformation matrix T is selected to minimize this MSE; the technique is denoted as $\mathrm{MMSE}_{\mathrm{K}}$, and T is a $k \times k$ matrix:

$$T = [mmse_1, mmse_2, \ldots mmse_k] \tag{4.35}$$

We estimate

$$\hat{b}_k = \operatorname{sgn}[MMSE_k^T y] \tag{4.36}$$

The spread spectrum sequence s_k of the matched filter output signal must meet the criterion [19]

$$s_k = \arg\min_{s_k} E\{[b_k - (s_k, y)]\}^2 \tag{4.37}$$

Then the output decision statistics is given by [19]

$$\hat{b}_k = \operatorname{sgn}(s_k, y) \tag{4.38}$$

If $b = [b_1, b_2, \ldots b_k]^T$, the linear transformation factor that minimizes the MSE is [19]

$$J(s_i) = E\{\|b_i - Ty\|\}^2 \tag{4.39}$$

The MSE is minimized, and the covariance matrix of the error vector is minimized to obtain [19]

$$\begin{aligned} \operatorname{cov}(b - Ty) &= E\{(b - Ty)(b - Ty)^T\} \\ &= E\{bb^T\} - E\{by^T\}T^T - TE\{yb^T\} + TE\{yy^T\}T^T \end{aligned} \tag{4.40}$$

Using (4.40) and the irrelevant uncorrelation between the noise and data vectors, and assuming $E\{b^2\} = 1$ (where a is the user signal amplitude, n is the noise, and σ^2 is the noise variance), we obtain [19]

$$E\{bb^T\} = 1 \tag{4.41}$$

$$E\{by^T\} = E\{bb^T AR\} = AR \tag{4.42}$$

$$E\{yb^T\} = E\{RAbb^T AR\} = RA \tag{4.43}$$

$$E\{yy^T\} = E\{RAbb^T AR\} + E\{nn^T\} = RA^2R + \sigma^2 R \tag{4.44}$$

By substituting these formulas into (4.40), the covariance matrix of the error vector can be expressed as [19]

$$\mathrm{cov}(b - Ty) = I + T(RA^2R + \sigma^2 R)T^T - ART^T - TRA \tag{4.45}$$

Because $\min\{\|x\|^2\} = \min\{tr(xx^T)\}$ [19],

$$\min J(T) = \min\{tr[cov(b - Ty)]\} \tag{4.46}$$

To find the minimum value problem, note that

$$\frac{d}{dM} tr\{\mathrm{cov}(b - Ty)\} = 0 \tag{4.47}$$

We substitute (4.45) into (4.47) to obtain

$$T_{MMSE}(RA^2R + \sigma^2 R) = AR \tag{4.48}$$

Assuming that the matrix R is nonsingular, the above formula is simplified as

$$T_{MMSE}(RA^2 + \sigma^2 I) = A \tag{4.49}$$

Because A is a diagonal matrix,

$$T_{MMSE} = A(RA^2 + \sigma^2 I)^{-1} \tag{4.50}$$

The output of the MMSE detector is [19]

$$\begin{aligned} \hat{b}_k &= \mathrm{sgn}[(T_{MMSE}y)_k] \\ &= \mathrm{sgn}\{([RA^2 + \sigma^2 I]^{-1} y)_k\} \end{aligned} \tag{4.51}$$

We multiply both sides of (4.50) by A^{-2} to obtain the second expression for the MMSE detector [19]:

$$T_{MMSE} = A^{-1}(R + \sigma^2 A^{-2})^{-1} \tag{4.52}$$

The corresponding detector is given by [19]

$$\begin{aligned} \hat{b}_k &= \mathrm{sgn}[(T_{MMSE}y)_k] \\ &= \mathrm{sgn}\{([R + \sigma^2 A^{-2}]^{-1} y)_k\} \end{aligned} \tag{4.53}$$

When σ tends to zero and $[R + \sigma^2 A^{-2}]^{-1}$ tends to R^{-1}, the MMSE detector converges to a decorrelator.

4.3.3 Optical Parallel Interference Cancellation (OPIC)

The parallel interference cancellation (PIC) detector has a multilevel structure. Each level estimates and removes the MAI caused by each user in parallel and then makes a data decision. Suppose the system has n users, and user 1 is the target user. Figure 4.17 shows a two-level PIC detector structure. The received signal is denoted as $r(t)$. The first stage of the detector is conventional correlation detection. It makes a decision based on the data signal of $n-1$ non-target users and spreads the determined jth non-target user signal $\hat{b}_i^{(j)}$ with the corresponding address code; that is, it regenerates the interference generated by each user to the target user and intercepts this regenerated interference in the received signal $r(t)$. In the second stage, the detector uses the correlation detector to detect and recover the target user signal. Figure 4.17 shows a block diagram of the PIC detector [20].

The PIC detector is resistant to the near/far effect; in particular, when the received user signal power difference is large, it can improve the detection of weak signals. Therefore, the system delay increases with increasing number of stages; however, among suboptimal multiuser detection technologies, PIC detection technology still has high practical value. To simplify the design and reduce the amount of calculation, the conventional correlation detector is selected as the first-stage detector for the general PIC detector.

The performance of a one-stage OPIC (OS-OPIC) detector in OCDMA has been evaluated [21]. The conventional PIC detector receiver signal is written as [21]

$$r(t) = \sum_{k=1}^{N} b_i s_k(t - \tau_k) \tag{4.54}$$

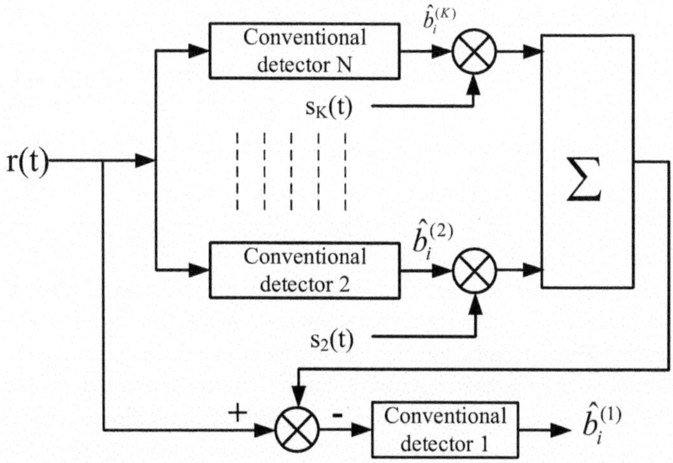

Fig. 4.17 Block diagram of PIC detector [20]

where τ_k is the delay of user k, and $0 \leq \tau_k \leq T_c$ and T_c are chip intervals. In addition, $b_i(0, 1)$ is the ith bit of user k. The signal entering the receiver of the target user is [21]

$$y(t) = r(t) - \sum_{j=2}^{N} \hat{b}_i^{(j)} s_j(t) \tag{4.55}$$

where $\hat{b}_i^{(j)}$ and $s_j(t)$ are the estimated value and address code, respectively, of user j. OS-OPIC is based on conventional OPIC but meets the requirements of the receiver with lower hardware complexity; it is very simple and effective. Unlike OPIC, it requires all non-target user information to recover the target user information. OS-OPIC detection needs only to provide an estimated $\hat{b}_i^{(k)}$ value for target user k; Use the corresponding address code $S_k(t)$ to spread the estimated valueand finally subtracts it from the received signal $r(t)$.Then the target user is detected by the conventional detector. The input signal of the conventional detector is [21]

$$y(t) = r(t) - \hat{b}_i^{(k)} s_k(t) \tag{4.56}$$

where $\hat{b}_i^{(k)}$ and $s_k(t)$ are the estimated value and address code, respectively, of non-target user k.

OS-OPIC can effectively eliminate MAI and reduce the required hardware complexity compared with conventional OPIC.

4.4 Multiuser Detection Technology for UV OCDMA Systems

4.4.1 Characteristics of UV Atmospheric Channel

When UV light propagates in the atmosphere, it has two typical characteristics:

① Exponential energy attenuation is caused by absorption and scattering by suspended particles in the atmosphere. The signal energy decays rapidly according to an exponential law. The light energy loss caused by atmospheric Rayleigh scattering is more than 1000 times that in the infrared band [22].

② UV light undergoes extremely strong scattering. Because there are many particles in the atmosphere, strong scattering occurs during atmospheric UV transmission. Consequently, UV communication systems can transmit signals in non-line-of-sight (NLOS) mode, and thus they can be useful in complex terrain. However, the greatest disadvantages of the scattering transmission mode are the low power efficiency and multipath delay caused by multipath propagation.

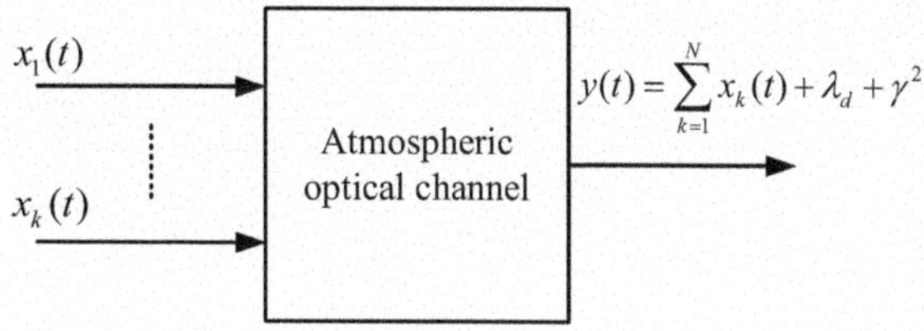

Fig. 4.18 Atmospheric OCDMA channel model [7]

The channel characteristics of NLOS UV scattering communication have been simulated and analyzed [23]; the results showed that signal transmission in UV communication is accompanied by multipath transmission, which greatly limits the system performance.

The main effects of atmospheric turbulence are beam broadening, scintillation, and angle-of-arrival fluctuation. When angle and position tracking are used, the drift and angle-of-arrival fluctuation can be greatly reduced. Atmospheric scintillation, which causes the received power to fluctuate, is one of the main problems in atmospheric wireless optical communication. The use of a large-aperture receiver can effectively suppress the effect of light scintillation on optical communication links.

At the receiving end, the variation in received power around the average power appears as additional noise; that is, the variance of the received power must be considered as detector noise. Figure 4.18 shows the atmospheric OCDMA channel model [7].

4.4.2 Decorrelation and Minimum Mean Square Error Detection (MMSE) Detection

Figure 4.19 shows a block diagram of delay attenuation in a multipath channel. To study the impact of time delay and multipath interference on the performance of multiuser detection algorithms in the UV atmospheric channel, we use this block diagram to simulate the energy attenuation and multipath delay in transmission according to the atmospheric channel analysis in the previous section.

In this section, we assume the following system in the simulation: the number of users k is 4; the average optical power of user k is λ_k, where [1, 1, 1, 3]; the average power of background light noise is λ_d; the mean value of thermal Gaussian noise is 1; and the variance is 0.001.

We selected a group of (30, 3, 1) OOCs as the address codes of the input signal. For user 1, the address code is (0, 1, 3); for user 2, the address code is (0, 8, 18); for user 3, the address code is (0, 4, 9); for user 4, the address code is (0, 6, 13).

The simulation was performed in MATLAB software. Taking user 1 as an example, the performance of decorrelation and MMSE detection under different conditions is analyzed.

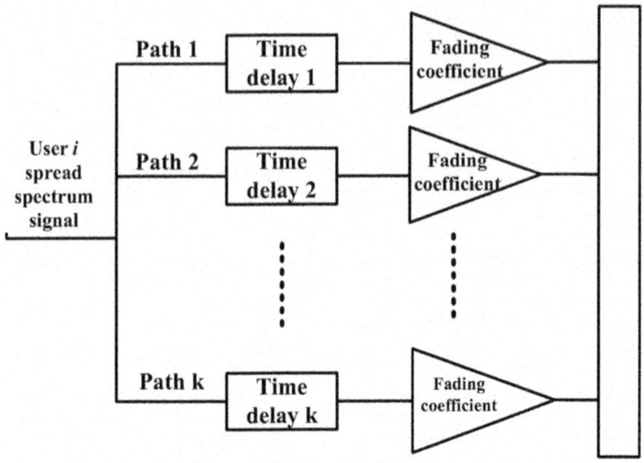

Fig. 4.19 Delay attenuation in multipath channel

(1) Background light noise λ_d

Figure 4.20 shows the BER of the system with conventional, decorrelation, and MMSE detection for background light noise with different average powers. Decorrelation and MMSE detection exhibit much better performance than conventional detection; they largely eliminate the impact of MAI, and the BER reaches 10^{-4}. Decorrelation and MMSE detection exhibit similar performance. Decorrelation detection outperforms MMSE detection when the background noise is low. When the average power of the background noise λ_d is greater than 0.7, MMSE detection outperforms decorrelation detection. When λ_d is greater than 1.1, the BER of decorrelation and MMSE detection is 10^{-1}, and the signal is buried in the background noise, making communication difficult.

If only the MAI is removed, the decorrelation detector exhibits optimal performance, but at the cost of noise enhancement. MMSE detection also exploits the

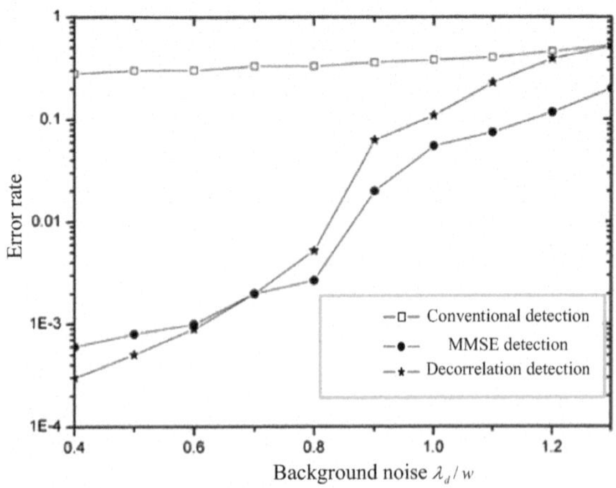

Fig. 4.20 BER of each detector versus background noise

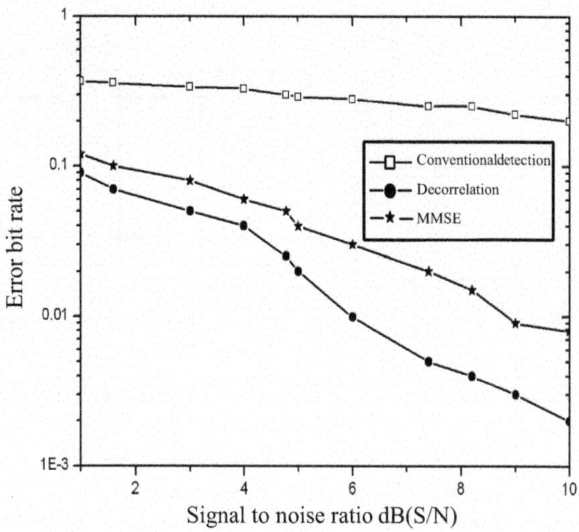

Fig. 4.21 BER of conventional, decorrelation, and MMSE detection for various signal-to-noise ratios in multipath channel

characteristics of the conventional decorrelation detector to solve the problem of correlation noise enhancement, but it cannot completely eliminate MAI.

(2) Multipath channel

The performance of conventional, decorrelation, and MMSE detection in multipath channels was studied. Four paths with attenuation coefficients of 0.9, 0.5, 0.2, and 0.1 were assumed. The paths with attenuation coefficients of 0.2 and 0.1 each have a delay of one codeword. Figure 4.21 shows the BER curves of conventional, decorrelation, and MMSE detection for different signal-to-noise ratios.

In a multipath channel, decorrelation and MMSE detection exhibit much better performance than conventional detection, and the system BER reaches 10^{-3}. In addition, decorrelation detection outperforms MMSE detection. With increasing signal-to-noise ratio, the performance advantage of decorrelation detection becomes more notable. The reason is that although MMSE detection can suppress MAI and background noise simultaneously, its performance depends on the power of interfering users, and multipath fading results in near/far effects. Decorrelation detection has the best ability to resist the near/far effect. By contrast, the simulation showed that MMSE detection could no longer resist the near/far effect at high signal-to-noise ratios.

4.4.3 MMSE-PIC Detection

In Sect. 4.3.2.2, we analyzed the MMSE detector, which exploits the characteristics of the conventional decorrelation detector, to solve the problem of understanding related noise enhancement issues. However, the effects of MAI and near/far effect

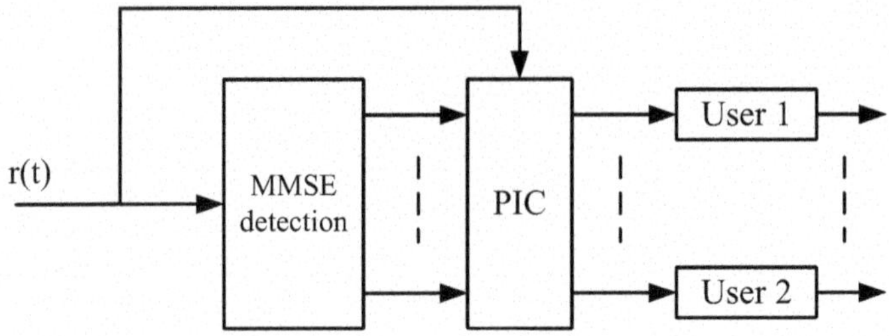

Fig. 4.22 MMSE_ PIC detector block diagram

cannot be completely eliminated. In particular, for the UV scattering channel, the near/far effect of the signal at the receiver becomes increasingly obvious. Therefore, joint detection by MMSE and PIC is used in OCDMA systems.

PIC detection is well able to resist the near/far effect, especially to improve weak power signals. Adding a PIC detector after the MMSE detector can further eliminate interference and reduce the BER of the system. Figure 4.22 shows a block diagram of the MMSE-PIC detector.

The linear transformation used in MMSE detection is given by

$$T_{MMSE}(RA^2R+\sigma^2R)=AR \tag{4.57}$$

In the first stage of MMSE-PIC detection, the output of MMSE detection is

$$\begin{aligned}\tilde{x}&=T_{MMSE}r\\&=[R+\sigma^2A^{-2}]^{-1}r\end{aligned} \tag{4.58}$$

For target user i and interfering user j in a system with n users, the PIC algorithm can be written as

$$\underline{\hat{x}}(i)=y-\sum_{\substack{j=1\\j\neq i}}^{K}R(j)\tilde{x}(j) \tag{4.59}$$

An OHL is introduced before the MMSE detector to limit the optical pulse of each chip before the receiver correlator performs correlation processing of the received signal. The OHL reduces the light intensity accumulation of multiple users in the time slot of the target user transmission signal; that is, it reduces the MAI. In this section, a dual-OHL structure is introduced to optimize the system transmission performance and further reduce the system BER. Figure 4.23 shows a block diagram of the MMSE-PIC detector based on optical hard limiting.

We assume the following system in the simulation: the number of users k is 4; the average optical power of user k is λ_k, where $\lambda_k=[1,4,1,3]$; and the UV channel is a

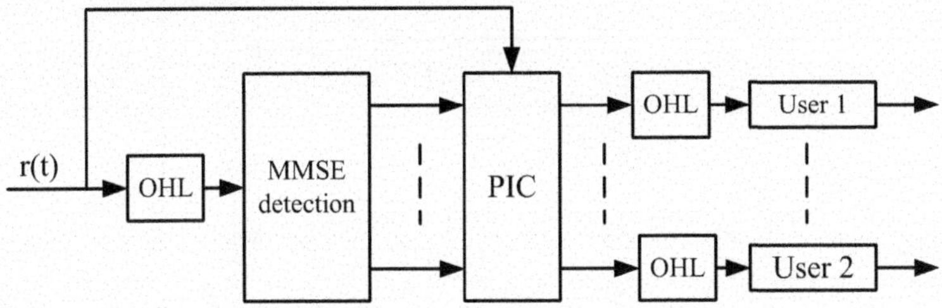

Fig. 4.23 Block diagram of MMSE-PIC detector based on optical hard limiting

typical Rayleigh fading channel in sunny weather (that is, the effects of atmospheric conditions such as rain and fog are ignored).

We select a group of (30, 3, 1) OOCs as the address codes of the input signal. For user 1, the address code is (0, 1, 3); for user 2, the address code is (0, 8, 18); for user 3, the address code is (0, 4, 9); for user 4, the address code is (0, 6, 13).

The simulation was performed in MATLAB software. Taking user 1 as an example, the performance of PIC detection and MMSE-PIC detection based on optical hard limiting was analyzed. Figure 4.24 shows the BER curve of each detector for different channels.

MMSE-PIC detection outperforms PIC detection in the multipath Rayleigh fading channel. After MMSE-PIC detection, the BER of user 1 reaches 10^{-3}. MMSE-PIC detection can improve weak power signals. However, the BER of the system increases with increasing number of Rayleigh fading channel paths.

Figures 4.25, 4.26 and 4.27 show schematic diagrams of the MATLAB simulation of the spread spectrum, receiving, and MMSE multiuser detection modules, respectively

MMSE-PIC detection can suppress MAI and improve weak power signals in optical multipath channels; it has a strong ability to eliminate the near/far effect and is suitable for atmospheric UV communication.

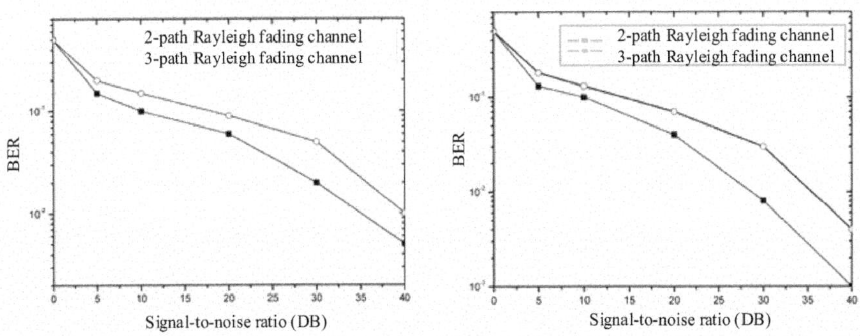

Fig. 4.24 BER curves of **a** PIC detection and **b** MMSE-PIC detection based on optical hard limiting for different channels

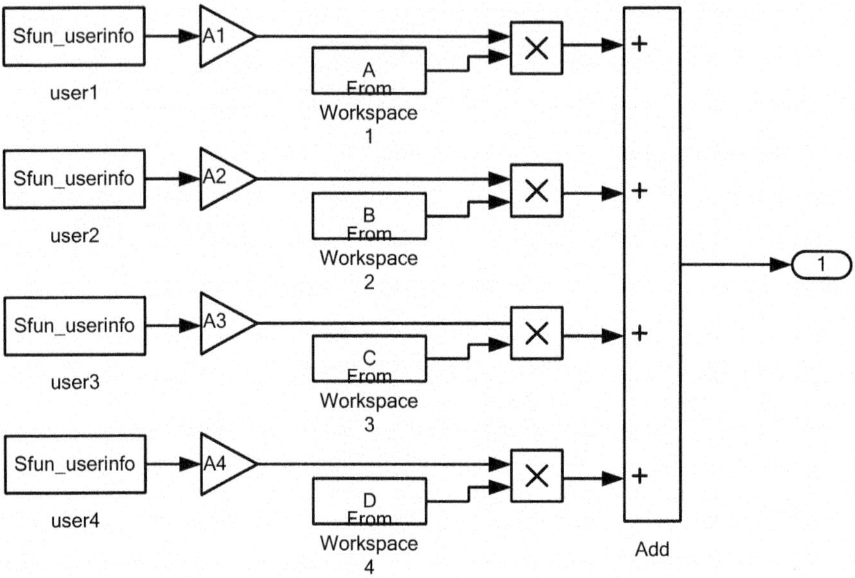

Fig. 4.25 Schematic diagram of MATLAB simulation of spread spectrum module

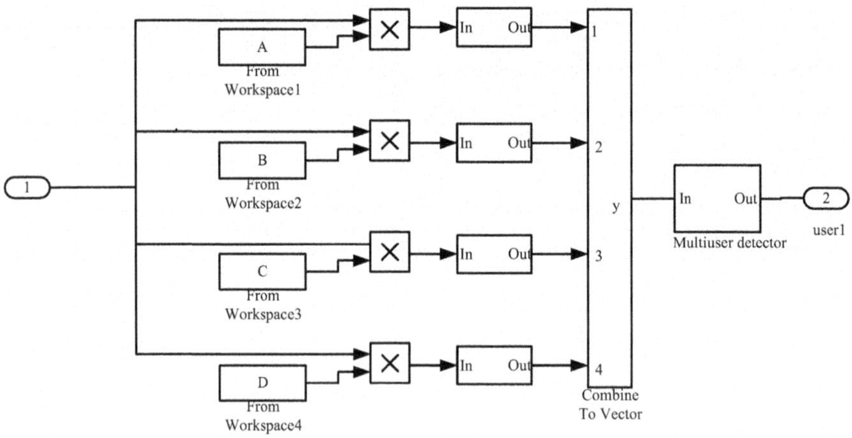

Fig. 4.26 Schematic diagram of MATLAB simulation of receiving module

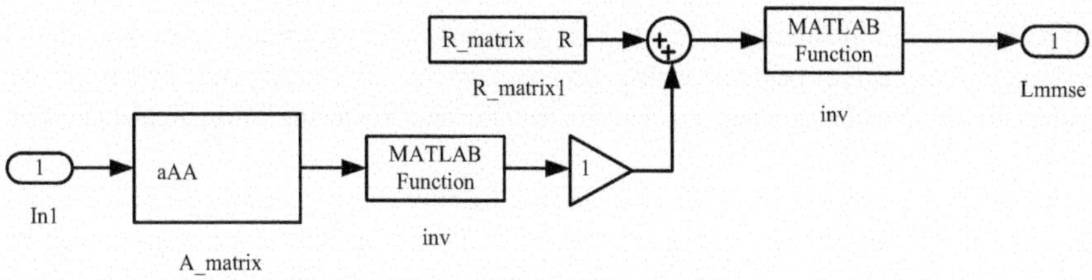

Fig. 4.27 Schematic diagram of MATLAB simulation of MMSE multiuser detection module

4.5 Multiuser Detection Technology for UV Ad Hoc Networks

In UV ad hoc networks based on OCDMA, the terminal typically uses a matched filter to detect the received data. Because of MAI in the network, the conventional detection method cannot meet the requirements of ad hoc networks. We use multiuser detection technology to improve the overall performance of the network.

Signal processing in networking requires the integration of signal processing technology into the design of the network structure and protocol to improve the performance of wireless networks. Specific research areas include network service modeling using signal processing methods, traffic-assisted multiuser detection, the adaptive prediction and tracking of network parameters, channel reservation based on conflict resolution, and network protocol design and performance analysis based on multipacket reception [24].

In multipacket reception [24], network nodes can use signal processing technology to distinguish and detect multiple data packets arriving simultaneously. In this case, conventional network nodes identify the data packets as conflicting and discard all the conflicting packets. Multipacket reception can greatly improve the throughput and delay of the network.

CDMA divides a shared channel into multiple subchannels. Each subchannel corresponds to a unique address code. If the signature codes are completely orthogonal and the node can operate in parallel on multiple subchannels (for example, if it is equipped with N matched filters to receive data on N subchannels simultaneously), for the node itself, the performance is equivalent to multipacket reception. In addition, compared with time-division multiple-access and frequency-division multiple-access systems, CDMA systems have the advantages of large capacity, anti-noise ability, and resistance to multipath effects., which make them suitable for wireless networks.

However, the number of available orthogonal feature codes is very limited. To obtain more feature codes, approximately orthogonal pseudo-random codes are generally used. In addition, considering the fading of the wireless channel, the movement of nodes, and other factors, even if an orthogonal code is used, it is difficult to maintain orthogonality in actual communication on ad hoc networks, and MAI occurs. Overall, MAI increases the BER of the receiver and reduces the network capacity. Locally, if the data packet to be received comes from a distant node and a node near the receiver is also sending data packets, strong interference from the near node will annihilate the weak signal from the far end, resulting in the near/far effect [25].

In addition, energy attenuation and multipath delay in the UV atmospheric channel increase the system noise and near/far effect. These problems must be solved to improve the performance of ad hoc networks based on UV OCDMA systems.

In OCDMA systems, the near/far effect can be avoided using power control. However, ad hoc networks have no special facility for centralized control, and the location of nodes changes constantly. Although the implementation of power control can be facilitated by the exchange of current power information before each data transmission and reception, this exchange occupies bandwidth and makes the system complex. Therefore, it is not feasible to use conventional power control methods to suppress near/far effects.

For OCDMA-based ad hoc networks, multiuser detection can be used to suppress MAI and avoid the near/far effect. When a multiuser detector is used, a node is much more likely to successfully extract its own packets from mixed signals, which greatly improves the efficiency of multipacket reception.

Most of the multiuser detectors used in previous studies of wireless ad hoc networks are classical MMSE detectors or nonlinear detectors such as PIC and continuous interference cancellation detectors. By contrast, multiuser detection technology for wireless optical ad hoc networks has not been reported. In this section, we analyze the effect of the blind adaptive algorithm on the throughput of ad hoc networks and propose to introduce the blind adaptive algorithm into OCDMA systems using OOCs.

4.5.1 Throughput Analysis of Ad Hoc Network Based on Multiuser Detection

The throughput of ad hoc networks has been analyzed [26] using a full-link finite-element ad hoc network as a model. The following assumptions were made regarding the ad hoc network [26].

(1) In each time slot, a terminal can only send or receive one packet. The packet length is fixed.
(2) The probability that any terminal sends packets in any time slot is p.
(3) When one terminal sends data, other terminals in the network become the target terminals of the data packet with equal probability.

The network is assumed to have m nodes (terminals), and the following random variables are defined:

L: The total number of packets sent in the network (that is, the number of terminals in the "send" state) in the current time slot.

X: In L packets, the total number of packets whose target node is in the "receive" state.

Y: The total number of packets among X packets that do not encounter conflicts.
N: The total number of packets successfully received in the current time slot.
In addition, $0 \leq N \leq Y \leq X \leq L \leq M$.

The probability that l nodes send data packets simultaneously in a time slot is [26]

$$p_l = C_M^l p^l (1-p)^{M-1}, l = 0, 1, 2, \ldots, M \tag{4.60}$$

The throughput T of the ad hoc network is defined as the average number of packets successfully received in the network within a time slot [26]:

$$T = \sum_{l=1}^{M} p_l \sum_{n=0}^{l} n r_{ln} \tag{4.61}$$

where r_{ln} is the corresponding element of the reception matrix R.

The reception matrix R is defined as [26]

$$R = \begin{pmatrix} r_{10} & r_{11} & 0 & & 0 \\ r_{20} & r_{21} & r_{22} & \cdots & 0 \\ \vdots & \vdots & & \ddots & \vdots \\ r_{M0} & r_{M1} & r_{M2} & \cdots & r_{MM} \end{pmatrix} \tag{4.62}$$

where r_{ij} represents the probability of the successful reception of j packets when the entire network sends i packets in one time slot. When $j > i$, $r_{ij} = 0$. In addition, considering the half-duplex characteristics of the terminal, when $i = m$, no node is in the receive state; therefore, $r_{M0} = 1$, and $r_{Mj} = 0 (j = 1, 2, \cdots M)$.

For nonzero elements in r, $r_{ln} (1 \leq l \leq M - 1, 0 \leq n \leq l)$. The probability can be expressed as [26]

$$\begin{aligned} r_{ln} &= \sum_{x=n}^{l} Prob\{\boldsymbol{N} = n, \boldsymbol{X} = x | \boldsymbol{L} = l\} \\ &= \sum_{x=n}^{l} Prob\{\boldsymbol{N} = n | \boldsymbol{L} = l, \boldsymbol{X} = x\} Prob\{\boldsymbol{X} = x | \boldsymbol{L} = l\} \end{aligned} \tag{4.63}$$

Suppose that when a terminal sends data, other terminals in the network become the target terminal of the data packet with equal probability; then [26]

$$Prob\{\boldsymbol{X} = x | \boldsymbol{L} = l\} = C_l^x \left(\frac{M-l}{M-1}\right)^x \left(\frac{l-1}{M-1}\right)^{l-x} \tag{4.64}$$

where $C_l^x = \frac{l!}{x!(l-x)!}$.

The spread spectrum protocol adopted here is based on the receiver. When the same terminal is the target node of two or more data packets, the target terminal cannot distinguish these packets, and they are lost. In the current time slot, the total number of packets discarded due to multipacket conflicts in the network is $(\boldsymbol{X}-\boldsymbol{Y})$, and $\boldsymbol{X}-\boldsymbol{Y}\geq 2$. Therefore [26],

$$\begin{aligned} & Prob\{\boldsymbol{N}=n,\boldsymbol{X}=x\} \\ &= \sum_{\substack{y=n \\ y\neq(x-1)}}^{x} Prob\{\boldsymbol{N}=n,\boldsymbol{Y}=y|\boldsymbol{L}=l,\boldsymbol{X}=x\} \\ &= \sum_{\substack{y=n \\ y\neq(x-1)}}^{x} Prob\{\boldsymbol{N}=n|\boldsymbol{L}=l,\boldsymbol{X}=x,\boldsymbol{Y}=y\}Prob\{\boldsymbol{Y}=y|\boldsymbol{X}=x,\boldsymbol{L}=l\} \end{aligned} \tag{4.65}$$

$Prob\{\boldsymbol{N}=n|\boldsymbol{L}=l,\boldsymbol{X}=x,\boldsymbol{Y}=y\}$ is related to the BER P_e of the multiuser detector in the physical layer.

The approximate solution of P_e is obtained by Gaussian approximation [26]:

$$P_e\approx Q\left(1\Big/\sqrt{\alpha_1}+\sum_{i=2}^{k}\beta_i\right) \tag{4.66}$$

where $\alpha_1\triangleq\left[\sigma^2\left(1+\|c_1-s_1\|^2\right)\right]/A_1^2$; $\beta_i\triangleq\left(A_i^2|s_i,c_1|^2\right)/A_1^2$, $i=2,3,\cdots,k$; s_i is the spread spectrum code of the interference data packet; A_1 and A_i are the amplitudes of the expected and interference data packets at the receiver input, respectively; c_1 is the linear multiuser detector of the target user; σ^2 is the noise power; and $Q(\cdot)$ is a function of Q. The parameters α_1 and β_i represent the contributions of noise and MAI, respectively, to P_e.

Equations (4.61)–(4.66) show that $P_e(k)$ is different for different multiuser detectors (i.e., different c_1); these differences result in difference in $Prob\{\boldsymbol{N}=n|\boldsymbol{L}=l,\boldsymbol{X}=x,\boldsymbol{Y}=y\}$ and ultimately in throughput. The following conclusions can be drawn [26].

(1) The environment affects the BER. With decreasing average noise power σ^2, the expected packet reception power A_1^2 increases, and $P_e(k)$ decreases.
(2) Multiuser detection affects the BER. To investigate these effects, the parameters σ and A_1 must be fixed, and the effect of x_1 or $c_1(c_1=s_1+x_1)$ on $P_e(k)$ must be considered.

   ① If $x_1=0(c_1=s_1)$, the detector degenerates into a conventional detector. In this case, $P_e(k)$ is determined by the cross-correlation coefficient between the spread spectrum codes corresponding to the transmission of the desired and interference data packets, and the ratio of their respective powers. In other words, the near/far effect caused by MAI causes the degradation of the receiver performance.

② If c_1 is selected so that $(s_i, c_1) = 0$ for each s_i, the detector becomes a correlation detector. In this case, MAI is completely eliminated, but $\|x_1\|^2$ plays a negative role because it amplifies the noise.

③ MMSE detectors make a reasonable trade-off in suppressing MAI and avoiding noise amplification. However, as the wireless channel is a time-varying system, it is difficult to recalculate the linear transformation matrix every time for linear multiuser detection, and the amount of computation is large.

The above analysis reveals that adaptive technology should be adopted for multiuser detection in ad hoc networks. Adaptive technology can not only automatically track channel changes, but also reduce the amount of computation. A typical adaptive linear detector does not need a matched filter. The received signal is first filtered by the chip and then sampled at the chip rate. The sampled signal is processed by the linear receiver, and the output signal is sampled at the symbol rate and finally evaluated. The linear receiver is calculated and adjusted by different algorithms. Adaptive algorithms include a general algorithm with a transmission sequence and a blind adaptive algorithm. Adaptive algorithms with a transmission sequence have been studied more extensively, in particular the least mean square (LMS) error random gradient algorithm and recursive least square (RLS) algorithm.

Multiuser detection technology for cellular CDMA communication systems has been proposed. Because the base station knows the address codes of all users in the cell, it is convenient to perform joint detection such as MMSE or decorrelation multiuser detection [25].

However, in distributed ad hoc networks, because the base station is not under centralized control, it is difficult for a node to obtain the address codes of adjacent nodes. It must use other means (such as the medium access control protocol) to obtain this information, which adds overhead. Therefore, it is not realistic to obtain the address codes of all interfering nodes and then perform multiuser detection. The multiuser detection algorithm proposed in Chapter 4 is not suitable for ad hoc networks.

Therefore, ad hoc networks should use a blind multiuser detection technology that needs only the address code of the target user. In addition, only the observation data are used instead of the training sequence. Sending a training sequence is not only a great waste of bandwidth resources, but also consumes more energy. Therefore, for ad hoc networks, where bandwidth and energy resources are extremely limited, blind multiuser detection has clear advantages. Moreover, the sending sequence not only occupies certain resources, but also improves the transmission power to ensure the reliability of the decision.

Energy attenuation and multipath delay in the UV atmospheric channel also increase the system noise and near/far effect. In the multipath channel, it is easy to produce deep fading, and training sequences must be sent frequently, which greatly degrades the system performance. Therefore, blind multiuser detection technology that does not interfere with user information is more suitable for UV ad hoc networks. Next, we analyze the blind adaptive RLS algorithm and introduce it into an OCDMA system to simulate and analyze its performance using OOCs.

4.5.2 Blind Adaptive Linear Detection Algorithm Based on Minimum-Output-Energy (MOE) Criterion

4.5.2.1 MOE Criterion

Honig proposed that a linear multiuser detector c can be decomposed into the sum of two mutually orthogonal components [27], one of which is the target characteristic sequence s of the user.

If user 1 is the target user, $\mathbf{c}_1$ can be represented as [27]

$$\mathbf{c}_1 = \mathbf{s}_1 + \mathbf{x}_1 \tag{4.67}$$

In addition, the conditions $\mathbf{s}_1^T\mathbf{x}_1 = \mathbf{0}$ and $<\mathbf{c}_1, \mathbf{s}_1> = <\mathbf{s}_1, \mathbf{s}_1> + <\mathbf{x}_1, \mathbf{s}_1> = \|\mathbf{s}_1\|^2 = 1$ are met.

Because $\mathbf{s}_1$ is the spread spectrum sequence of user 1, the linear multiuser detector $\mathbf{c}_1$ can also directly describe the characteristic waveform of user 1 using the orthogonal component $\mathbf{x}_1$, which is invariant in the adaptive algorithm; the second component $\mathbf{x}_1$ must be updated. Therefore, the updated design of the detector is an update of the filter coefficient $\mathbf{c}_1$, which is equivalent to an update of $\mathbf{x}_1$.

The minimum-output-energy detector can be seen as adding a quadrature receiving component to the conventional detector, and this vector can be adjusted adaptively by the output value. The control criterion is the constrained minimum-output-energy (CMOE) criterion, where [27]

$$CMOE(\mathbf{x}_1) = E\{< \mathbf{s}_1 + \mathbf{x}_1, \mathbf{r} >^2\} = E\{((\mathbf{s}_1 + \mathbf{x}_1)^T \mathbf{r})^2\} \tag{4.68}$$

We take the minimum value. That is, the energy output by the detector can be expressed as the sum of the energies of the desired signal and the interference signal (background noise and MAI), and the energy of the desired signal is independent of the selection of $\mathbf{x}_1$. Therefore, by minimizing the output energy, the energy of the interference component is minimized, and thus the interference is reduced.

According to the CMOE criterion, the blind multiuser detector $\mathbf{c}_1$ shall satisfy the following constrained optimization problem [27]:

$$\begin{cases} \min E\{(c_1^T r)^2\} \\ s.t.\ s_1^T c_1 = 1 \end{cases} \tag{4.69}$$

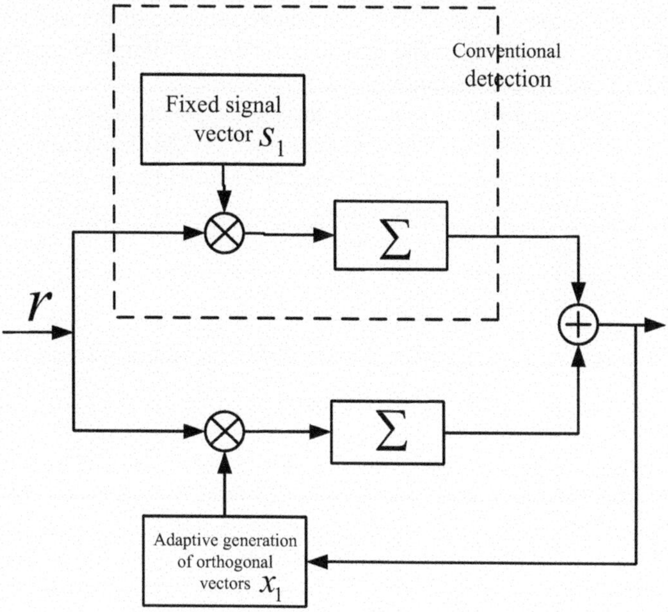

Fig. 4.28 Block diagram of minimum-output-energy multiuser detector [29]

The function $CMOE(\mathbf{x}_1)$ is a convex function in the space orthogonal to $\mathbf{x}_1$, with no local or global minimum. It is easy to adopt a gradient descent algorithm, including multiuser detection algorithms such as LMS and RLS [28]. Figure 4.28 shows the minimum-output-energy multiuser detector [29].

4.5.2.2 Blind Adaptive RLS Algorithm

The LMS and RLS algorithms are used in blind adaptive multiuser detection. The LMS algorithm has the advantages of a simple structure and strong robustness; however, the convergence speed is slow. The convergence performance of the algorithm depends on the dispersion of the eigenvalues of the input signal autocorrelation matrix. With increasing dispersion, the convergence is degraded. The convergence of the RLS algorithm is independent of the input signal spectrum; the algorithm has good convergence and numerical characteristics, has good adaptability to nonstationary signals, and is convenient for modular implementation [30]. Because it converges rapidly, the RLS algorithm is more suitable for wireless channels in complex and changeable environments. This section analyzes the performance of the RLS adaptive algorithm in UV ad hoc networks.

The key to the RLS algorithm is the replacement of the LMS criterion used in the LMS algorithm [31] with the quadratic time-average minimization criterion, that is, the averaging and minimization of the square of all the errors from the initial time to the current time, and the use of the least-square estimate of the filter tap weight vector in the $n-1$th iteration. The new data for the nth iteration can be used to calculate the updated vector estimation.

The RLS adaptive algorithm extends the least-square method to an adaptive algorithm. An exponential weighted least-square method is considered here instead of a general least-square method. That is, the weighted sum of the squares of the errors is used as the cost function, and the exponential weighted sum is used to replace $CMOE(\mathbf{c}_1) = E\{(\boldsymbol{c}_1^T\boldsymbol{r})^2\}$ to obtain the following constrained optimization problem [32]:

$$\begin{cases} \min \sum_{i=1}^{n} \lambda^{n-i}[\mathbf{c}_1^T(n)\mathbf{r}(i)]^2 \\ \boldsymbol{s.t.}\ \boldsymbol{s}_1^T\boldsymbol{c}_1(n) = 1 \end{cases} \tag{4.70}$$

The weighting factor λ is called the forgetting factor and satisfies $0 < \lambda < 1,\ 1-\lambda < 1$; its function is to add a relatively large weight to the error closer to time n, and a relatively small weight to the errors farther from time n. In other words, λ has a certain forgetting effect on the errors at each time. By using the Lagrange multiplier method, the closed-form solution of the optimization problem can be obtained as [33]

$$\mathbf{c}_1(n) = \frac{\mathbf{R}^{-1}(n)\mathbf{s}_1}{\mathbf{s}_1^T\mathbf{R}^{-1}(n)\mathbf{s}_1} \tag{4.71}$$

where $\mathbf{R}(n) = \sum_{i=1}^{n} \lambda^{n-1}[\mathbf{r}(i)\mathbf{r}^T(i)]^2$ represents the autocorrelation matrix of the observed signal.

The MOE detection algorithm based on the constrained RLS method is thus given by [33]

$$\boldsymbol{k}(n) = \frac{\boldsymbol{R}^{-1}(n-1)\boldsymbol{r}(n)}{\lambda + \boldsymbol{r}(n)^T\boldsymbol{R}^{-1}(n-1)\boldsymbol{r}(n)} \tag{4.72}$$

$$\mathbf{h}(n) = \mathbf{R}^{-1}(n)\mathbf{s}_1 = \frac{1}{\lambda}[\mathbf{h}(n-1) - \mathbf{k}(n)\mathbf{r}(n)^T\mathbf{h}(n-1)] \tag{4.73}$$

$$\mathbf{c}_1(n) = \frac{1}{\mathbf{s}_1^T\mathbf{h}(n)}\mathbf{h}(n) \tag{4.74}$$

The inverse matrix of the correlation matrix is updated as follows [33]:

$$\mathbf{R}^{-1}(n) = \frac{1}{\lambda}[\mathbf{R}^{-1}(n-1) - \mathbf{k}(n)\mathbf{r}(n)^T\mathbf{R}^{-1}(n-1)] \tag{4.75}$$

The blind RLS detector output is

$$\tilde{b}(n) = < \mathbf{c}_1, \mathbf{r} > = \mathbf{c}_1^T(n)\mathbf{r}(n) \tag{4.76}$$

4.5.2.3 Simulation Analysis

The performance of multiuser detection algorithms can be evaluated measuring the output-signal-to-interference ratio (SINR). At time n, the SINR of detector output is defined as [33]

$$SINR_n = \frac{[E(c_n^T r_n)]^2}{\text{var}[c_n^T r_n]} \tag{4.77}$$

We assume a synchronous OCDMA system under additive Gaussian white noise, and we expect user 1 to have unit energy, that is, $A_1^2 = 1$, with a signal-to-noise ratio of 20 dB and a processing gain or spread spectrum gain of 31. There are three interfering users, two of which have a signal-to-noise ratio of 30 dB, whereas the other has a signal-to-noise ratio of 40 dB. The blind adaptive multiuser detector using the RLS algorithm is simulated, and the BER curves are shown in Figs. 4.29, 4.30, and 4.31.

Figure 4.29 shows the BER obtained in 100 independent simulations with $\lambda = 0.99$. The tracking of the expected user signal is relatively good, and the estimated BER of the information characters transmitted by the expected user is low; thus, the

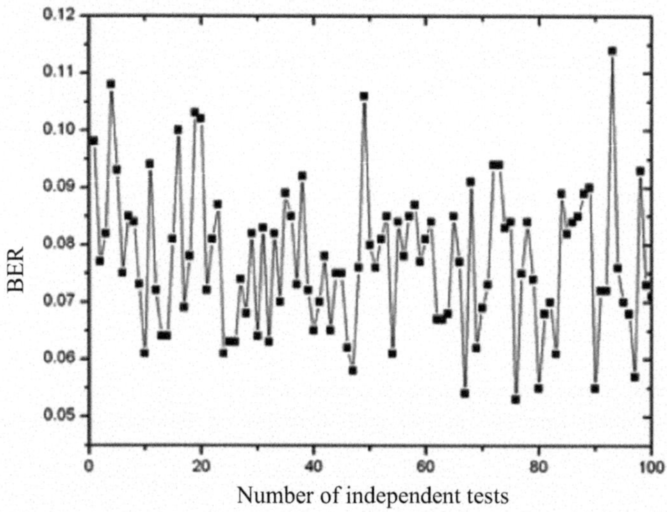

Fig. 4.29 BER curve of RLS algorithm with forgetting factor $\lambda = 0.99$

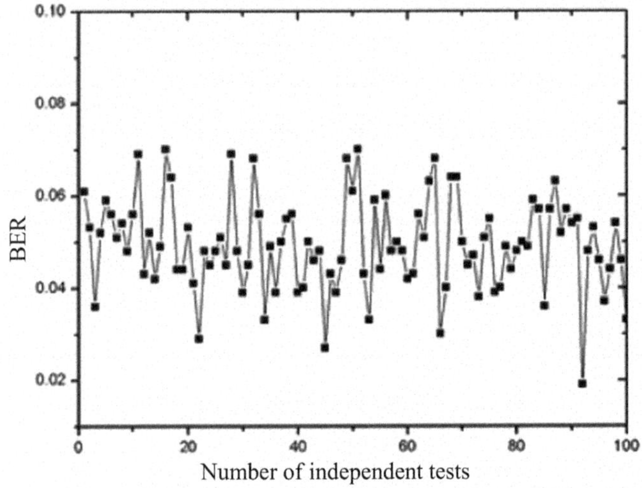

Fig. 4.30 BER curve of RLS algorithm with forgetting factor $\lambda = 0.997$

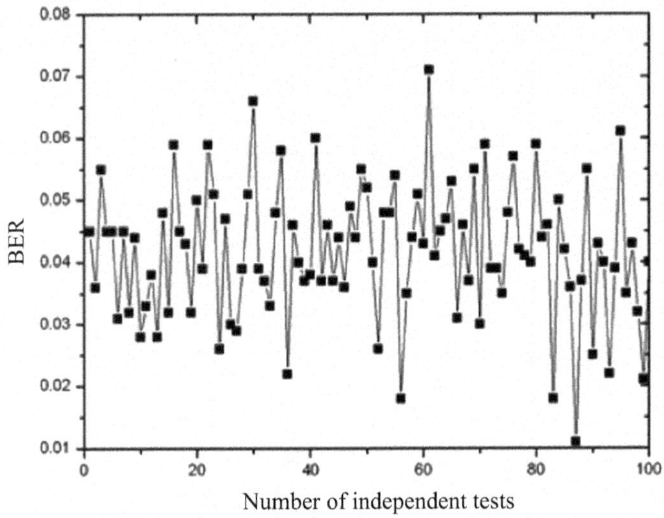

Fig. 4.31 BER curve of RLS algorithm with forgetting factor $\lambda = 0.9997$

algorithm can achieve blind multiuser detection in an OCDMA system. In addition, the RLS algorithm can converge well without a training sequence and suppress MAI; the system BER reaches 10^{-2}.

Figures 4.30 and 4.31 show the BER curves for $\lambda = 0.997$ and $\lambda = 0.9997$, respectively. When λ is closer to 1, the performance of the algorithm improves, and the BER decreases. Next, strong MAI with unit energy is added, and 1000

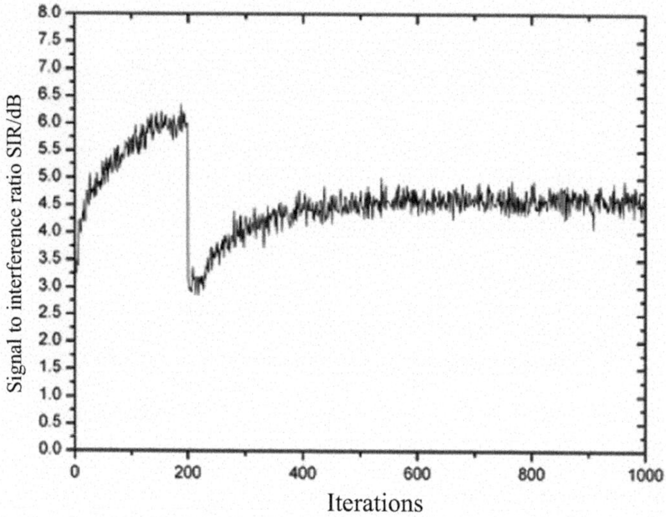

Fig. 4.32 SINR curve of RLS algorithm for $\lambda = 0.99$

independent experiments are performed in which the algorithm is iterated 200 times. Figure 4.32 shows the convergence of the algorithm for $\lambda = 0.99$.

When strong interference suddenly occurs in the system, the algorithm can quickly converge to the optimal value, and the SINR of convergence in this case does not differ greatly from that before the strong interference appeared. The algorithm can not only rapidly converge and track the dynamic environment at any time, but also ensures a relatively high output SINR.

Appendices B-1, B-2, and B-3 show the MATLAB data generation and multiuser detection algorithm programs. The blind adaptive RLS algorithm has a fast convergence rate and good performance. It has good ability to suppress MAI and eliminate the near/far effect in OCDMA systems. It does not need to send training sequences and thus conserves valuable bandwidth resources and reduces energy consumption. Therefore, in terms of performance and complexity, it is highly suitable for use in UV ad hoc networks.

References

1. Ma F (2008) Research and implementation of spread spectrum technology. Master's thesis of Xi'an University of Electronic Science and Technology, Xi'an, 5
2. Caixiaozhe (2006) Research on detection method of direct sequence spread spectrum signal. Master's thesis of Xi'an University of Electronic Science and Technology, Xi'an, pp 7–8
3. Fuxiaomei, Yu J, Wangwenrui et al (2005) Research on two dimensional optical orthogonal code coded optical code division multiple access system. Optoelectron Laser 16(12):1459–1462
4. Lin P, Hu Y, Zhouxiuli (2005) 2D w_multiuser detection technology in OCDMA system. Opt Commun Res 2:50–52

5. Lixiaobin (2007) Research on optical code division multiple access multiuser interference suppression method. Semiconductor Optoelectron 28(1):116–130
6. Wangfuchang (2006) Optical code division multiple access_OCDMA_research and implementation of communication system. Master's thesis of Zhejiang University, Hangzhou, pp 9–10
7. Wang Y, Wan S, Hu Y (1999) Research on coding technology of wireless optical code division multiple access communication system. J Univer Electron Sci technol 28(4):415–418
8. Fuxiaomei (2006) Research on key technologies of optical code division multiple access. Master's thesis of Tianjin University, Tianjin, pp 13–21
9. Zhu Y, Zhangbaofu, Cheng S, et al (1997) Optical fiber CDMA communication technology. J Commun Eng College 11(2):13–18
10. Chung FRK, Salehi JA, Wei VK (1989) Optical orthogonal codes: design, analysis, and applications. IEEE Trans Inf Theor 35(3):595–604
11. Johnson S (1962) A new upper bound for error-correcting codes. IEEE Trans Inf Theory 8(3):203–207
12. Liangyuhui (2003) Research on optical orthogonal code in OCDMA system. Master's thesis of University of Electronic Science and Technology, Chengdu, pp 13–17
13. Yinxiaoli, Zhang Q, Yu C, et al (2008) Optical orthogonal codes are generated by genetic algorithm. Opt Commun Res 6:24–26
14. Pan Y, Sun Q (2009) Research on frequency hopping scheme of OCDMA orthogonal code selection based on energy efficiency ratio. Opt Fiber Cable Appl Technol 3:38–41
15. Chu W, Colbourn CJ (2004) Optimal (n, 4, 2)-OOC of small orders. Discret Math 279(1–3):163–172
16. Verdu S (1986) Minimum probability of error for asynchronous Gaussian multiple-access channels. IEEE Trans Inf Theory 32(1):85–96
17. On Y (2002) Research on multiuser detection technology for large capacity wireless optical CDMA communication system. Master's thesis of University of Electronic Science and Technology, Chengdu, pp 20–44
18. Ruan C (2007) Research on multi-user detection algorithm based on WCDMA. Master's thesis of Wuhan University of Technology, Wuhan, pp 21–26
19. Huang K (2008) Research on multiuser detection technology in CDMA system. Master's thesis of Wuhan University of Technology, Wuhan, pp 32–34
20. Elfade N, Aziz AA, Yousif H, et al (2007) Multi-user detection for the OCDMA: optical parallel interference cancellation with optical limiter. In: International conference on intelligent and advanced systems, Kuala Lumpur, pp 444–446
21. Elfade N, Aziz AA, ldriss E (2008) New receiver architecture based on optical parallel interference cancellation for the optical CDMA. J Commun 3(1):64–70
22. Blue sky, Niguoqiang (2003) Simulation of atmospheric transmission characteristics of UV communication. J Beijing Univer Technol 23(4):419–423
23. Feng T, Chen G, Fang Z (2006) Channel characteristics of non line of sight ultraviolet scattering communication. Infrared Laser Eng 35(supply):226–230
24. Swami A, Tong L (2004) Signal Processing for networking: an integrated approach. IEEE Sig Proc Mag 21(9):18–19
25. Qianxiaocong (2006) Research on wireless ad+hoc networks based on multi-user detection. Doctoral Dissertation of Shanghai Jiaotong University, Shanghai, pp 6–58
26. Qianxiaocong, Zhengbaoyu, Zhaoxianjing (2006) Research on throughput of ad hoc networks based on multi-user detection. J Electron Inf 28(8):1429–1434
27. Honig M, Madhow U, Verdu S (1995) Blind adaptive multiuser detection. IEEE Trans Inf Theory 41(4):944–960
28. Zhangxianda (20002) Modern signal processing. Tsinghua University Press, Beijing
29. Songchunling (2004) Research on the application of blind multiuser detection in CDMA. Master's thesis of University of Electronic Science and Technology, Chengdu, pp 28–35
30. Li Y (2003) Simulation comparison of LMS and RLS adaptive algorithms China radar 04:35–38

31. Zvonar Z, Brady D (1996) Linear multipath-decorrelating receivers for CDMA frequency–selective fading channels. IEEE Trans Commun 44(6):650–653
32. Tian J (2007) Research on blind multiuser detection technology in CDMA systems. Master's thesis of Xi'an University of Electronic Science and Technology, Xi'an
33. Poor HV, Wang X (1997) Code-aided interference suppression for DS/CDMA communications—part II: parallel blind adaptive implementations. IEEE Trans Commun 45(9):1112–1122
34. Liang W (2011) Research on multi-user detection technology in UV ad hoc networks. Master's thesis of Xi'an University of technology, Xi'an

Chapter 5
MAC Layer Protocol Fairness in Ultraviolet Self-organizing Networks

The media access control (MAC) protocol is an important part of the wireless ad hoc network protocol stack. It defines how messages are transmitted over the medium and directly controls how messages are sent and received over the wireless channel. The MAC protocol provides unified services to the network by channel segmentation, allocation, and masking of the underlying channel control methods to achieve functions such as error, congestion, and traffic control. The ability of MAC protocols to use limited wireless resources fairly and efficiently is crucial to the performance of ad hoc networks. The IEEE 802.11 distributed coordination function (DCF) was originally designed for applications in wireless local area networks; when it is applied to wireless ad hoc networks, inequities exist in the access channels of nodes or data streams. When the unfairness is extreme, a single node or data stream can monopolize the entire channel for a period of time. In this case, other nodes or data streams in the network cannot compete for the channel and are starved. This chapter discusses fairness issues in IEEE 802.11 DCF-based wireless ad hoc networks.

5.1 MAC Protocol for Wireless Ad Hoc Networks

5.1.1 Types of MAC Protocol for Wireless Ad Hoc Networks

After years of research, dozens of channel access protocols have been proposed for ad hoc networks. As MAC protocols are highly channel-dependent, the design of channel access protocols should consider channel multiplexing techniques, various application environments and design goals, and other factors. According to the number of channels in an ad hoc network, we classify classical MAC protocols as single-channel, dual-channel, or multichannel protocols [1], as shown in Table 5.1.

X. Ke, *UV Light Self Organizing Network Theory*, Optical Wireless Communication Theory and Technology, https://doi.org/10.1007/978-981-97-7854-6_5

Table 5.1 Types of MAC protocols

Single-channel-based channel access protocols	Carrier-sense multiple access (CSMA) series
	Multiple access collision avoidance (MACA)
	Floor acquisition multiple access
Dual channel based channel access protocol	Dual-busy-tone multiple access
	Basic access protocol solutions for wireless (BAPU)
	Carrier listening dual-channel multiple access (DCMA)
Multichannel based channel access protocols	Hop-reservation multiple access
	Dynamic channel assignment
	Multichannel MAC

(1) Single-channel-based channel access protocols

As the name implies, single-channel-based ad hoc network channel access protocols are used for networks where only one common channel is used between communicating nodes. In this case, all control and data messages must be sent and received on the same channel. As messages are subject to propagation delays, hidden terminals, exposed terminals, and the random movement of nodes during channel transmission, conflicts may occur between control messages, between control and data messages, and between data messages in ad hoc networks based on single-channel MAC protocols. Data messages are usually much longer than control messages; consequently, data message conflicts are a major factor affecting channel utilization. Therefore, a main objective of single-channel-based access protocols is to minimize or even eliminate data message conflicts by designing suitable mechanisms using control messages. Typical single-channel-based channel access protocols for ad hoc networks are the carrier sense multiple access (CSMA) family [2], multiple access collision avoidance (MACA) [3], and floor acquisition multiple access [4].

(2) Dual-channel-based channel access protocols

Dual-channel-based network channel access protocols are used for ad hoc networks where control and data messages can be transmitted on separate channels. These protocols specify that the control channel can transmit only control messages and the data channel can transmit only data messages. Therefore, control messages do not conflict with data messages when a dual-channel-based channel access protocol is used. Dual-channel access has a unique advantage over single-channel access in solving the problem of hidden and exposed terminals. By designing a reasonable control mechanism, the effects of hidden and exposed terminals on the performance of the ad hoc network can be completely eliminated, and data message conflicts can be avoided. Typical channel access protocols for dual-channel-based ad hoc networks are dual-busy-tone multiple access [5], basic access protocol solutions for wireless (BAPU) [6], and carrier listening dual-channel multiple access (DCMA) [7].

(3) Multichannel-based channel access protocols

Multichannel-based channel access protocols are used in ad hoc networks with three or more channels. Because multiple channels are available, the access control mechanism can be more flexible; for example, different channels can be used for simultaneous communication between adjacent nodes, a single channel can be used as a common control channel, or both control and data messages can be transmitted on a single channel. Channel allocation algorithms allocate channels to communication nodes to maximize the number of network nodes that can communicate simultaneously. Access control functions schedule the appropriate channel access for network nodes to avoid or reduce conflicts. Typical multichannel-based channel access protocols for ad hoc networks are hop-reservation multiple access [8], dynamic channel assignment [9], and multichannel MAC [10].

As single-channel-based access protocols do not require complex channel allocation and because wireless media resources are limited, most MAC protocols in use are based on single channels. Typical single-channel-based access protocols are described below.

(1) CSMA series

The CSMA series is the first packet radio network channel access protocol to use the carrier listening mechanism, which can be used in mobile networks. If the channel is free, a message is sent immediately; otherwise, a random delay is selected, and the message is retransmitted at the end of the delay period.

CSMA protocols can be divided into three types according to the listening strategy: non-adherent, I-adherent, and P-adherent protocols. (a) Non-adherent CSMA does not continuously listen to the channel; when the channel is busy or in conflict, a network node that is attempting to send a frame waits for a random period of time and then listens again. These protocols have good channel utilization, but at the cost of longer latency and less throughput. (b) For I-adherent CSMA protocols, when the channel is busy or in conflict, a network node that is attempting to send a frame listens continuously and sends the frame as soon as the channel becomes available. These protocols have low latency and good throughput when the network load is low. However, when the network load is high, their mechanism results in multiple collision conflicts when multiple nodes send frames simultaneously, degrading system performance. (c) P-adherent CSMA combines the features of non-adherent and I-adherent protocols. When the channel is busy or in conflict, a network node that is attempting to send a frame continues to listen with probability P, waits for a random period of time with probability $1 - P$, and then listens again. The probability P can be reasonably adapted to the requirements of the network. The CSMA family includes CSMA with collision detection (CSMA/CD) and CSMA with collision avoidance (CSMA/CA), which is the best known of the IEEE 802.11 DCF protocols. CSMA/CA is used by Ethernet. Ethernet also uses CSMA/CD, which saves time and bandwidth by immediately terminating the frame currently in transmission as soon as a conflict is detected and waiting for a period of time to retry.

(2) MACA

MACA was the first wireless network channel access protocol to use the RTS/CTS handshake mechanism to solve the problem of hidden and exposed terminals. First, the source node sends an RTS frame. The destination node replies with a CTS frame to confirm that it is ready to receive the data frame; the source node sends the data frame only after receiving the CTS frame. The MACA partially solves the problem of hidden sending terminals. The reason is that if the data frame is sent only after a hidden terminal receives the CTS frame correctly, transmission is delayed, whereas a hidden terminal that does not receive the CTS frame correctly may still send the message, causing a message collision conflict. When the destination node receives a data frame, its neighboring nodes will not send messages; that is, the reception of data frames by the destination node will not experience interference. When a collision occurs, the MACA fallback strategy uses the binary exponential backoff (BEB) algorithm.

(3) MACAW

The MACA wireless (MACAW) protocol improves on the RTS/CTS handshake mechanism by adding a data sending (DS) control frame before the data frame to inform the node that it is exposed to the source node and an ACK control frame after the data frame to confirm the successful reception of the data frame. Specifically, after the source and destination nodes complete the RTS/CTS handshake mechanism, the source node sends a DS frame to inform its neighbors of the time to send subsequent data and ACK frames. These neighboring nodes wait for the specified time. Although the performance of the MACAW protocol has been improved somewhat compared to that of the MACA protocol, it has not completely solved the problem of hidden and exposed terminals [11].

5.1.2 IEEE 802.11 DCF Protocol

IEEE 802.11 DCF is based on and extends the CSMA/CA approach; it has three mechanisms: a basic DCF approach with two handshakes, another RTS/CTS approach with four handshakes, and a hybrid approach.

5.1.2.1 Basic DCF Approach with Two Handshakes

(1) Interframe gap

CSMA/CA requires a certain time interval between frame transmissions; when the channel is idle during this interval, the node can access the channel. IEEE 802.11 strictly defines four types of interframe gaps [12]. The short interframe space (SIFS) is used for MAC protocol data units other than the first one in an ACK frame,

CTS frame, or segmented burst transmission. The centralized coordination function interframe space [point coordination function interframe space (PIFS)] is used in centralized coordination function mode. The DCF interframe space (DIFS) is used to transmit data and management frames from nodes in DCF mode. The extended interframe space (EIFS) is used in DCF mode to provide sufficient time for a node to answer if a frame is not received correctly. In the IEEE 802.11 standard, the access priority is determined by specifying the length of the interframe gap, where the access priority is higher for shorter interframe gaps. The four types of interframe gaps are available in the order SIFS, PIFS, DIFS, EIFS.

(2) Principle of basic access methods

Before a node can send data, it must listen to the channel. If the channel is busy, the node does not send data and freezes its avoidance counter; if the channel is idle, it decreases the avoidance counter. The value of the avoidance counter is a randomly selected integer between 0 and the current contention window (CW), which is determined by the avoidance policy adopted by the node since the last transmission was completed. The avoidance counter is decremented to 0. During this period, if the channel becomes busy, the node freezes the avoidance counter for the DIFS time, after which the channel becomes idle again, and the node completes the rest of the avoidance process. Figure 5.1 shows the basic DCF data transmission process.

(3) Carrier listening mechanism

IEEE 802.11 DCF uses physical and virtual carrier listening mechanisms. When either type of listening mechanism detects that the channel is busy, the MAC layer of the node informs its higher-level protocol stack that the channel is busy and cannot be occupied and that the node must wait to access the channel. The carrier listening range of a node is generally larger than the transmission range.

The physical carrier listening mechanism is provided by the physical layer to determine whether another node in the network is using the channel to send data

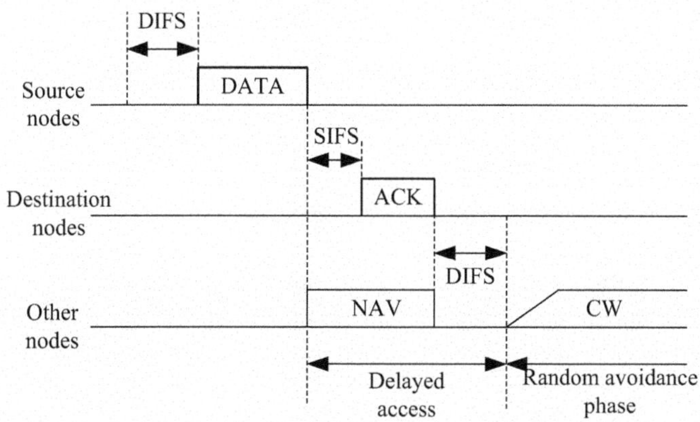

Fig. 5.1 Basic data transmission process in IEEE 802.11 DCF method

according to whether the relative received signal strength exceeds a set threshold and to inform the MAC layer of the result in a timely manner. Its performance depends on the channel medium and modulation method.

The virtual carrier listening mechanism is provided by the MAC layer and is implemented by introducing the network allocation vector (NAV), which is the equivalent of a subtraction counter. When the NAV is greater than 0, the channel is busy, and other nodes are occupying the channel. The node that wants to send data estimates how long it will need to occupy the channel to complete the entire transmission process and puts the time value into the Duration/Identification Code (ID) field of the MAC frame. The sending node that is occupying the channel notifies the other nodes in its coverage area via the MAC frame carrying this information, and the other nodes (except the destination node) extract the Duration/ID field value to update the NAV in time. The update rule is as follows. When the node receives a Duration/ID field value greater than its own NAV, it updates the NAV with the Duration/ID field value. When the NAV is not 0, it indicates that the channel is busy, and the NAV continues to count in decreasing units. When the NAV is decremented to 0, it indicates that the channel is idle.

(4) BEB fallback algorithm

The main concept of the BEB algorithm is that the CW decreases to the minimum value, CW_{min}, if the data frame interaction is successful and doubles to the maximum value if the interaction fails. As shown in (5.1), the size of the CW indirectly reflects the ability of the node to compete for the channel.

$$\begin{cases} CW = \min(CW \times 2, CW_{max}) & \text{Interaction failure} \\ CW = CW_{min} & \text{Interaction success} \\ \quad BackoffTime(BO) = Random(0, CW) \times aSlotTime \end{cases} \tag{5.1}$$

where $Random(0, CW)$ is a pseudo-random number equally distributed over the $[0, CW]$ segment, and *aSlotTime* is the unit time interval. The setting of the CW size is the key to the fallback algorithm. The reason is that the value of the fallback counter is a randomly determined integer of the node between 0 and the currently set CW. Setting a larger CW may increase the avoidance counter, which may reduce the probability of collision, but also increase the time required for the avoidance process.

5.1.2.2 RTS/CTS Method with Four Handshakes

The RTS/CTS method is effective in mitigating the adverse effects of collisions caused by hidden terminals. Figure 5.2 shows the data transmission process for this method.

The RTS/CTS method replaces collisions that occur when a data frame is transmitted with collisions that occur when an RTS or CTS frame is transmitted, at a

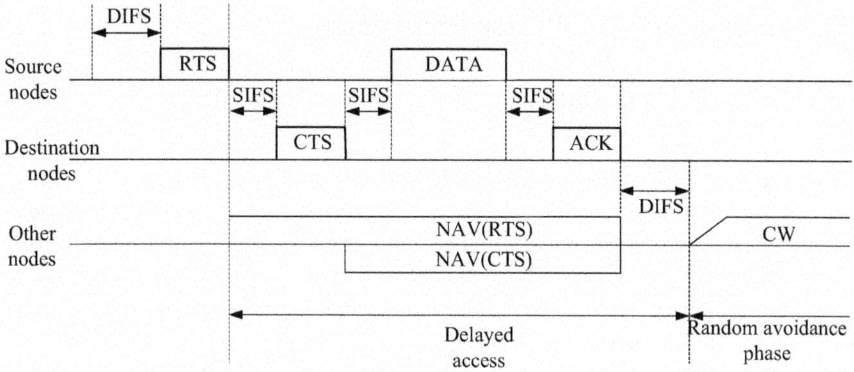

Fig. 5.2 Data transmission process in RTS/CTS method

small cost. Because the RTS and CTS frames are short (20 and 14 bytes, respectively) compared to the maximum length of the data frame, which can be up to 2346 bytes, the retransmission overhead is minimal.

In contrast to the basic method, the source node sends an RTS frame to reserve the channel before sending a data frame. Both the RTS and CTS frame contain a Duration/ID field that indicates the time required for transmission from the RTS or CTS frame to the ACK frame. All nodes in the network, except the source and destination nodes, update their NAV based on the Duration/ID field in the message instead of sending the frame, and the NAV indicates the length of time that other nodes in the network will occupy the channel. The CTS frame reserves the channel for the source node to send the data frame and notifies all other stations in the communication range, thus essentially solving the hidden terminal problem.

5.1.2.3 Hybrid Mode

The RTS/CTS method performs well under heavier network loads; however, when the network is lightly loaded, it is not ideal because of the overhead of its two additional handshake mechanisms, that is, the sum of the time spent transmitting RTS and CTS frames and the two SIFS. This mechanism serves to reduce collisions under heavier network loads but increases the transmission delay at lighter loads. It also affects the network throughput because of the introduction of RTS and CTS frames, which reduces the proportion of valid data frames to successfully transmitted frames.

Therefore, in simulation experiments and practical applications, a hybrid approach that combines the basic and RTS/CTS methods is typically used. In the hybrid approach, the network is considered to be lightly loaded when the length of the data frame is less than the RTS threshold, and the network node uses the basic DCF method, that is, the two-handshake mechanism. When the length of the data frame is greater than the RTS threshold, the network is considered to be heavily loaded, and the network node uses the RTS/CTS method.

5.2 Analysis of MAC Protocol Unfairness

5.2.1 Fairness Objectives

Fairness at the MAC layer means that individual network nodes sharing the medium have equal ability to use resources. In a wireless ad hoc network, the network provides an end-to-end flow of services to users. The service flow is generally transmitted along a route consisting of multiple links in the network, that is, by multihop transmission. Therefore, the performance of each link segment can have a cascading effect on the performance of the entire route. In severe cases, if a segment of a link in a route fails completely, the entire route fails, and the service flow in transmission is lost. In this case the route must be re-established, and the overall network overhead is quite high. Therefore, the fair use of shared media on each link segment may appear to be a local performance issue, but it affects the overall performance of the network. Typically, network throughput and fairness are in a reciprocal relationship; that is, increasing network throughput tends to result in unfairness between nodes, and reduced network throughput is required to improve fairness. Because wireless ad hoc networks are self-organizing, considerable experimentation is involved in practical performance testing. Therefore, it is important to find suitable fairness criteria in theory to design fairness algorithms at the MAC layer. The three main fairness indices (FIs) are as follows.

5.2.1.1 Max–Min Fairness Index (FI)

The max–min FI [13] is the ratio of the throughput of the largest link to the throughput of the smallest link in the network. It is expressed as follows:

$$FI = \frac{Throughput_{\max}}{Throughput_{\min}} \tag{5.2}$$

where $\mathrm{Throughput}_{\max}$ and $\mathrm{Throughput}_{\min}$ are the throughputs of the maximum and minimum links, respectively; $FI \in (0, \infty)$. When the FI is 1, that is, the throughput of the maximum link is equal to that of the minimum link, nodes in the network have equal probability of accessing the channel, and the fairness is optimal. As the FI increases, the difference between the throughputs of the maximum and minimum link increases, and the fairness is degraded. Therefore, FIs close to 1 are desirable.

5.2.1.2 Improving the FI

The improved FI [14] is the ratio of the difference between the maximum link and minimum link throughputs to the total network throughput. It is expressed as follows:

$$FI = \frac{Throughput_{\max} - Throughput_{\min}}{Throughput_{all}} \tag{5.3}$$

where $\text{Throughput}_{\text{all}}$ is the total network throughput. $FI \in \left(0,\ 1\right)$; when FI is 0, the throughput of the maximum link is equal to that of the minimum link, indicating that the nodes in the network have equal probability of accessing the channel, and the fairness is optimal. The least fair case is when only one link in the network occupies the shared channel, and the FI is 1. Therefore, FIs close to 0 are desirable.

5.2.1.3 Weighted FI

The weighted FI [15] incorporates weights based on the importance of business flows. We define the weighted FI as

$$FI = \frac{\left(\sum_i T_i/\phi_i\right)^2}{n \times \sum_i \left(T_i/\phi_i\right)^2} \tag{5.4}$$

where T_i and ϕ_i denote the throughput and weight of the first service flow, respectively, and n denotes the number of service flows in the network. $FI \in \left(0,\ 1\right)$, and the fairness increases as the FI approaches 1.

Of these three FIs, the max–min FI is the simplest, but it has significant limitations. If only one of multiple business streams has a relatively abnormal throughput (much higher or lower than the others), and the remaining business streams have similar throughput, the FI calculated using (5.2) is very large and reflects a fairness that is consistent with the circumstances. The improved FI overcomes the limitations of the max–min FI to a certain extent. By contrast, the weighted FI addresses the limitations of the max–min FI by considering the relative importance of business flows. Consequently, it is the most widely used FI, and we used it in our simulation analysis.

5.2.2 *Causes of Fairness Issues*

An analysis of IEEE 802.11 DCF showed that the main causes of fairness problems in wireless ad hoc networks are the following [16].

(1) Inequities arising from the backoff mechanism

The IEEE 802.11 DCF protocol uses the BEB, an algorithm that always favors nodes that have successfully transmitted in future contention for the channel. If a node transmits successfully, its CW decreases to the minimum value, $CW_{\min}$. If transmission fails, the CW takes the smaller of 2CW and $CW_{\max}$. However, a single communication process generally does not accurately reflect the actual competition for the channel, especially when the network is busy and there are many conflicts, and thus each node has a large CW after multiple instances of conflict avoidance. At this point, if a node successfully sends a message, its CW is reduced to CW_{min}, whereas the CWs of other nodes remain large, and thus differ greatly from that of the successful node. In subsequent competitions for channel resources, the successful node always has the advantage, whereas other nodes are starved because they cannot access the channel, resulting in serious unfairness.

(2) Inequities arising from the short-term random selection of counters

Short-term random selection of the avoidance counter values is not reasonable when the CWs are equal. Theoretically, when the number of random picks is infinite, the values of the selected fallback counters tend to be evenly distributed in the range of values. However, communication between nodes in an ad hoc network is often short-lived or even bursty. Consequently, short-term avoidance counter values are selected for a given time period, where some nodes select large values, and others select small values. For example, for a short period of time, when the CWs are equal, the avoidance counters of some nodes in the ad hoc network have values between 0 and CW/4 several times, whereas others have avoidance counter values of 3CW/4 to CW several times. Consequently, the CWs are the same, but the avoidance counter values are very different, resulting in different ability to compete, and thus unfairness.

(3) Inequities arising from asymmetry in the information on internode channels

In the IEEE 802.11 DCF protocol, channel access is initiated by the sender. Certain nodes in the network cause asymmetry in the channel information between the sender and receiver nodes because of their geographical location, resulting in unfairness. An example is illustrated Fig. 5.3. When node 1 sends data stream f1 to node 2, node 3 sends data stream f2 to node 2, and node 4 sends data stream f3 to node 3. For the area in which data stream f1 is transmitted, node 3 is within the coverage of receiving node 2 but outside the coverage of sending node 1; it is thus a hidden terminal. The information asymmetry appears in two ways. First, when node 3 sends data stream f2 to node 2, it does not known that node 1 is also sending data stream f1 to node 2 because it is outside the coverage of sending node 1; therefore, it causes a collision between data streams f1 and f2. Second, nodes 2 and 3, at the center of the network topology, are subject to interference from two neighboring nodes. For successful data transmission, nodes 1 and 3 must not send data. By contrast, nodes 1 and 4, at the edge of the network topology, experience interference from only one neighboring node; when node 4 is sending data, only node 3 must not send data.

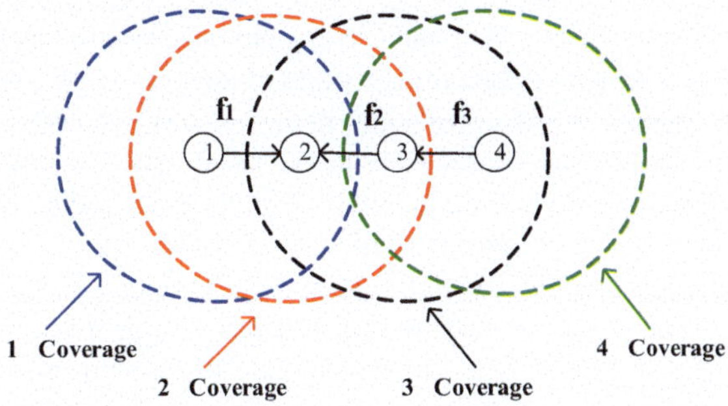

Fig. 5.3 Hidden terminal topology

5.3 Improving the Fairness of the MAC Protocol

As an important topic in research on wireless ad hoc networks, the fairness of MAC protocols has received considerable attention. Many MAC protocol improvement schemes have been proposed to address the causes of unfairness. Algorithms for improving fairness fall into two main categories [17].

5.3.1 Improved Update Rules for Contention Window

In the multiplicative increase–linear decrease (MILD) algorithm. when a node fails to transmit, the size of the CW is the smaller of α times the original CW and CW_{max} (the maximum CW); if a node successfully transmits, the CW is the larger of (CW – Contention Window). The MILD algorithm can cause the CW to change smoothly so that most nodes have similar CWs, which improves the fairness of channel access by network nodes. For networks with many nodes, the throughput of the MILD algorithm is slightly better than that of the BEB algorithm. However, for networks with fewer nodes, because the CW decreases linearly, the CW changes more slowly than the network conditions do; consequently, the CWs of network nodes are often larger than the reasonable CW value. In this case, the throughput performance of the MILD algorithm is worse than that of the BEB algorithm. Because MILD uses a smaller CW_{min} value than the BEB algorithm, it can outperform the BEB algorithm in terms of throughput when the number of network nodes is very small.

The linear MILD (LMILD) algorithm [18] is an improved version of the MILD algorithm. If a node fails to send, the CW takes a small value between αCW and CW_{max}, and neighboring nodes in the network that listen to the collision increase their CWs by a fixed value L until it is maximum. The LMILD algorithm reduces the probability of the next collision by increasing the CWs of neighboring nodes that can listen to the collision when a node fails to send and by decreasing the CWs of all nodes in the network by a fixed value β.

In gentle DCF (GDCF) [19], if a node fails to send, its CW is set to a smaller value between 2CW and CW_{max}. If the node succeeds, it must set its CW to a larger value between CW/2 and CW_{min}, but only after it has successfully sent C messages in a row, where the optimal value of the consecutive success parameter C is related to the number of active nodes in the network. Therefore, the GDCF algorithm can be improved by combining it with a series of neighborhood node information acquisition algorithms. When the number of nodes in the network is unknown, a parameter C between 4 and 8 ensures a high level of network performance. The GDCF algorithm is fairer than the BEB algorithm for networks with more nodes.

The conic increase–multiplicative/linear decrease (CIMLD) [20] algorithm uses the threshold CW_{basic} to divide the CW domain into two regions and dynamically adjusts the value of the avoidance counter by calculating the multiplicative factor for avoidance according to a segmented quadratic curve. When a message collision occurs, if the CW is between CW_{min} and CW_{basic}, the competition is more intense, and message collisions are relatively frequent because the CW is small. Therefore, a larger multiplication factor (trending to curve C1) must be used so that the CW of the node increases quickly to near CW_{basic}, reducing the probability of further message collisions. If the CW is between CW_{basic} and CW_{max}, the probability of message collision is relatively small because of the large CW; in this case, a smaller multiplication factor (trending to curve C2) is used to balance the variation in the CW of the node and thus improve its throughput. CW_{basic} the smoother the curve is. This algorithm is fairer than the BEB algorithm.

5.3.2 *Predicting or Monitoring Network Status to Dynamically Adjust Channel Access*

The distributed foundation wireless MAC [21] algorithm is implemented as follows. (1) To establish the listening policy, count values are set according to the type of data frames received. When an RTS frame is received, if the destination address is the node itself, the value of T_i is predicted as the sum $T_{RTS} + T_{CTS}$ for the time the node occupies the channel; otherwise, the value of T_o is predicted as the sum T_{RTS} for the time the neighboring node occupies the channel. When a CTS frame is received, if the destination address is the node itself, the value of T_i is predicted to accumulate $T_{RTS} + T_{CTS} + T_{data}$; otherwise, the value of T_o is predicted to accumulate $T_{RTS} + T_{CTS}$. When a data frame is received, if the destination address is the node itself, the value of T_i is added to $T_{RTS} + T_{CTS} + T_{data} + T_{ACK}$; otherwise, the value of T_o is added to $T_{RTS} + T_{CTS} + T_{DATA}$. When an ACK frame is received, if the destination address is itself, the value of T_i is added to $T_{RTS} + T_{CTS} + T_{data} + T_{ACK}$; otherwise, the value of T_o is added to $T_{RTS} + T_{CTS} + T_{data} + T_{ACK}$. A CW selection mechanism is applied, where the FI, $F = (T_i/W_i)/(T_o/W_o)$, is estimated from the count value obtained from the listening policy. W_i and W_o ($W_o = 1 - W_i$) are the weights of the channels occupied by this node and neighboring nodes, respectively. When $F > C$, the node is considered to have a large capacity to occupy the channel, and the CW is taken as a small value between 2CW and CW_{max} to reduce its capacity to occupy the channel appropriately. When $C \geq F > 1/C$, the node is considered to have a moderate capacity to occupy the channel, and the CW remains unchanged. When $F \leq 1/C$, the node is considered to have a small capacity to occupy the channel, and its CW is the larger of CW/2 and CW_{min}.

In the new self-adaptive DCF (NSAD) [22] algorithm, when successful transmission occurs, the CW of the node is reset, not to CW_{min}, as in the BEB algorithm, but to the initial CW, CW_{init}, which is dynamically updated according to network conditions. Each node in the network calculates the value of the current network load parameter L according to the lengths of the currently monitored network conflict and network idle time; it then instantly adjusts CW_{init} according the predicted L relative to the optimal load parameter $L_{opt.}$ In addition, because it adds an optimized CW field to the data frame structure, the NSAD algorithm can inform all neighboring nodes in the network of this value when the data frame is transmitted, allowing all neighboring nodes to use the same CW to ensure fairness.

The adaptive efficiency fairness tradeoff algorithm [23] is a fast conflict resolution strategy that uses a rapid avoidance counter update mechanism called fast collision resolution (FCR) [24]. FCR reduces the time wasted by the node in the idle time between two consecutive transmissions, which improves network throughput. However, it also increases the possibility that a node successfully acquires a channel multiple times in a row over a period of time, which reduces network fairness. A fairness mechanism is used to control the number of consecutive successful and failed transmissions by a node. After a successful transmission, the CW is set to the larger of CW/2 and CW_{min}; in addition, the continuous transmission success counter is incremented by 1, and the continuous transmission failure counter is cleared. When a transmission fails, the CW is set to the smaller of 2CW and CW_{max}; the continuous transmission failure counter is incremented by 1, and the continuous transmission success counter is cleared. To ensure fairness, the CW is set to the maximum value if the continuous transmission success counter reaches the threshold N_{s_TH}. If the continuous transmission failure counter reaches the threshold N_{f_TH}, the CW is set to the minimum value.

In the dynamic prediction backoff algorithm [25], the node starts listening to the channel when the backoff process begins to predict the current contention for the channel. The time slots are divided into three states (idle, busy, and conflicting), and the number of time slots in each state is counted. The sum of the number of slots in the busy and conflicting states over the number of slots in all states is defined as the conflict probability P. If the node transmits successfully, the time slot counter is reset to zero before the next transmission. If the node fails to transmit, P is calculated, and the CW is adjusted according to the value of P. The time slot counters continue to increase with the next transmission. Network conditions are predicted according to the calculated P, and various methods are used to adjust the CW. If $0 \leq P < P_1$, the network load is considered light. When a collision occurs, the CW is simply increased to $1/(1 - P)$ times the original value; when a successful transmission occurs, the CW is reset to the initial value, CW_{min}. If $P_1 \leq P < P_2$, the network load is considered moderate. When a collision occurs, the CW is increased to $1/(1 - P)$ times the original value. When a successful transmission occurs, the CW is reduced to the larger value of P times the initial value and CW_{min}. If $P_2 \leq P < 1$, the network load is considered to be light, and collisions are very frequent. When a collision occurs, the CW is increased to a value between twice the initial value and CW_{max}; when a successful transmission occurs, CW is decreased to a value between P times the initial value and CW_{min}.

5.4 Avoidance Algorithm with Improved Fairness

The history-based dynamic fair backoff (HDFB) algorithm dynamically adjusts the CW by predicting the network load according to the past success or failure of a node in sending and past CW values. The new DCF (NDCF) algorithm is based on the size of the last successful transmission and takes a different approach to the selection of the CW. We analyze the causes of fairness problems in IEEE 802.11 DCF and the improved algorithms mentioned in other studies. We then discuss two improved fallback algorithms [26–28]; the main implementation program code list is shown in Appendix C.

5.4.1 HDFB Algorithm

The HDFB algorithm first divides the CW from 32 to 1024 into four levels according to (5.5), where higher levels indicate smaller values. Then, the selection mechanism for the CW is divided into three parts: coarse, fine, and additional tuning. Figure 5.4 shows a flow chart of the mechanism.

$$\begin{cases} CW[i]\_low = (CW_{\min} + 1) \times 2^{i-1} \\ CW[i]\_high = (CW_{\min} + 1) \times 2^{i+1} \end{cases} \quad i = 1, 2, 3, 4. \tag{5.5}$$

where $CW[i]\_low$ and $CW[i]\_high$ denote the lower and upper bounds of the CW at level i, respectively.

(1) Coarse tuning (to avoid exclusive channel use)

The status of two complete transmissions will have four cases: 00, 01, 10, and 11 (success: 1, failure: 0). The CW level is determined by the status of the last two complete messages sent. For 00, two consecutive transmissions have failed, indicating that the node cannot compete successfully for media resources. Consequently, the CW level is increased by 1, which improves the competitiveness of the node and helps it to compete for the channel the next time. For 01 and 10, two consecutive messages have been transmitted successfully after one failure each, indicating that the node can compete appropriately for the channel in the current network environment. The CW level is maintained, and the corresponding CW is unchanged. For 11, two successive message transmissions have been successful, indicating that the node

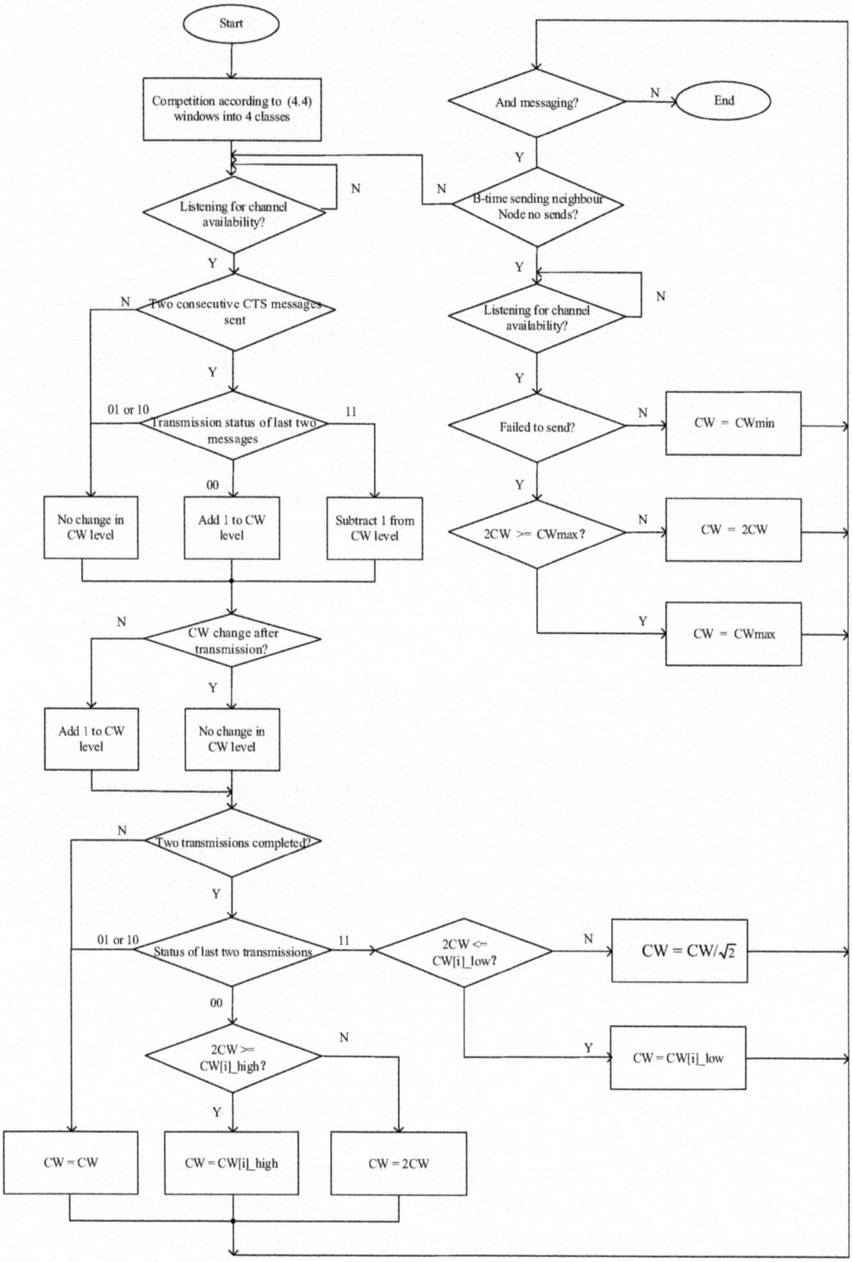

Fig. 5.4 HDFB algorithm flow chart

is highly competitive and continues to occupy the channel. To prevent it from monopolizing the channel, the CW level is decreased by 1; as a result, the CW increases, limiting its ability to compete.

(2) Fine tuning (to avoid re-collisions)

Under a heavy network load, collisions generally occur during the four steps of complete message transmission: RTS, CTS, data, and ACK. Consequently, the CW is selected several times. The two CW selections have four possible states: 00, 01, 10, and 11 (call to subroutine to decrease the CW: 1, call to subroutine to increase the CW: 0). The CW is dynamically adjusted according to the history of the last two CW selections. For 00, the increase-CW subroutine has been called twice in a row, indicating that the network is heavily loaded and prone to collisions if data are sent. To avoid another collision, the CW is doubled. For 01 and 10, the increase-CW subroutine and decrease-CW subroutine have each been called once, indicating a moderate channel load; the CW remains unchanged. For 11, the decrease-CW subroutine has been called twice in a row, indicating a light channel load and less likelihood collision if data are sent. Therefore, to increase the network throughput, the CW is set to $CW \big/ \sqrt{2}$. In fine tuning, the CW is between CW[i]_low and CW[i]_high, where CW level i was determined by coarse tuning.

(3) Additional tuning

When the network load is high, owing to the coarse tuning strategy of the HDFB algorithm, some nodes have a small CW level. After a period of time, the network load is reduced, and these nodes are likely to be stable for a long period of time, with two consecutive messages sent (01 and 10). In addition, when there is only one data stream in the network, with no contention, again because of the coarse tuning strategy, the CW level is kept at a minimum, and the CW is maximum. In both cases, the throughput of the network is reduced.

A monitoring strategy is added in each of these situations to eliminate their adverse effects. Specifically, the CSMA/CA carrier listening mechanism is used to count messages sent (either successfully or unsuccessfully) and add 1 whenever a message is sent. For the former case, when the count value 1 reaches the threshold A, a determination is made once. If the CW level remains the same during this process, the CW level is increased by 1 to make it easy to access the channel. If the CW level changes, nothing is done. For the latter case, when the count value 2 reaches the threshold B, adjudication is performed once. If no neighboring node has been listening for messages during this process, it is determined that only one data stream in the network is not contested, and the BEB algorithm is adopted to ensure network throughput. Otherwise, the HDFB algorithm is used. If an active neighbor node is listened to during the BEB algorithm, the BEB algorithm mechanism is immediately withdrawn, and the HDFB algorithm is used again.

5.4.2 New DCF (NDCF) Algorithm

When the CWs are the same, it can be assumed that network nodes should be equally competitive. When transmission fails, as in the BEB algorithm, the CW is doubled; when transmission succeeds, the next action depends on the value of the randomly generated avoidance counter. If the value of the randomly generated fallback counter BO is less than CW/2, the node can compete one level higher than its own rank. To enhance fairness, the node is penalized accordingly; the rank of the CW is unchanged, inhibiting the ability of the node to compete for the channel again. In addition, when BO $\geq$ CW/2, because the node sets the avoidance counter according to the mechanism in line with its own competitive ability, it is rewarded by having its competitive ability increased by keeping the value of the CW at CW/2 and increasing the rank of the CW. Under certain circumstances (e.g., when hidden terminals expose terminal effects), the competitive ability of some nodes can easily reach the maximum. Thus, when the CW is set to CW_{min}, these competitive nodes occupy the channel for a long time and cause neighboring nodes in the network to starve, which creates unfairness. To address this problem, if the value of the avoidance counter BO is $<$ CW/2 when the node with the strongest competitive ability successfully transmits, to improve the fairness, the node is penalized by having its ability to occupy the channel again reduced. This reduction decreases the CW level so that CW = 2CW. If BO $\geq$ CW/2, the CW value is unchanged. Adopting this strategy reduces the impact of unreasonable short-term random selection of the avoidance counter value. Figure 5.5 shows a flowchart of the algorithm.

5.5 Performance Analysis of BEB and NDCF Algorithms

5.5.1 Performance Analysis of BEB Algorithm

The MAC protocol of IEEE 802.11 is a random access mechanism where the system throughput increases with increasing load until a critical point (i.e., the maximum throughput), at which the input load to the network is low. As the input load increases, the system throughput decreases; the decrease is dramatic in severe cases. In practice, it is difficult to control the network input load to ensure that the system throughput is maintained at its maximum. Therefore, maximum throughput is not a suitable metric for evaluating system performance. The concept of “saturation throughput” has been proposed as an alternative metric. By saturation, we mean that the sending queue of a node is always non-empty; that is, there are always data packets in the queue waiting to be sent. We use the two-dimensional Markov chain model proposed by Bianchi [29] to analyze the saturation throughput performance of IEEE 802.11. We denote the probability of collision each time a node sends a data frame as p. In addition, we assume that all nodes have the same information about the channels.

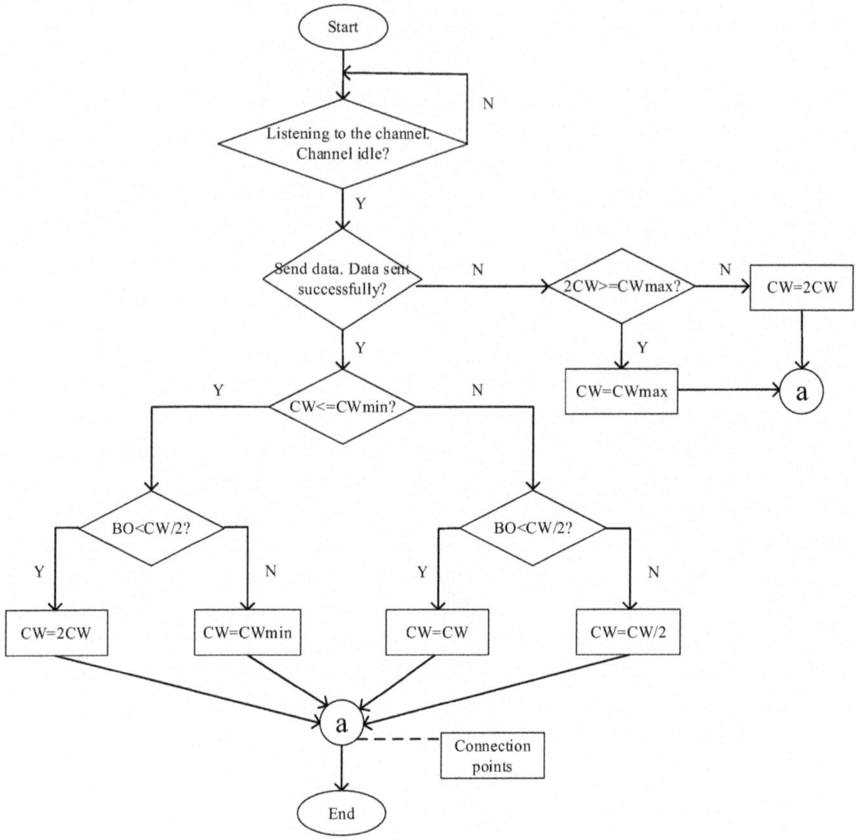

Fig. 5.5 Flow chart of NDCF algorithm

5.5.1.1 Analytical Model of Saturated State

IEEE 802.11 DCF uses a binary exponential fallback rule. Let $b(t)$ represent the statistical process of taking the value of the node fallback counter at time t, where the count is reduced by 1 at the beginning of each time slot. Let $s(t)$ represent the stochastic process of taking the values of the node fallback level. We built a two-dimensional stochastic process model $\{s(t), b(t)\}$ to analyze the performance of IEEE 802.11 DCF; it is a two-dimensional discrete-time Markov chain with the state transfer diagram shown in Fig. 5.6. This model is valid for the basic, RTS/CTS, and hybrid DCF modes.

Using the Markov chain model, we obtain a system of one-step transfer probability equations:

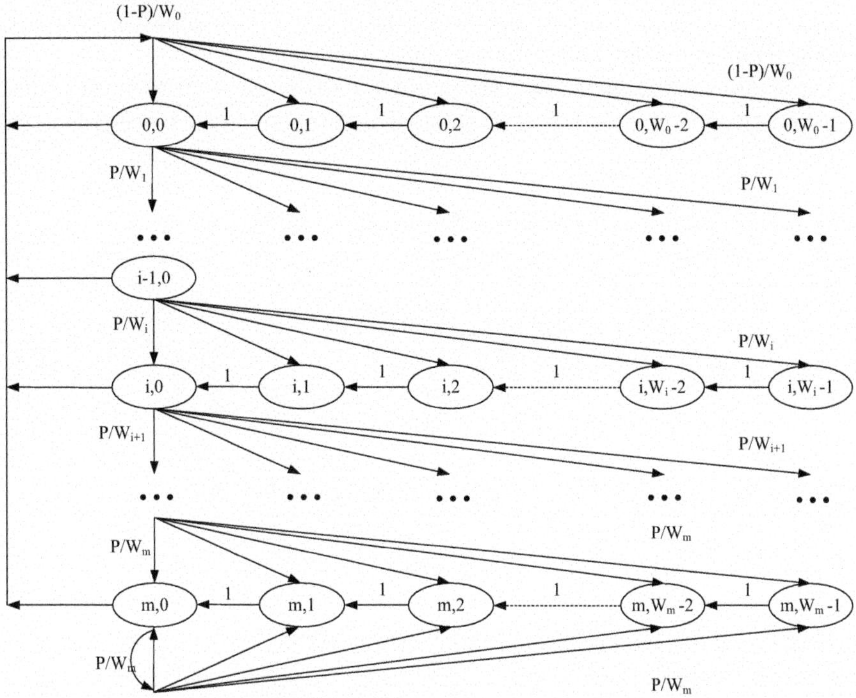

Fig. 5.6 Markov model of IEEE 802.11 DCF

$$\begin{cases} P\{i,k|i,k+1\} = 1 & k \in [0, W_i - 2], \quad i \in [0, m] \\ P\{0,k|i,0\} = (1-p)/W_0 & k \in [0, W_0 - 1], \quad i \in [0, m] \\ P\{i,k|i-1,0\} = p/W_i & k \in [0, W_i - 1], \quad i \in [1, m] \\ P\{m,k|m,0\} = p/W_m & k \in [0, W_m - 1] \end{cases} \tag{5.6}$$

where P is the probability of collision for each data frame sent by the node, W_0 is the minimum CW, W_i is the CW in the ith avoidance (according to the binary exponential avoidance rule, $W_i = 2^i \times W_0$), K is the value of the avoidance counter of the node, i is the order of the avoidance counter, and m is the maximum value of i.

The four equations in (5.6) represent the four main events in the IEEE 802.11 channel access process. (1) At the beginning of each time slot, the fallback counter decreases by 1. (2) When the node transmits successfully, the fallback counter is set to order 0. The value of the fallback counter is chosen randomly with medium probability in $(0, W_0 - 1)$. (3) When the node transmits unsuccessfully at order $i - 1$ of the avoidance counter, the order of the avoidance counter is increased by 1, and the value of the avoidance counter is chosen at random with medium probability in $(0, W_i - 1)$. (4) When the order of the avoidance counter reaches the maximum order m, if the node still fails in the next transmission, the order of the avoidance counter is not increased, and the value of the avoidance counter is chosen randomly with medium probability in $(0, W_{\text{m}} - 1)$.

Let $b_{i,k} = \lim_{t\to\infty} p\{s(t) = i, b(t) = k\}, i \in (0, m), k \in (0, W_i - 1)$ be the stable distribution of a Markov chain. It is inferred from the state transfer diagram in Fig. 5.6 that

$$\begin{cases} b_{i-1,0} \times p = b_i,\ 0 \to b_i,\ 0 = p^i \times b_0,\ 00 < i < m \\ b_{\text{m}-1,0} \times p = (1-p) \times b_{\text{m},0} \to b_{\text{m},0} = \frac{p^m}{1-p} \times b_0,\ 0 \end{cases} \tag{5.7}$$

$$b_i, k = \frac{W_i - k}{W_i} \times \begin{cases} (1-p)\sum\limits_{j=0}^{m} b_{\text{j},0} & i = 0 \\ p \times b_{i-1,0} & 0 < i < m, k \in (1, W_i - 1) \\ p \times \left(b_{\text{m}-1,0} + b_{\text{m},0}\right) & i = m \end{cases} \tag{5.8}$$

In addition, $\sum_{j=0}^{m} b_{\text{j},0} = b_{\text{j},0} / (1-p)$, (5.8) can be reduced to

$$b_{i,k} = \frac{W_i - k}{W_i} \times b_{i,0} i \in (0, m), k \in (0, W_i - 1) \tag{5.9}$$

By the normalization principle of probability, (5.7) and (5.9) can be combined in (5.10) to find $b_{0,0}$.

$$\begin{aligned} 1 &= \sum_{i=0}^{m}\sum_{k=0}^{W_i-1} b_{i,k} = \sum_{i=0}^{m} b_{i,0} \sum_{k=0}^{W_i-1} \tfrac{W_i-k}{W_i} = \sum_{i=0}^{m} b_{i,0} \times \tfrac{W_i+1}{2} \\ &= \tfrac{b_{0,0}}{2}\left[W_0\left(\sum_{i=0}^{m-1} (2p)^i + \tfrac{(2p)^m}{1-p}\right) + \tfrac{1}{1-p}\right] \end{aligned} \tag{5.10}$$

$$b_{0,0} = \frac{2 \times (1-2p) \times (1-p)}{(1-2p) \times (W_0 + 1) + p \times W_0 \times (1 - (2p)^m)} \tag{5.11}$$

Let the probability that a node sends a data frame at any moment be τ, which is the sum of the probabilities when the avoidance counter of any order decreases to 0. Then we have

$$\tau = \sum_{i=0}^{m} b_{i,0} = \frac{b_{0,0}}{1-p} = \frac{2 \times (1-2p)}{(1-2p) \times (W_0 + 1) + p \times W_0 \times (1 - (2p)^m)} \tag{5.12}$$

From (5.12), the value of τ depends on p. If $m = 0$, τ is independent of p; that is,

$$\tau = \frac{2}{W_0 + 1} \tag{5.13}$$

The probability that only one node in the network sends a data frame is then $(1-\tau)^{n-1}$, and we have

$$p = 1 - (1 - \tau)^{n-1} \tag{5.14}$$

The solution of the nonlinear system of equations in (5.12) and (5.14) can be found numerically. It can be concluded that p is determined only by the number of network nodes n, minimum CW W_0, and maximum order of avoidance m.

5.5.1.2 Throughput Analysis

We analyze the saturation throughput of the system. We define P_{tr} as the probability that at least one node in each time slot sends a data frame, P_s as the probability of successful data frame transmission, τ as the probability that a node sends a data frame at a given moment, and n as the number of nodes. Then we have

$$P_{tr} = 1 - (1 - \tau)^n \tag{5.15}$$

$$P_s = \frac{n \times \tau \times (1 - \tau)^{n-1}}{P_{tr}} = \frac{n \times \tau \times (1 - \tau)^{n-1}}{1 - (1 - \tau)^n} \tag{5.16}$$

The normalized saturation throughput S is equal to the ratio of the effective data frame transmission time to the total transmission time during successful transmission. In addition, the total transmission time is the sum of the channel idle time, data frame transmission time, and data frame collision time. Then we have

$$\begin{aligned} S &= \frac{P_s \times P_{tr} \times E[MSDU]}{(1 - P_{tr}) \times \sigma + P_s \times P_{tr} \times T_s + P_{tr} \times (1 - P_s) \times T_c} \\ &= \frac{E[MSDU]}{T_s - T_c + \frac{\sigma \times (1 - P_{tr})/P_{tr} + T_c}{P_s}} \end{aligned} \tag{5.17}$$

where E[MSDU] is the average length of the transmitted MAC service data unit (MSDU), σ is the time slot width, T_s is the time required to send one data frame, and T_c is the time required for a data frame collision. Equation (5.17) is a general formula where only the values of T_s and T_c must be substituted for the basic, RTS/CTS, and hybrid approaches. We analyze only the hybrid method, as it is the most widely used. The other two methods can be analyzed similarly by drawing on the analysis below. When the length of the data frame is less than the RTS threshold in the RTS/CTS method, the basic method is used to send the data frame, and vice versa, the RTS/CTS method is used when the length exceeds the RTS threshold. The probabilities that the data frame is sent using the basic and RTS/CTS methods are p(TH) and 1 – P(TH), respectively. T_s for the hybrid method is obtained accordingly.

$$T_s = T_s^{BAS} \times P(TH) + T_s^{RTS} \times (1 - P(TH)) \tag{5.18}$$

where T_s^{BAS} and T_s^{RTS} are obtained from (5.19) and (5.20), respectively, and δ is the average propagation delay.

$$T_s^{BAS} = PHY_{hdr} + MAC_{hdr} + E[MSDU] + SIFS + \delta + ACK + \delta + DIFS \quad (5.19)$$

$$T_s^{RTS} = RTS + SIFS + \delta + CTS + SIFS + \delta + T_s^{BAS} \quad (5.20)$$

When T_c is found to simplify the analysis, only the case of two data frames colliding within the same time slot is considered. In the hybrid approach, there are three cases. (1) Both nodes sending data frames use the basic approach; the collision time is represented by $T_c^{BAS+BAS}$, and the probability of occurrence is $P^2(TH)$. (2) Both nodes use the RTS/CTS approach. Collisions can occur only between two RTS frames; the collision time is $T_c^{RTS+RTS}$, and the probability of occurrence is $[1 - P(TH)]^2$. (3) One node uses the basic approach, and the other uses the RTS/CTS approach. The collision time is $T_c^{BAS+RTS}$, and the probability of occurrence is $2P(TH) \times [1 - P(TH)]$.

$$\begin{aligned} T_c^{RTS} &= P^2(TH) \times T_c^{BAS+BAS} + (1 - P(TH))^2 \times T_c^{RTS+RTS} \\ &\quad + 2 \times P(TH) \times (1 - P(TH)) \times T_c^{BAS+RTS} \end{aligned} \quad (5.21)$$

where $T_c^{BAS+BAS}$, $T_c^{RTS+RTS}$, and $T_c^{BAS+RTS}$ are given by (5.22), (5.23), and (5.24), respectively, and $E[MSDU^*]$ is the average length of the MSDU for the long transmission in which the collision occurred.

$$\begin{cases} T_c^{BAS+BAS} = PHY_{hdr} + MAC_{hdr} + E[MSDU^*] + DIFS + \delta & \text{Uncollided nodes} \\ T_c^{BAS+BAS} = PHY_{hdr} + MAC_{hdr} + E[MSDU^*] + DIFS + \delta & \\ \quad +SIFS + ACK + \delta & \text{Collidingnodes} \end{cases} \quad (5.22)$$

$$\begin{cases} T_c^{BAS+RTS} = RTS + DIFS + \delta & \text{Uncollided nodes} \\ T_c^{BAS+RTS} = RTS + DIFS + \delta & \\ \quad +SIFS + CTS + SIFS + \delta & \text{Collision Nodes} \end{cases} \quad (5.23)$$

$$\begin{cases} T_c^{BAS+RTS} = PHY_{hdr} + MAC_{hdr} + E[MSDU] + DIFS + \delta & \text{Uncollided nodes} \\ T_c^{BAS+RTS} = PHY_{hdr} + MAC_{hdr} + E[MSDU] + DIFS + \delta & \\ \quad +SIFS + ACK + \delta & \text{Collision Nodes} \end{cases} \quad (5.24)$$

The above analysis shows that both T_s and T_c are approximately fixed values. In addition, in (5.17) the normalized saturation throughput S is maximum if (5.25) is taken to be minimum.

$$\frac{(1-P_{tr})/P_{tr}+T_c/\sigma}{P_s}=\frac{\overline{T_c}-(1-\tau)^n\times\left(\overline{T_c}-1\right)}{n\times\tau\times(1-\tau)^{n-1}} \tag{5.25}$$

where $\overline{T_c}=T_c/\sigma$, and σ is the time slot width. Taking the derivative of (5.25) with respect to τ and setting it to zero yields

$$(1-\tau)^n-\overline{T_c}\times\left\{n\times\tau-\left[1-(1-\tau)^n\right]\right\}=0 \tag{5.26}$$

For $\tau << 1$,

$$(1-\tau)^n\approx 1-n\times\tau+\frac{n\times(n-1)}{2}\times\tau^2 \tag{5.27}$$

Substituting (5.27) into (5.26) gives an approximation of τ:

$$\tau=\frac{\frac{\sqrt{n+2\times(n-1)\times\left(\overline{T_c}-1\right)}}{n}-1}{(n-1)\times\left(\overline{T_c}-1\right)}\approx\frac{1}{n\times\sqrt{\overline{T_c}/2}} \tag{5.28}$$

By substituting (5.28) into (5.13), when $m=0$ (the CW value is constant), the optimal CW for obtaining the maximum throughput is given by

$$W_{opt}\approx n\sqrt{2\overline{T_c}} \tag{5.29}$$

5.5.2 *Performance Analysis of NDCF Algorithm*

A process is said to be Markovian if the future of the process depends only on the present and not on the past. Because the fallback avoidance strategy of the HDFB algorithm depends not only on the present but also on the past, it cannot be analyzed in the same way as the BEB algorithm in terms of Markov modeling. In the following, we perform a Markov modeling analysis of the performance of the NDCF algorithm.

Referring to the flow chart of the NDCF algorithm in Fig. 5.6, let $B(t)$ represent the statistical process of determining the value of the avoidance counter of the node at time t, which is decremented by 1 at the beginning of each time slot, and let $S(t)$ represent the stochastic process of taking the value of the avoidance level of the node. We built a two-dimensional stochastic process model $\{S(t), B(t)\}$ to analyze the NDCF algorithm. Figure 5.7 shows the Markov state transfer diagram. W_0 is the minimum CW, W_i is the size of the CW in the ith avoidance, K is the value of the avoidance counter of the node, i is the order of the avoidance counter, m is the maximum value of i, P is the probability of collision each time a node sends a data frame, and P_0 is the probability that the avoidance counter takes the value in the first half of the CW.

Using the Markov chain model, we can obtain a system of one-step transfer probability equations:

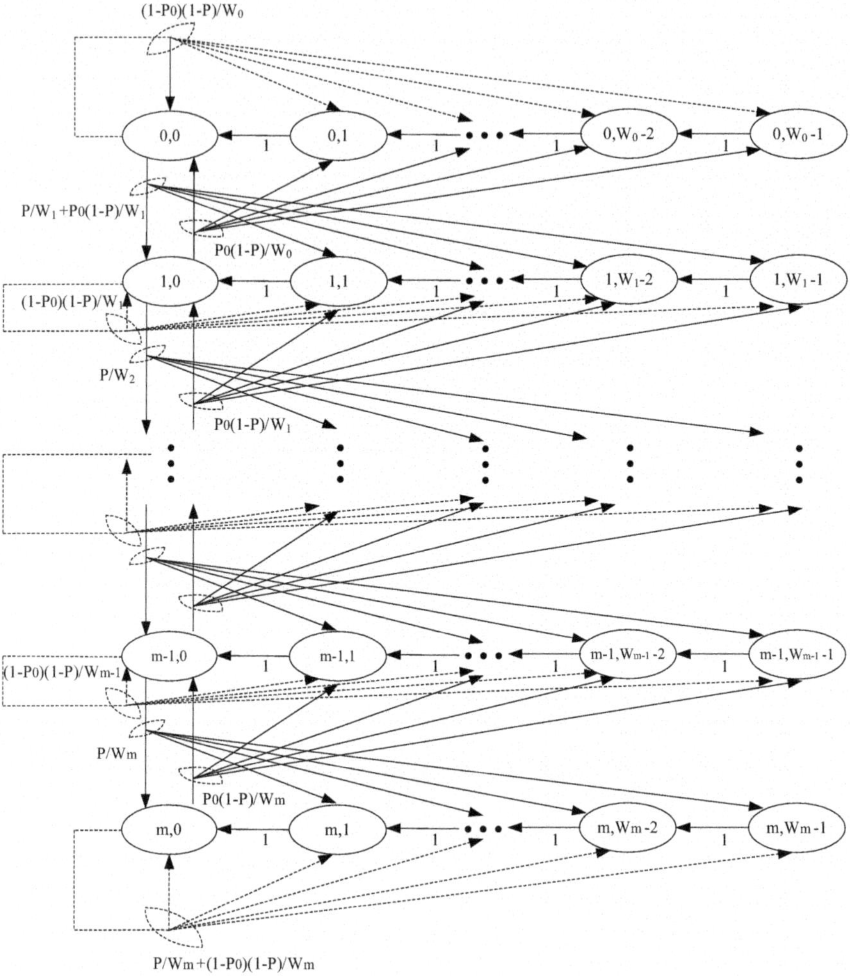

Fig. 5.7 State diagram of NDCF Markov model

$$\begin{cases} P\{i,k|i,k+1\}=1 & k\in[0,W_i-2], i\in[0,m] \\ P\{0,k|0,0\}=\frac{(1-P_0)\times(1-P)}{W_0} & k\in[0,W_0-1] \\ P\{1,k|0,0\}=\frac{P}{W_1}+\frac{P_0\times(1-P)}{W_1} & k\in[0,W_1-1] \\ P\{i,k|i-1,0\}=p/W_i & k\in[0,W_i-1], i\in[2,m] \\ P\{m,k|m,0\}=\frac{P}{W_m}+\frac{(1-P_0)\times(1-P)}{W_m} & k\in[0,W_m-1] \\ P\{i-1,k|i,0\}=\frac{P_0\times(1-P)}{W_{i-1}} & k\in\left[0,W_{i-1}-1\right], i\in[1,m-1] \\ P\{i,k|i,0\}=\frac{(1-P_0)\times(1-P)}{W_i} & k\in[0,W_i-1], i\in[1,m-1] \end{cases} \tag{5.30}$$

These equations represent the seven main events in the NDCF channel access process. (1) At the beginning of each time slot, the fallback counter is decremented by one count. (2) When the fallback counter is of order 0, and the last fallback counter value is taken in the second half of the CW, the fallback counter remains of order 0 if the transmission is successful, and the value of the fallback counter is chosen randomly with medium probability in $(0, W_0-1)$. (3) There are two cases in which the avoidance counter order is changed from order 0 to order 1: when transmission succeeds and the last avoidance counter value is taken in the first half of the CW, and when transmission fails. (4) When a node fails to send with an avoidance counter of order $i-1$, the order of the avoidance counter is increased by 1, and the value of the avoidance counter is chosen at random with medium probability in $(0, W_i-1)$. (5) There are two cases in which the avoidance counter order is kept at order m: when transmission succeeds and the last avoidance counter value is taken in the second half of the CW, and when transmission fails. (6) When the avoidance counter is between order 1 and order $m-1$ and the last avoidance counter value is taken in the first half of the CW, transmission succeeds; the order of the avoidance counter is increased by 1, and the value of the avoidance counter is chosen at random with medium probability in $(0, W_i-2$. (7) When the avoidance counter is between order 1 and order $m-1$ and the last avoidance counter value is taken in the second half of the CW, transmission succeeds; the order of the avoidance counter is unchanged, and the value of the avoidance counter is chosen at random with medium probability in $(0, W_i-1)$.

Let $b_{i,k}=\lim_{t\to\infty}\mathrm{P}\{\mathrm{S}(t)=i,\mathrm{B}(t)=k\}, i\in(0,m), k\in(0,W_i-1)$ be the stable distribution of a Markov chain. According to the state transfer diagram in Fig. 5.7,

$$b_{i,k}=\frac{W_i-k}{W_i}\times b_{i,0}=\frac{W_i-k}{W_i} \times\begin{cases} (1-P_0)\times(1-P)\times b_{0,0}+P_0\times(1-P)\times b_{i+1,0} & i=0 \\ P\times b_{0,0}+p_0\times(1-P)\times b_{0,0}+ \\ (1-P_0)\times(1-P)\times b_{1,0}+(1-P_0)\times(1-P)\times b_{i+1,0} & i=1 \\ p\times b_{i-1,0}+(1-P_0)\times(1-P)\times b_{i,0}+P_0\times(1-P)\times b_{i+1,0} & 1<i<m \\ p\times b_{i-1,0}+(1-P_0)\times(1-P)\times b_{i,0} & i=m \end{cases} \tag{5.31}$$

where $b_{i,K}$ is the probability that the node has avoidance order i and the avoidance counter takes the value K. In general, $P_0 = 0.5$, and according to the normalization principle of probability, we can substitute (5.31) into (5.32) and simplify to find $b_{0,0}$:

$$1 = \sum_{i=0}^{m}\sum_{k=0}^{W_i-1} b_{i,k} = \sum_{i=0}^{m} b_{i,0} \sum_{k=0}^{W_i-1} \frac{W_i - k}{W_i} = \sum_{i=0}^{m} b_{i,0} \times \frac{W_i + 1}{2}$$
$$= \frac{b_{0,0}}{2} \times W_0 \times \left\{1 + 2 \times \frac{1+P}{1-P}\left[1 + \frac{4P}{1-P} + \left(\frac{4P}{1-P}\right)^2 + \ominus + \left(\frac{4P}{1-P}\right)^{m-2}\right]\right.$$
$$\left. + \left(\frac{4P}{1-P}\right)^{m-1}\right\} = \frac{b_{0,0}}{2} \times W_0 \times \frac{3(1-P)^m - (1+7P) \times (4P)^{m-1}}{(1-5P) \times (1-P)^{m-1}} \tag{5.32}$$

$$b_{0,0} = \frac{2 \times (1-5P) \times (1-P)^{m-1}}{W_0 \times \left[3(1-P)^m - (1+7P) \times (4P)^{m-1}\right]} \tag{5.33}$$

Let the probability that a node sends a data frame at any moment be τ, which is the sum of the probabilities when the avoidance counter of any avoidance order decreases to 0. Then we have

$$\tau = \sum_{i=0}^{m} b_{i,0} = b_{0,0} \times \left\{1 + \tfrac{1+P}{1-P}\left[1 + \tfrac{2P}{1-P} + \left(\tfrac{2P}{1-P}\right)^2 + \ominus + \left(\tfrac{2P}{1-P}\right)^{m-2}\right] + \left(\tfrac{2P}{1-P}\right)^{m-1}\right\}$$
$$= \frac{4\times(1-5P)\times\left[(1-P)^m-(2P)^{m-1}\right]}{W_0\times(1-3P)\times\left[3(1-P)^m-(1+7P)\times(4P)^{m-1}\right]} \tag{5.34}$$

The probability that only one node in the network sends a data frame is then $(1-\tau)^{n-1}$, and we have

$$P = 1 - (1-\tau)^{n-1} \tag{5.35}$$

The relationship between P and the number of network nodes n, minimum CW W_0, and maximum avoidance order m is obtained from (5.34) and (5.35).

The relationship between the saturation throughput and fairness of the NDCF and BEB algorithms is analyzed qualitatively. The NDCF algorithm is fairer than the BEB algorithm because of its avoidance backoff strategy. Fairness refers to the equal ability of each node to access a channel, which can be expressed as a similar difference in avoidance wait time between nodes. The BEB algorithm reduces the CW to a minimum after a successful transmission and doubles it to a maximum after a failed transmission, resulting in a large variation in the differences between the avoidance waiting times of nodes. The NDCF algorithm, by contrast, makes small changes in the CW. In addition, the HDFB algorithm must select the avoidance window twice before the avoidance rule is changed, whereas the BEB algorithm selects it only once. The HDFB algorithm uses the same strategy as the BEB algorithm after transmission failure; however, when transmission succeeds, the CW is divided only by $\sqrt{2}$. Thus, the change in the CW is not as large as that in the BEB algorithm, which reduces the CW to the minimum value.

The inference process for the saturation throughput is illustrated in Fig. 5.8. The throughput performance of the NDCF algorithm is inferior to that of the BEB algorithm. The difference in avoidance waiting time is smaller for the NDCF algorithm; thus, the total avoidance waiting time is longer. The avoidance wait time is proportional to the average propagation delay δ, which is larger for the NDCF algorithm than for the BEB algorithm. Consequently, the time T_s required for the transmission of one data frame and the time T_c required for one data frame collision are larger for the NDCF algorithm. The inverse relationship between T_c and the probability τ that a node sends a data frame at any moment is obtained from (5.28), which shows that the τ value of the NDCF algorithm is small. Equations (5.15) and (5.16) also indicate that τ is small. Consequently, the probability P_{tr} that at least one node in each time slot sends a data frame is also small, and $1 - P_{tr}$ is large, whereas the probability P_s that only one data frame is successfully transmitted remains essentially the same. Finally, according to the formula for the normalized saturation throughput S in (5.17), the average length of the transmitted MSDU, E[MSDU], remains the same; $T_s - T_c$ and P_s are essentially unchanged, whereas $\sigma \times (1 - P_{tr}) / P_{tr}$ and T_c are large. Therefore, S is small; that is, the saturation throughput of the NDCF algorithm is lower than that of the BEB algorithm.

$$S = \frac{E[MSDU]}{T_s - T_c + \frac{\sigma \times (1 - P_{tr}) / P_{tr} + T_c}{P_s}} \tag{5.36}$$

Both improved algorithms proposed above, the HDFB and NDCF algorithms, adopt the strategy of adaptively setting the avoidance according to the state of certain indicators. Because of the directional nature of ultraviolet (UV) light transmission, the directional limitation is exacerbated by hidden terminals and the asymmetry of node access to channel information. Therefore, the concept of improved fairness based on the MAC layer of ad hoc networks for radio-frequency communication differs slightly from that proposed in the papers cited in Sect. 5.3. Therefore, the HDFB algorithm provides a multilevel prediction of the channel state based on the status of the last two packet transmissions from a node and the last two avoidance window selections. The channel busy and idle states are predicted indirectly using the transmit/receive and collision states of the node, rather than by directly monitoring the channel using the carrier listening mechanism. The NDCF algorithm determines the suitability of a node's access to a channel according the value of its last randomly generated fallback counter.

Large δ → large T_c and large T_s ($T_s - T_c$ essentially constant) → small τ → small P_{tr}, P_s essentially constant → large $(1 - P_{tr})/P_{tr} + T_c$ → small S.

Fig. 5.8 Inference process for saturation throughput

5.6 Simulation and Analysis of NDCF Algorithm

5.6.1 Experimental Environment and Parameters

Simulation experiments based on IEEE 802.11 DCF were performed. The performance of the BEB algorithm and the improved algorithms (the HDFB and NDCF algorithms) for the MAC layer protocol were simulated and analyzed. The UV optical communication model is used instead of the default two-ray ground reflection model in the NS2 network simulation software.

(1) UV optical communication model

Figure 5.9 shows a single-scattering model of non-line-of-sight (NLOS) UV communication.

The estimated power of the received signal in the single-scattering model of NLOS UV communication is defined as [30]

$$\mathrm{E_r} \approx \frac{E_t K_s P(\theta_s) A_R V \sin^4\theta_s \left[-\frac{K_e r}{\sin\theta_s}(\sin\theta_T + \sin\theta_R) \right]}{2\pi r^4 \sin^2\theta_T \sin^2\theta_R \left(1 - \cos\frac{\phi_T}{2}\right)} \tag{5.37}$$

where E_r is the total received energy, E_t is the total transmitted energy, θ_T is the transmitter elevation angle, θ_R is the receiver elevation angle, ϕ_T is the aperture angle of the transmitted beam, ϕ_R is the field of view of the receiver, and r is the distance from the transmitter (Tx) to the detection receiver (Rx). V is the volume of the overlapping region of the transmitter and receiver elevation angles, and $r1$ and $r2$ are the distances of the transmitter and receiver, respectively, from V. θ_s is the sum of θ_T and θ_R ($\theta_s = \theta_T + \theta_R$). A_R is the receiver photosensitive surface area, $P(\theta_s)$ is the single-scattering phase function (the entire sphere is normalized to 4π), and Ke is the atmospheric attenuation coefficient (km^{-1}), which is the sum of the atmospheric scattering coefficient Ks and atmospheric absorption coefficient Ka ($Ke = Ks + Ka$).

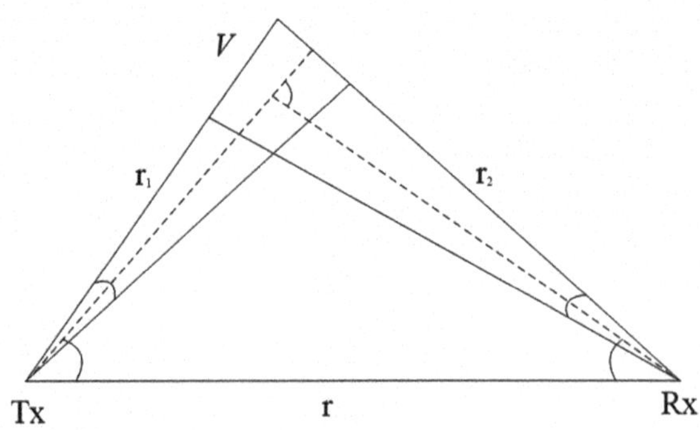

Fig. 5.9 Single-scattering model of NLOS UV communication

Table 5.2 Simulation parameter settings

Channel bit rate	2 Mbps	Slot time	20 μs
DIFS	50 μs	SIFS	10 μs
Preamble length	144 bits	PLCP header length	48 bits
RTS	20 bits	CTS/ACK	14 bits
CW_{max}	1023	CW_{min}	31
ShortRetryLimit	7	LongRetryLimit	4

(2) Program implementation of NDCF and HDFB algorithms

(a) For the HDFB algorithm, the main modifications were made to the mac-802_11.h and mac-802_11.cc files. The modified program modules are as follows: mac802_11::set_cw(): performs the coarse tuning strategy, grading the CW as well as supplementing it; mac802_11::inc_cw(): is called when transmission fails in the fine tuning strategy, and mac802_11::rst_cw(): is called when transmission succeeds in the fine tuning strategy. In addition, mac802_11::RetransmitDATA(), mac802_11::recv_timer(), mac802_11::recvACK (Packet *p), and other program modules must be modified to complete the count of the sent messages in the two monitoring mechanisms in the supplementary policy.

(b) For the NDCF algorithm, most of the modifications are made in mac-802_11.h and mac-802_11.cc. The modified program modules are as follows: mac802_11::inc_cw(): selects the fallback counter value to be set when transmission fails. mac802_11::rst_cw(): selects the avoidance counter value to be set when transmission succeeds.

(c) The other network parameters in the simulation are listed in Table 5.2.

5.6.2 Simulation Results and Performance Analysis

5.6.2.1 Chain Topology

We used a very common four-node chain topology for an ad hoc network in the simulation (Fig. 5.10). Arrows indicate data transmission between nodes, and dashed lines indicate that nodes are within communication range of each other but there is no service transmission between them. Node C sends data stream f2 to node D, and node A sends data stream f1 to node B. Node B is within range of node C and outside the range of node D (the receiving node); it is therefore an exposed receiving terminal. For the four-node exposed terminal topology shown in Fig. 5.10, to avoid the effects of the congestion control mechanism of the TCP, the service type uses a constant bit rate (CBR), where data are sent continuously at a fixed rate during the simulation. The CBR is set to 800 kbps, the data packet length is set to 1000 bytes, and the simulation time is 200 s. The simulation throughput and fairness are shown in Figs. 5.11 and 5.12, respectively.

For the HDFB algorithm, the throughputs of nodes C and A are 365.1 and 345.0 kbps, respectively; the values differ by approximately 20 kbps, indicating the best fairness. For the NDCF algorithm, the throughputs of nodes C and A are 537.8 and 248.2 kbps, respectively, which differ by approximately 290 kbps, indicating adequate fairness. For the BEB algorithm, nodes C and A have throughputs of 664.6 and 128.3 kbps, respectively, which differ by 536 kbps; here the fairness is poor, and node C has almost exclusive access to the channel. The BEB algorithm introduces

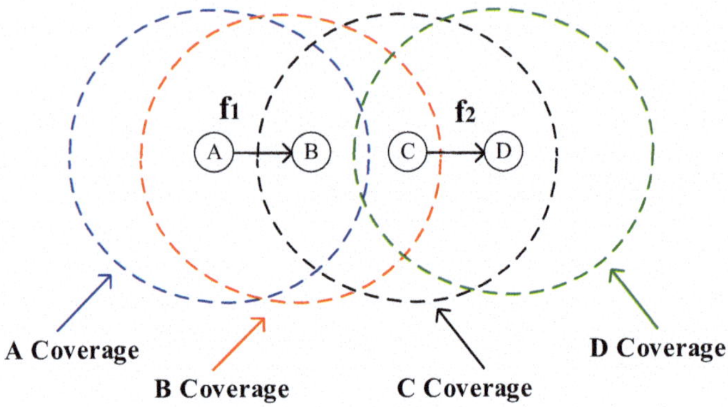

Fig. 5.10 Four-node topology (exposed terminals)

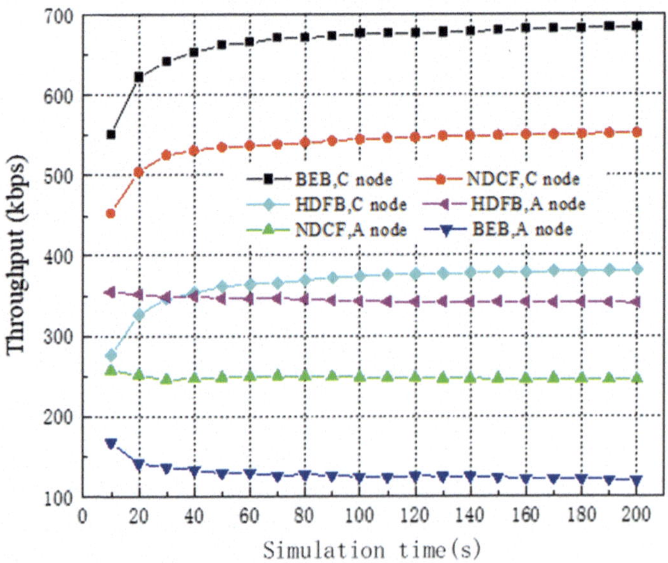

Fig. 5.11 Sending node throughput in four-node topology [26]

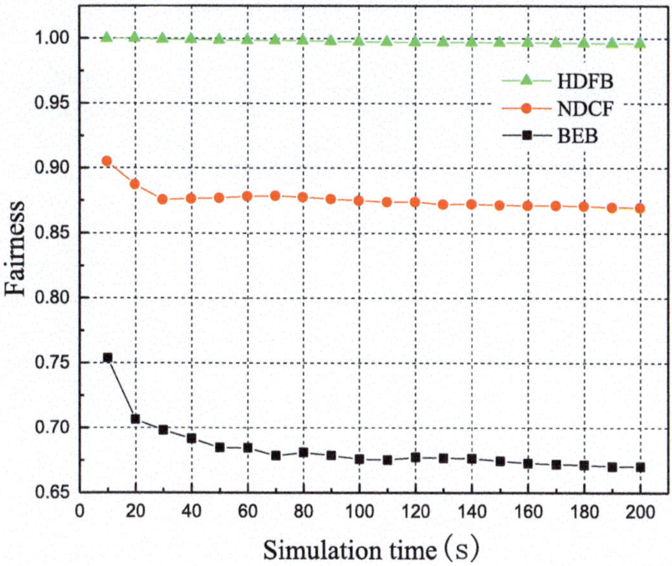

Fig. 5.12 FI versus time in four-node topology [26]

a serious inequity for the following reasons. Node B receives an RTS data packet from node C, which interferes with the reception of the RTS data packet from node A (causing a conflict and forcing A to retransmit) or delays the transmission of the CTS ACK packet (generating unnecessary transmission delay), which weakens the channel occupation of data stream f1. This mechanism in the BEB algorithm always favors the node that transmitted successfully, enabling it to occupy the channel again; consequently, node C again has priority in sending data. This behavior is repeated several times, and eventually the data stream f2 has exclusive access to the channel.

5.6.2.2 Hybrid Topology

We simulate the 12-node hybrid topology in Fig. 5.13, where arrows indicate data transmission between nodes, and dashed lines indicate that nodes are within communication range of each other but there is no internode service transmission. Nodes 0, 1, and 3 simultaneously send data streams F_1, F_2, and F_3, respectively, to node 2. The three nodes are mutually hidden terminals. and F_1, F_2, and F_3 form contention area A. Node 4 sends data stream F_4 to node 6. In addition, F_5 is a data stream from node 5 to node 7, which is an exposed terminal with respect to F_3. F_3, F_4, and F_5 form contention area B, and F_6, F_7, and F_8 form contention area C. The service type used in the simulation is CBR, the data packet length is set to 512 bytes, and the simulation time is 300 s. The simulated throughput and fairness are shown in Figs. 5.14 and 5.15, respectively.

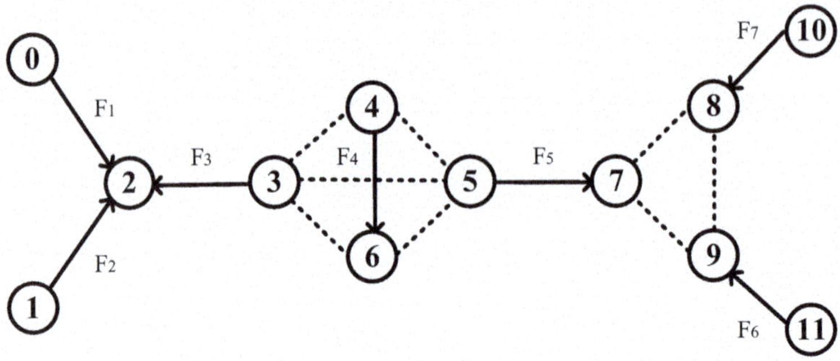

Fig. 5.13 Hybrid network topology with 12 nodes

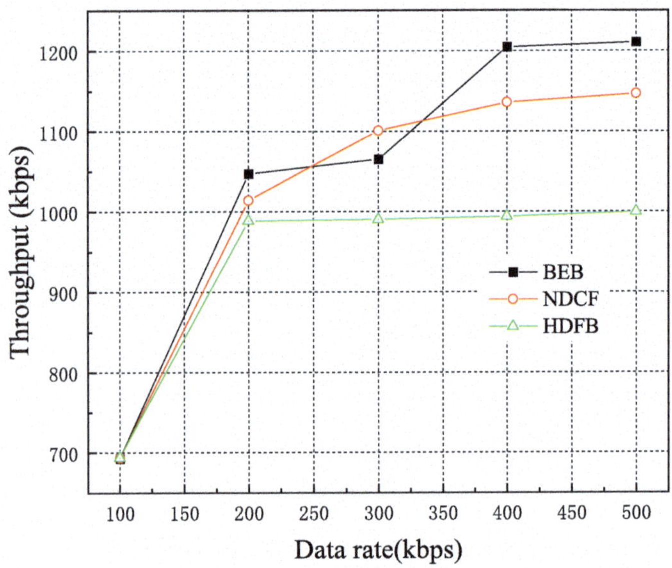

Fig. 5.14 Simulated total throughput versus CBR [26]

As the CBR increases, the network load gradually increases, resulting in more collisions and lower fairness. The analysis considers three CBR ranges. (1) When the CBR is between 100 and 200 kbps, increasing the CBR results in higher network throughput and lower fairness, almost regardless of the algorithm adopted for backoff avoidance. The FI decreases from 0.999 to approximately 0.903, and the throughput increases from 693.8 kbps to approximately 1014.5 kbps. The HDFB algorithm exhibits the worst performance. The reason is that under light loads, this algorithm has too much idle time due to waiting because of the reliance on historical data. (2) When the CBR is between 200 and 400 kbps, the network throughput increases, and the fairness decreases, but not as much as in the first situation. The idle time due to waiting determined by the HDFB algorithm decreases with increasing CBR,

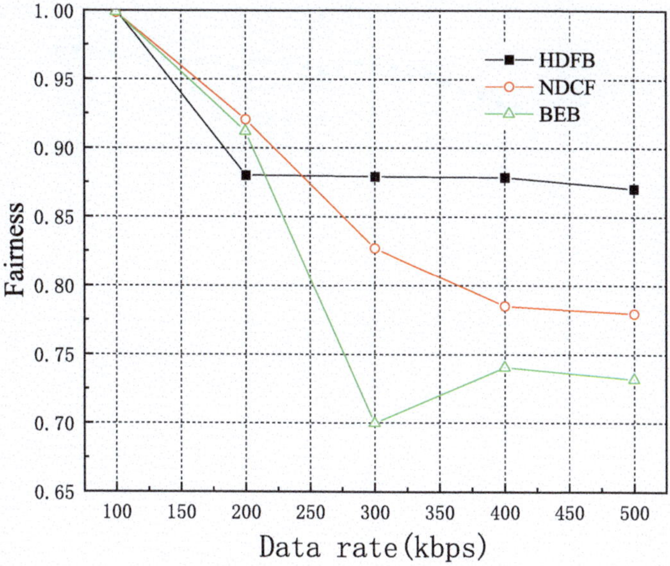

Fig. 5.15 Simulated FI versus CBR [26]

and its dynamically adjusted avoidance strategy for predicting channel conditions begins to make a significant difference, where the FI remains at approximately 0.877. The NDCF algorithm adjusts the avoidance strategy based on the value of the last randomly generated avoidance counter, which reduces the unfairness caused by the unreasonable short-term random selection of avoidance counter values. Its FI is approximately 0.823 on average. (3) When the CBR is between 400 and 500 kbps, increasing the CBR has minimal impact on the network throughput and fairness. The reason is that the network is overloaded, and the number of collisions between data flows exceeds the ability of the three backoff avoidance algorithms to mitigate collisions. Overall, the HDFB algorithm has the best fairness and smoother fairness, with the lowest total throughput; the NDCF algorithm has better fairness and total throughput, and the BEB algorithm has the worst fairness and highest total throughput (Figs. 5.16 and 5.17).

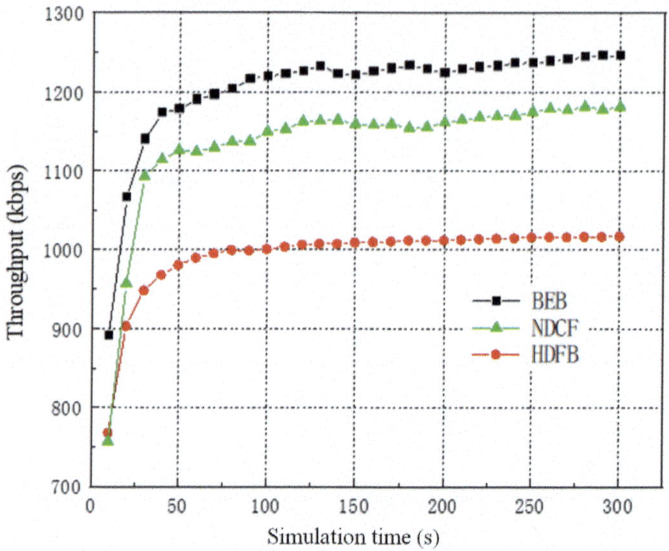

Fig. 5.16 Simulated total throughput versus time (CBR: 400 kbps) [26]

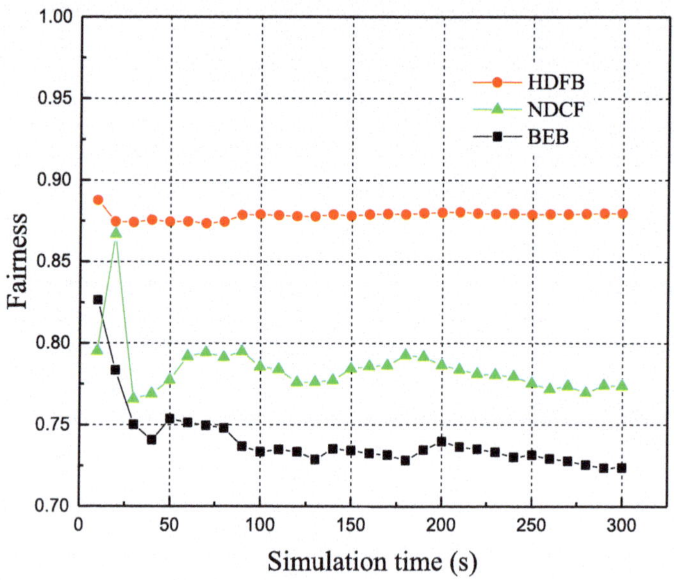

Fig. 5.17 Simulated FI versus time (CBR: 400 kbps) [26]

References

1. Zh M (2005) Proposed and Implementation of a New MAC Protocol for Ad Hoc Networks. Zhejiang University, Hangzhou, pp 32–39
2. Huang DJ (2006) Simulation study based on improved CSMA/CD protocol. Master's thesis, Hefei University of Technology, Hefei, pp 38–41
3. Karn P (1990) MACA-a new channel access method for packet radio. In: ARRL/CRRL amaterur radio 9th computer networking conference, London, Ontario, Canada, pp 134–140
4. Fullmer CL, Carcia-Luna-Aceves JJ (1995) Floor acquisition multiple access (FAMA) for packer-radio networks. Conf Appl Technol Arch Protoc Comput Communication Camb 25(4):262–273
5. Haas J, Deng J (2002) Dual busy tone multiple access (DBTMA): a new medium access control for multihop networks. IEEE Trans Commun 50(6):975–985
6. Bianchi G (2000) Performance analysis of the IEEE 802.11 distributed coordination function. IEEE J Sel Areas Commun 18(3):535–547
7. Acharya A, Ganu S, Misra A (2006) DCMA: a label switching MAC for efficient packet forwarding in multihop wireless networks. IEEE J Sel Areas Commun 24(11):1995–2004
8. Yang ZY, Garcia-Luna-Aceves JJ (1999) Hop-reservation multiple access (HRMA) for ad-hoc networks, vol 1. INFOCOM'99, New York, pp 194–201
9. Hiraga K, Akabane K, Shiba H et al (2008) Channel assignment and reallocation algorithms for cognitive radio systems. In: Proceeding of APCC 2008, Tokyo, pp 1–4
10. Wang M, Ci L, Zhan P et al (2008) Multi-channel MAC protocols in wireless ad hoc and sensor networks. Comput Commun Control Manag Guangzhou 2:562–566
11. Xu CW, Wu XY, Liu HJ (2008) Ad Hoc network topology and hidden terminal/exposed terminal research. Commun Technol 41(2):103–105
12. IEEE Std.802.11 (1999) Wireless LAN medium access control (MAC) and physical layer (PHY) specifictions, pp 65–97
13. Huang LF, Huang Y, Li YD (2001) Effectiveness max-min fairness criterion and its application to ABR services. J Commun 22(07):10–17
14. Fan Y (2007) Design of MAC protocol for Ad Hoc networks based on CSMA/CA improvement. Master's thesis, University of Electronic Science and Technology, Chengdu, pp 34–47
15. Jain RK, Chiu DM, Hawe WR (2009) A quantitative measure of fairness and discrimination for resource allocation in shared computer systems. http://www.cs.wustl.edu/~jain/papers/ftp/fairness.pdf
16. Tao L (2008) Research and improvement of MAC fairness in wireless self-organizing networks. Master's thesis, Beijing University of Posts and Telecommunications, Beijing, pp 29–34
17. Li N, Han L (2005) A review of fallback algorithms for wireless self-organizing networks. Comput Appl 25(6):1244–1250
18. Deng J, Varshney PK, Haas ZJ (2004) A new backoff algorithm for the IEEE802.11 distributed coordination function.Commun Netw Distrib Syst Model Simul San Diego 215–225
19. Wang C, Li B, Li L (2004) A new collision resolution mechanism to enhance the performance of IEEE802.11DCF. IEEE Trans Veh Technol 53(4):1235–1246
20. Kui XY, Du HK(2009) CIMLD: an adaptive MAC fallback algorithm in multi-hop Ad Hoc networks. Small Microcomput Syst 30(4):679–682
21. Bensaou B, Wang Y, Ko CC (2000) Fair medium access in 802.11 based wireless ad-hoc networks. IEEE/ACM MobiHoc Workshop, Boston, pp 99–106
22. Peng Y, Cheng SD (2004) An adaptive wireless local area network protocol. J Software 15(4):604–615
23. Cui HX, Wei G (2009) A novel backoff algorithm based on the tradeoff of efficiency and fairness for Ad Hoc networks. Int Conf Commun Soc Yunnan 2:81–86
24. Kwon YY, Fang YY, Latchman H (2003) A novel MAC protocol with fast collision resolution for wireless LANs, vol 2, pp 853–862. IEEE INFOCOM 2003, San Francisco
25. Qin YW (2007) Research and improvement of MAC layer protocols in Ad Hoc networks. Master's thesis, Central South University, Changsha, pp 21–29

26. Chen X (2010) Research on the fairness of MAC layer protocols in UV Ad Hoc networks. Master's thesis, Xi'an University of Technology, Xi'an
27. Ke X-Z, He H, Chen X (2010) History-based dynamic fairness backoff avoidance algorithm in UV Ad hoc networks. Laser J 31(3):32–34
28. Ke X-Z, He H, Chen X (2010) A new MAC layer fallback avoidance algorithm for UV self-organized communication networks. Optoelectron-Laser 21(7):1002–1006
29. Bianchi G (2001) Performance analysis of the IEEE802.11 distributed coordination function. IEEE J Sel Area Commun 18(3):535–547
30. Xu ZY et al (2008) Analytical performance study of solar blind non-line-of-sight ultraviolet short-range communication links. Optics Lett 33(16):1860–1862

Chapter 6
Ultraviolet Wireless Optical Mesh Network Access Protocol

Wireless mesh networks (WMNs) are multihop wireless networks. They combine the advantages of wireless local area networks (WLANs) and wireless ad hoc networks. The WMN is a network technology with large capacity, high speed, and wide coverage [1]. The combination of ultraviolet (UV) communication and a WMN can not only fully exploit the advantages of UV non-line-of-sight (NLOS) communication, including its strong anti-interference ability, but also extend the effective communication distance because of the multihop communication capability of WMNs, which compensates for the short UV communication distance. Therefore, UV wireless optical mesh networks have potential application prospects. The key technologies of UV wireless optical mesh networks include the following [2].

(1) Network configuration and deployment

To obtain a UV wireless optical mesh network with good scalability, fault tolerance, self-regulation ability, and large coverage and network capacity, it is necessary to study the network configuration and deployment. First, we need to plan the deployment locations of mesh routers, which are less mobile and form the backbone network. To reduce costs, as few mesh routers as possible should be used while ensuring that there are no gaps in wireless signal coverage. However, in hotspot areas, multiple paths are provided to increase the number of users who can access the network. These paths can be combined with multiple input–multiple output and directional antenna technology to further improve the network transmission capacity. Mesh routers can provide access services for various heterogeneous networks. Therefore, WMNs should be capable of allocating multiple wireless interfaces for mesh routers to ensure connectivity between networks. Multichannel methods can be used to increase the throughput of WMNs. However, in wireless multihop environments, the multichannel communication mode also faces many problems that urgently require solution.

X. Ke, *UV Light Self Organizing Network Theory*, Optical Wireless Communication Theory and Technology, https://doi.org/10.1007/978-981-97-7854-6_6

(2) Media access control (MAC) protocol design

The MAC protocol determines how to allocate wireless channel resources among competing users, that is, how to control access to the channel to send data. The quality of MAC protocol design directly affects the network throughput, fairness, delay, channel spatial multiplexing rate, and other performance metrics. Unlike radio-frequency communication, UV communication uses a type of scattering channel. The coverage of UV NLOS communication has a certain directionality. Only nodes within range of transmitting nodes can receive information. Consequently, conventional MAC protocols may not be directly applicable to UV WMNs. MAC protocols based on directional antenna access can control the spatial multiplexing rate and communication distance of the channel, but they introduce the problems of directional hidden terminals, deafness, and the need to determine node location The multichannel, multi-interface MAC protocol uses multiple channels so that a single node can use different channels to communicate with different nodes simultaneously, which improves the network performance. Our goal is to analyze the conventional, directional-antenna-based, and multichannel MAC protocols and design an access protocol suitable for UV WMNs.

(3) Routing protocol design

WMNs use multihop relay to provide network access. Therefore, routing protocol design can be based on routing protocols for ad hoc networks, such as ad hoc on-demand distance vector (AODV) [3], dynamic source routing [4], destination-sequenced distance vector [5], optimized link state routing (OLSR) [6], and other methods. However, protocols for WMNs and ad hoc networks differ. For example, WMNs are composed of access points, mesh routers, mesh gateways, etc., and nodes are relatively fixed. In addition, most data services in WMNs are sent from user nodes to gateway nodes, whereas ad hoc networks provide mainly node-to-node transmission. Moreover, WMNs have relatively low power requirements. The study of routing protocols for WMNs can also refer to routing technologies used on the Internet. For example, protocols such as the zone routing protocol [7] are used within and outside of the autonomous region.

(4) Quality of service (QoS)

There are two reasons for QoS-related problems in WMNs: (1) the admission control and differentiated service mechanism for mixed services transmitted on the access and backbone networks to ensure the appropriate service level and (2) the need for a conventional two-layer mechanism to obtain end-to-end traffic information and guarantee services on multihop networks. The QoS guarantee mechanism in the MAC layer (for example, IEEE 802.11e service differentiation) can provide the appropriate QoS only on single-hop links. To support end-to-end real-time services, it is also necessary to design a reasonable QoS routing mechanism at the network layer.

6.1 UV NLOS Communication Networks

Research on UV NLOS communication has focused mainly on single- and multiple-scattering atmospheric channel models. In particular, it has considered the relationship between the path loss and the elevation, divergence, and field of view of the UV transceiver, and the relationship between the bit error rate (BER) and data transmission rate and distance [8, 9] and the effects of modulation and coding [10] and atmospheric conditions [11] on the transmission distance of UV communication. However, research on UV communication networks has not been reported.

6.1.1 UV Communication Channel Bandwidth

Table 6.1 shows the data transmission rate [12] of line-of-sight (LOS) and NLOS UV communication using 10 UV light-emitting diode light sources in a 24-cell array. The luminous power of each light source was 0.5 mW, the receiving antenna gain was 100, and on–off keying modulation was used. The UV NLOS communication mode used a divergence angle of 1°, field of view of 60°, transmitter elevation angle of 45°, and receiver elevation angle of 60°. The data transmission rate of the short-range LOS mode can reach gigabytes per second, whereas the data transmission rate of the NLOS mode is much lower.

Reference [13] analyzed the relationship between the NLOS channel bandwidth and geometric angles. The channel bandwidth decreases with increasing communication distance. It also decreases with increasing UV transmitter and receiver elevation angles, divergence angle, and field of view. The channel bandwidth depends mainly on the elevation angles of the transmitter and receiver during communication, whereas the divergence angle and field of view have relatively little effect on channel bandwidth. When the transmitter and receiver elevation angles are relatively small, the channel bandwidth is affected more strongly by the transmitter elevation angle than by the receiver elevation angle. When these angles are relatively large, the channel bandwidth does not depend significantly on them. Because the divergence angle and field of view affect the channel bandwidth, it is more effective in this case to adjust the divergence angle. The formula for channel bandwidth estimation [13] is as follows:

Table 6.1 Data transmission rate of UV communication for various distances and BERs [12]

Range	LOS		NLOS	
	BER = 10^{-3}	BER = 10^{-6}	BER = 10^{-3}	BER = 10^{-6}
10 m	6 Gbps	2 Gbps	8 Mbps	3 Mbps
100 m	50 Mbps	20 Mbps	700 kbps	300 kbps
1000 m	300 kbps	100 kbps	20 kbps	9 kbps

$$B = \frac{\sqrt{(2^{1/\alpha} - 1)}}{2\pi\beta} \tag{6.1}$$

where α and β are the parameters of the model, which can be estimated by numerically finding the minimum value of the nonlinear least-mean-square criterion using the channel model.

6.1.2 Received Optical Power of UV NLOS Communication

The received optical power [12] of UV NLOS communication is calculated as follows:

$$P_{r,NLOS} = \frac{P_t \cdot A_r \cdot K_s \cdot P_s \cdot \phi_2 \cdot \phi_1^2 \cdot \sin(\theta_1 + \theta_2)}{32\pi^3 \cdot r \cdot \sin(\theta_1) \cdot (1 - \cos\frac{\phi_1}{2})} \cdot e^{-\frac{Ke \cdot r \cdot (\sin\theta_1 + \sin\theta_2)}{\sin(\theta_1 + \theta_2)}} \tag{6.2}$$

The values of each parameter in the formula are as follows: the atmospheric scattering coefficient K_s is 0.49 km^{-1}, and the atmospheric absorption coefficient K_e is 1.23 km^{-1} [14], the Phase function of scattering angle $P_s = 1$ and the Receiving aperture area $A_r = 1.8\,cm^2$ [12], and the transmission power $P_t = 1$ W. According to reference [15], the single-scattering atmospheric channel model for UV NLOS communication is relatively accurate when the communication distance r is less than 200 m. Because the simulation in this chapter is based on the single-scattering atmospheric channel model, r is set to < 200 m in the simulation. We set the divergence angle, field of view, and transmitter and receiver elevation angles in the three UV NLOS communication modes according to (6.2). The received power is shown in Table 6.2, where the divergence angle and field of view are both 60°. For NLOS(A), the transmitter and receiver elevation angles of the node are both 89°. Omnidirectional transmission and reception are approximated by using these angles. For NLOS(B), the node has a transmitter elevation angle of 30° and a receiver elevation angle of 90°. For NLOS(C), the node has a transmitter elevation angle of 20° and a receiver elevation angle of 60°.

For NLOS(A) and NLOS(B), the communication distances at which each mode achieves the same received power differ greatly. For example, in NLOS(B) mode, the received optical power is 4.3094×10^{-12} W at a communication distance of 200 m, whereas in NLOS(A) mode, a similar optical power is received at a communication distance of less than 5 m. In NLOS(A) mode, the power attenuates very rapidly, and the communication distance is short. The power attenuation in the NLOS(B) and NLOS(C) modes is of the same order of magnitude. However, the power attenuation in the NLOS(B) mode is larger than that in NLOS(C) mode. If the source and destination nodes in a communication network are widely separated and the UV NLOS(A) mode is used, the transmitter must transmit at higher power or use multihop communication. When NLOS(B) and NLOS(C) modes are used, the source and destination nodes may communicate through single-hop communication.

Table 6.2 Received optical power in UV NLOS communication

Distance	NLOS(A) (89°, 89°)	NLOS(B) (30°, 90°)	NLOS(C) (20°, 60°)
1 m	2.47862×10^{-11}	1.316952×10^{-9}	2.19067×10^{-9}
5 m	3.73946×10^{-12}	2.61154×10^{-10}	4.35498×10^{-10}
10 m	1.31444×10^{-12}	1.29194×10^{-10}	2.16112×10^{-10}
50 m	1.56839×10^{-14}	2.3728×10^{-11}	4.0691×10^{-11}
100 m	2.31222×10^{-16}	1.06652×10^{-11}	1.88671×10^{-11}
150 m	4.5451×10^{-18}	6.39171×10^{-12}	1.16641×10^{-11}
200 m	1.0051×10^{-19}	4.3094×10^{-12}	8.11237×10^{-12}

Unit *W* angles represent transmitter and receiver elevation angles

6.1.3 Forwarding Structure of UV NLOS Communication Node

Because UV light is affected by atmospheric absorption, scattering, and turbulence, its power decays rapidly, and the UV communication distance is short. When the distance between the source and destination nodes is small, communication can occur directly by a single hop. Otherwise, multihop communication must be used, which requires forwarding by intermediate nodes. According to the analysis of UV NLOS communication coverage, the UV NLOS(A) mode is omnidirectional, and thus forwarding is easy. The UV NLOS(B) mode uses directional transmission and omnidirectional reception, and the UV NLOS(C) mode uses directional transmission and reception. Consequently, only nodes within the coverage of the transmitter can receive information. Therefore, we designed a UV forwarding structure with multiple transceivers, as shown in Fig. 6.1.

In this structure, there are multiple UV transceivers on the forwarding node, which are located in the circular area of the device. To enable nodes to transmit information in UV communication networks, devices with this type of forwarding structure can

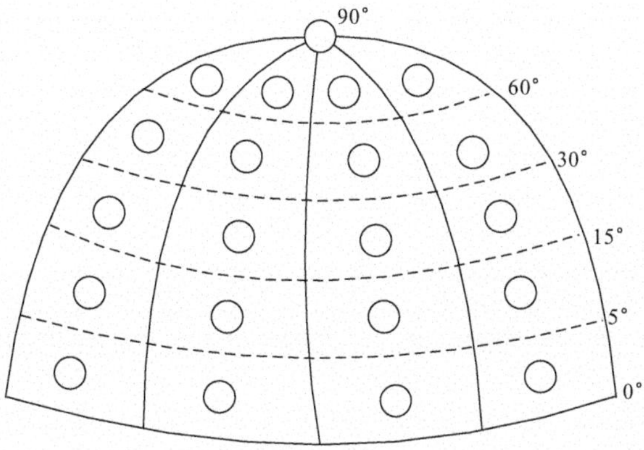

Fig. 6.1 Schematic diagram of structure of forwarding device

adjust the direction and elevation angle of nodes during directional transmission by selecting UV transceivers in different sectors and levels. For example, the forwarding node communicates with the source node by selecting the UV transceiver at level 3 of sector 2 and communicates with the destination node through the UV transceiver at level 2 of sector 4. For a single-frequency UV light source, the UV transceiver operates in only one direction at a certain time and stops working in other directions. As shown in Fig. 6.1, the structure of the forwarding device is divided into eight sectors and six levels. The reason is that with increasing divergence angle in UV communication, the dispersion of the beam energy increases, directional transmission becomes less advantageous, and the spatial multiplexing rate of the channel decreases. UV communication changes obviously with elevation angle at small elevation angles; therefore, the level interval is small at small elevation angles and large at large elevation angles.

6.2 MAC Protocols

Many MAC protocols based on wireless ad hoc networks have been proposed. According to the access mode, MAC protocols can be divided into two categories: contention-based channel access protocols and non-contention-based channel differentiation protocols. Contention-based channel access protocols assume that there is no central entity in the network to allocate channel resources, and each node must compete for media resources to transmit. When multiple nodes try to transmit simultaneously, collisions occur. Competitive channel access protocols include ALOHA, carrier sense multiple access (CSMA), CSMA with collision avoidance (CSMA/CA), IEEE 802.11 distributed coordination function (DCF), and IEEE 802.11e. By contrast, non-competitive channel discrimination protocols allocate dedicated channel resources to each node that needs to communicate in advance. They include time-division multiple access (TDMA), frequency-division multiple access, code-division multiple access, and token-passing schemes. To improve network performance, multichannel and directional antenna technology has been introduced into the design of MAC protocols, and multichannel-based MAC protocols [16] and directional-antenna-based MAC protocols have been proposed [17]. When multiple channels are used, a single node can use different channels to communicate with different nodes simultaneously, which reduces interference and conflicts in the network and improves the network capacity. Multichannel-based MAC protocols include dual-busy-tone multiple access (DBTMA) [18]. MAC protocols based on directional antennas [directional MAC (DMAC) protocols] increase both the transmission distance of node communication and the spatial multiplexing rate of the channel because the interference area becomes smaller. To fully exploit the advantages of multichannel networks and directional antennas, it has been suggested that these two technologies be combined to improve network performance [19]. Typical omnidirectional and DMAC protocols are introduced below.

6.2.1 Omnidirectional MAC Protocols

The first contention-based MAC protocol, ALOHA [20], was proposed for single-hop wireless networks by researchers including Abramson at the University of Hawaii in the 1970s. Subsequently, other protocols such as the slot ALOHA, CSMA [21], CSMA/CA, IEEE 802.11 DCF [22], and IEEE 802.1le MAC [23] protocols were developed to solve the problem of competitive media sharing.

6.2.1.1 IEEE 802.11 DCF Protocol

The IEEE 802.11 standard specifies two media access methods: the DCF in CSMA/CA, which is based on competitive channel access, and the point coordination function (PCF), which provides non-competitive access.

Figure 6.2 shows the data transmission process of the IEEE 802.11 DCF protocol. Before sending data, each node first uses the CMSA/CA mechanism to identify busy or idle channels. Carriers can be monitored physically or virtually. In physical carrier monitoring, channel occupation is determined by receiving information on surrounding nodes In virtual carrier monitoring, the occupation time of the channel is obtained by analyzing the received information; that is, a node receiving request-to-send (RTS) and clear-to-send (CTS) packets sets the directional network allocation vector (DNAV) according to the value of the duration field in the packet. A channel is identified as idle only when both physical and virtual carrier monitoring indicate that it is not occupied. If the channel is idle, the node waits for the distributed interframe space (DIFS) time. If the channel remains idle during this period, the node starts sending data; if the channel becomes busy during this period, the backoff algorithm is executed, and a backoff time between 0 and the minimum backoff value is randomly selected. When the channel is idle for another DIFS time, the backoff value is decreased by one time slot until it reaches 0, and the node starts to send data. If the channel becomes busy during the decrement process, the node freezes the backoff counter and continues to decrement the backoff value after the channel is idle for a DIFS time. After receiving the data from the sending node, the receiving node sends a confirmation frame to the sending node. The sending node waits for a short interframe space (SIFS) time after sending the message. If the sending node receives an acknowledgement (ACK) frame from the receiving node within the SIFS time, data transmission is successful. Otherwise, the sender resends the packet. If the retransmission failure threshold is exceeded, the packet is discarded.

The IEEE 802.11 DCF protocol uses the binary exponential backoff algorithm. After each node fails to send, the contention window (CW) is doubled up to the maximum value, CW_{max}, which is specified in the standard as 1. When transmission succeeds, the CW is reduced to the minimum value, CW_{min}.

To further reduce transmission failure caused by collision during packet transmission, the IEEE 802.11 standard introduced a channel reservation mechanism. That is, before sending a data message, the sending node first sends an RTS frame, and

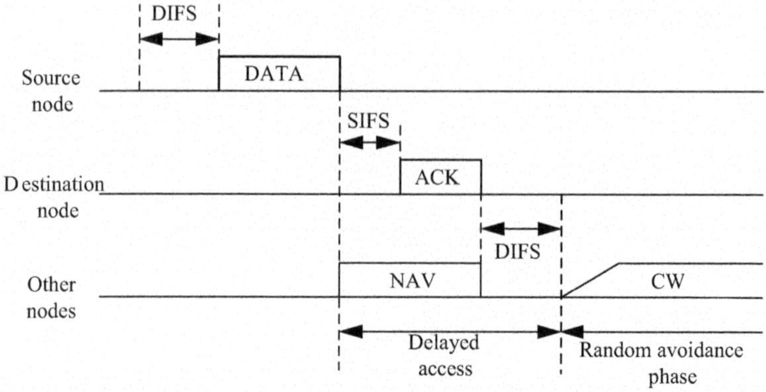

Fig. 6.2 Basic IEEE 802.11 DCF data transmission process

the receiving node replies with a CTS frame. After receiving the CTS, the sending node indicates that channel reservation was successful and sends the data. Because RTS and CTS frames are 20 and 14 bytes long, respectively, and thus are small compared with data packets (0–2312 bytes), this mechanism avoids the monopolization of the channel for a long period when the transmission of a large packet fails. An RTS threshold value is set in the protocol. When this value is exceeded, RTS/CTS mode is used; otherwise, data are sent directly. The protocol also specifies that the maximum numbers of retransmissions of short frames (RTS frames) and data frames are 4 and 7, respectively. Figure 6.3 shows the data transmission process based on RTS/CTS mode.

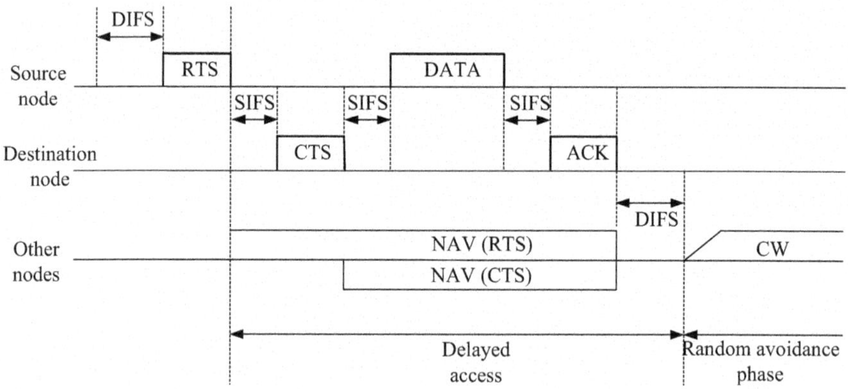

Fig. 6.3 RTS/CTS data transmission process

6.2.1.2 IEEE 802.11e

Because IEEE 802.11 supports only best-effort service, it does not provide any priority mechanism or service differentiation features and cannot guarantee QoS. To compensate for this deficiency and ensure communication quality for high-bandwidth applications such as voice and video, the IEEE Standards Organization proposed an enhanced version of the IEEE 802.11 standard, IEEE 802.11e, which defines the QoS of WLANs, such as support for voice over IP. The IEEE 802.11e standard defines the hybrid coordination function, where the DCF and PCF are replaced by new access methods to provide improved access bandwidth and reduce the delay of high-priority communication. Enhanced distributed coordinated access (EDCA) and hybrid control channel access are used to extend the functions of the DCF and PCF, respectively, in the IEEE 802.11 standard.

EDCA specifies four access types, each of which corresponds to a type of data. Each access category has four parameters that are configured: (1) the minimum CW, CW_{min}; (2) the maximum CW, CW_{max}; (3) the transmission opportunity limit, TXOPlimit; and (4) the arbitration interframe space number (AIFSN). By setting these parameters for each type of data, the network administrator can adjust the network according to the application mix and traffic. Table 6.3 shows the parameter configuration of 802.11e EDCA.

The priority of service quality from high to low follows the order voice, video, best-effort service, and background service. The AIFS, CW_{min}, and CW_{max} values for voice service are relatively small, because voice communication requires a short delay time. TXOPlimit specifies the maximum duration that each access category (AC) can continuously occupy a channel. According to the definition of TXOP, ACs with larger TXOPlimit values can occupy the channel for longer and thus have higher priority.

However, the IEEE 802.11e standard based on service differentiation cannot really guarantee QoS. When a network is overloaded, although high-priority services still obtain bandwidth first, the QoS is significantly reduced, and low-priority services are

Table 6.3 EDCA default parameter configuration [23]

Access cate-gory (AC)	CW_{min}	CW_{max}	AIFSN	TXOP limit		
				For PHYs defined in clauses 15 and 18	For PHYs defined in clauses 17 and 19	For other PHYs
AC_BK	aCW_{min}	aCW_{max}	7	0	0	0
AC_BE	aCW_{min}	aCW_{max}	3	0	0	0
AC_VI	$(aCW_{min} + 1)/2 - 1$	aCW_{min}	2	6.016 ms	3.008 ms	0
AC_VO	$(aCW_{min} + 1)/4 - 1$	$(aCW_{min} + 1)/2 - 1$	2	3.264 ms	1.504 ms	0

easily starved. To provide better QoS guarantee, the admission control mechanism [24] was introduced to control the admission of newly added services; it ensures the QoS of real-time services and prevents QoS degradation caused by increases in the number of real-time services. Two types of admission control algorithms are used [25]: model-based and measurement-based algorithms. Model-based admission control algorithms require the network to know the traffic model in advance before accepting or rejecting a connection request from a new flow. Measurement-based admission control algorithms do not need to know the traffic model in advance, but rather determine whether to accept new flows by measuring the network load.

6.2.1.3 DBTMA

Hidden and exposed terminal problems greatly decrease the MAC layer capacity of ad hoc networks. Even when the network is fully connected, the RTS/CTS switching method cannot solve these problems. Moreover, the condition that all relevant nodes can receive RTS/CTS frames is often difficult to meet. Some nodes in the network may be outside the transmission range of RTS/CTS frames, but they can interfere with the sending nodes. In addition, when the network load is heavy, the probability of RTS/CTS frame collision is high. To solve these problems, a channel separation method is proposed, which divides the channel into data and control channels. The control channel sends control frames, and the data channel is used to send data packets.

The DBTMA protocol uses the control and data channels to send control and data frames, respectively. In addition, it introduces two outgoing busy-tone signals for the control channel, the receive busy tone (BTr) and send busy tone (BTt), which are used to indicate that a node is receiving and sending data in the data channel, respectively. The two busy-tone signals are separated in frequency to avoid mutual interference [26].

A node operating in the DBTMA protocol can be in one of the following states: (1) IDLE, (2) CONTEND, (3) sending RTS (S_RTS), (4) sending data (S_DATA), (5) waiting for BTr (WF_BTR), (6) waiting for data (WF_DATA), or (7) waiting (WAIT). Figure 6.4 shows the state diagram of the DBTMA protocol. Theoretical analysis and simulation [18] have shown that the DBTMA protocol is superior to MAC protocolsusing only the RTS/CTS mechanism. The network utilization of DBTMA is greatly improved compared to those of multiple access with collision avoidance (MACA) and MACA for wireless networks. Because the busy-tone signal is always present during communication, the absence of conflicts between user data frames is ensured. However, this access method results in the waste of channel resources. The control channel is not used when data messages are transmitted. During the control of information exchange, the data channel is idle. Although the DBTMA protocol avoids the problems of hidden and exposed terminals, it reduces channel utilization.

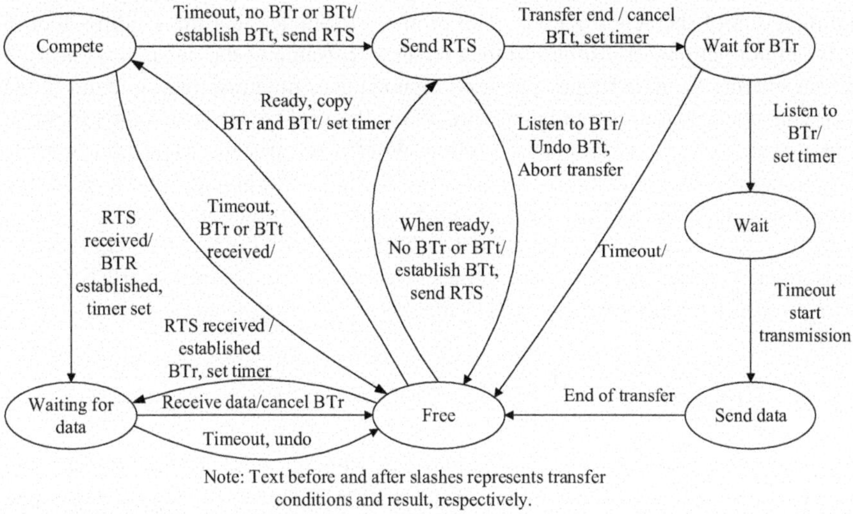

Fig. 6.4 State diagram of DBTMA [18]

6.2.2 Directional MAC Protocols

Omnidirectional MAC protocols are used in networks containing omnidirectional antennas. In 1990, Zander proposed the use of directional antennas in slotted ALOHA multihop packet radio networks [27]. For multihop networks such as mesh networks, the use of directional antennas can solve the problem of exposed terminals and greatly reduce the impact of hidden terminals [28]. Directional antennas provide a spatial separation method that can separate transmissions that may conflict in space in omnidirectional mode, thus improving the network throughput. In addition, omnidirectional antennas send energy in all directions during transmission; most of this energy is wasted, and only a small part can be received. DMAC protocols must use directional antenna mode when sending data and ACK frames. Many combinations may be used to transmit RTS and CTS frames, for example, omnidirectional RTS (ORTS)/omnidirectional CTS (OCTS), directional RTS (DRTS)/OCTS, DRTS/DRTS, and multihop RTS [29].

6.2.2.1 ORTS/OCTS

The ORTS/OCTS protocol proposed by Nasipuri et al. [30] is an improved version of IEEE 802.11 DCF. It uses beam-switching antennas, assuming that the direction of each antenna element remains unchanged when moving, and the idle node uses all its directional antennas to listen to the channel in all directions. Because directional antennas are used, the transmitting node must determine the direction of the receiving node before sending a data frame. Consequently, the receiving node also must know the direction of the sending node. The protocol stipulates that the sending node S uses omnidirectional transmission to send an RTS frame to the destination node D before sending the data frame. The sending node does not know the direction of the receiving node, so it must send in all directions. If node D receives this RTS frame, it replies with an omnidirectional CTS frame. When it receives the RTS frame, node D determines the direction of node S from the direction of the antenna that receives the strongest signal. Similarly, node S also determines the direction of the destination node from the received omnidirectional CTS signal. After the RTS/CTS interaction ends, the two nodes can send data frames directionally. In contrast to IEEE 802.11 DCF, this protocol does not use ACK frames.

6.2.2.2 DRTS/OCTS

The size of directional antennas limits their application in handheld devices, but they are very practical for vehicle-mounted devices. In addition, the higher frequency band reduces the physical size of the directional antenna, and the conventional MAC protocol is not suitable for directional antennas. Reference [31] introduced a dynamic slot assignment (DSA) protocol for directional antennas; it requires each node to determine the location information of its neighbor nodes and itself. Each node uses multiple sector antennas, and each antenna corresponds to an angle, for example, 90°. These antennas are used together to obtain omnidirectional communication. If the physical location of a node is unknown, it must be obtained using GPS. According to the location information, the node selects the directional antenna in the direction of the appropriate destination node to send the packet. Two options are discussed in the DSA protocol.

DMAC-1: When the transmitter is sending data packets, it is assumed that no other node nearby is transmitting data; that is, all antennas at the transmitter are not blocked. The transmitter uses a directional antenna to send a DRTS, which contains the location information of the transmitter. After receiving the RTS, the receiver replies by sending a CTS in all directions. The CTS contains the location information of the transmitting and receiving nodes. After receiving the CTS packet, the sending end sends the data by directional transmission. If the receiving end correctly receives the data, it responds with a directional ACK packet If other nodes listen to the RTS or CTS packets from a certain direction, the directional antenna in that direction is blocked until the current communication ends.

DMAC-2: Because DMAC-1 may sometimes increase the conflict of control packets, DMAC-2 modifies the sending rules for DRTS and ORTS at the sending end. That is, if all the directional antennas at the transmitting end are not blocked, an ORTS is sent; otherwise, if the transmitter expects that the directional antenna in the transmission direction is not blocked, it sends a DRTS in that direction. If the directional antenna in this direction is blocked, it delays until this antenna is idle.

However, there is a problem in both DMAC-1 and DMAC-2. The directional antenna of a node in one direction may be blocked, and thus it cannot reply with an OCTS to RTS packets sent from other nodes in the other direction. To solve this problem, directional wait-to-send packets are introduced so that the node receiving the RTS packet can reply to the node sending the RTS packet with a directional request to wait for transmission; as a result, the node receiving the packet avoids unnecessary backoff. Reference [31] reported the simulation of 5×5, 3×3, and 6×6 grid topologies in different traffic scenarios. The simulation results showed that DMAC outperforms IEEE 802.11, DMAC-1 is better in some scenarios, and DMAC-2 is better in some scenarios. However, the protocol does not consider the deafness caused by directional antennas and directional hidden terminals.

6.2.2.3 DRTS/DRTS

Reference [32] proposed a MAC protocol for sending circular RTS and DRTS frames. The sending node sends circular and DRTS frames in each beam direction. The coverage in each beam direction does not overlap. The receiving node selects the appropriate beam direction to receive and uses this direction to reply with CTS frames. The sending node receives an OCTS and determines the beam direction used to receive the OCTS; it then performs directional transmission of data and ACK frames. The directional transmission of each frame in this protocol takes full advantage of the directional antenna to improve the transmission distance. The transmission of a circular RTS reduces the problem of hidden terminals and their deafness to some extent. The transmitting and receiving nodes do not need to know their respective geographical location information in advance. However, it does not completely solve the problem of deafness and conflict; it only avoids the deafness of neighbors around the transmitting node. When the receiving node does not reply to the CTS, neighbor nodes around the sending node set their DNAV values because they received RTS frames. Therefore, no packets can be transmitted in the direction of the DNAV, which reduces the network capacity. In addition, it is unnecessary to send RTS frames in an empty sector without neighbors. As the number of antenna beams increases, this effect becomes more severe. Therefore, the circular DRTS transmission strategy may not be highly effective.

The circular RTS and CTS MAC (CRCM) protocol is based the circular and directional transmission of RTS frames [33]. It also adopts a circular directional mode for the transmission of CTS frames, which reduces deafness and hidden terminals around the transmitting and receiving nodes. However, if there is no neighbor node in the appropriate direction, the circular RTS sent in this direction is unnecessary and increases the packet delay. The DNAV setting is also unreasonable. If a node fails to send a packet, according to the DNAV value, the nodes around the sending and receiving nodes continue to wait for the appropriate time to compete for channel access. This waiting is unnecessary.

The demand-assigned multiple access (DAMA) [34] protocol uses a new adaptive mechanism to overcome the limitations of DMAC and CRCM. To exploit the high

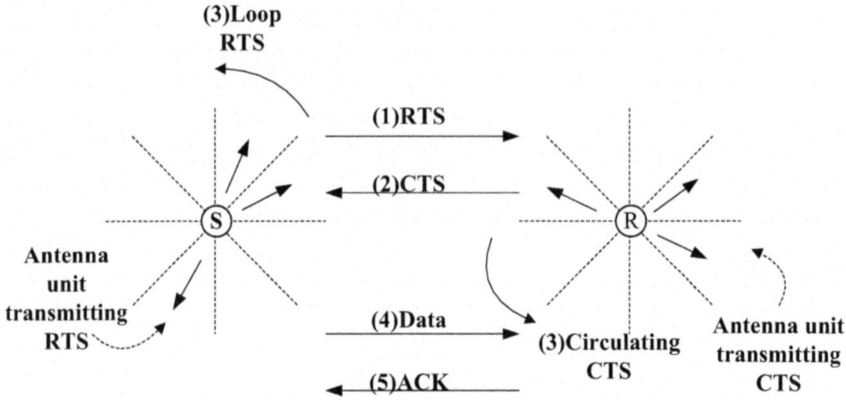

Fig. 6.5 RTS/CTS/data/ACK interaction in DAMA [34]

gain of directional antennas, all transmissions in DAMA are directional; a node does not depend on information on high-priority neighbors. But identifies its neighbors by internode communication. To avoid deafness and new hidden terminal problems, the DAMA protocol introduces circular directional RTS and CTS. However, in contrast to CRCM, DAMA transmits RTS and CTS only on the beam direction toward neighbors (as shown in Fig. 6.5). To complete this transmission, DAMA introduces an adaptive mechanism to learn and cache the neighbor information according to sectors. During network initialization, DAMA scans all antenna beam directions, like CRCM, and then collects and caches neighbor information according to the received responses. To make the protocol easier to add, DAMA was modified according to IEEE 802.11 MAC. Simulations of the DAMA protocol in different topologies [34] showed that the use of directional antenna can improve the channel spatial multiplexing rate and increase the transmission distance to reduce the number of hops between transmission node pairs. The results show that the DAMA protocol outperforms the CRCM protocol.

However, self-induced blocking may occur in directional antennas in networks using the MAC protocol because a single MAC is used to buffer data packets in the direction of all antenna beams. That is, when a packet is sent at the network layer, its next hop is determined, and the packet is placed in the link layer queue. When the MAC layer is idle, the packets queued from the link layer are marked and put into the buffer for processing. If the DNAV value of the beam direction of the packets to be transmitted is set, the buffer waits for the medium to become idle before transmission. Other packets queued at the link layer can be transmitted in the idle beam direction. In this scenario, the packet waits for the medium to become idle, reducing the overall throughput of the system. To address this problem, a cross-layer design method was introduced, and an enhanced DAMA (EDAMA) protocol was proposed [34]. The network layer can sense the MAC layers of different antenna beams. Each antenna beam has a single MAC layer buffer, and each buffer corresponds to a specific queue of the link layer. In addition, EDAMA introduces a separate backoff time for each antenna beam to allow simultaneous data transmission. Simulation results [34] showed that for different traffic types and numbers of antenna beams, the performance of EDAMA always exceeds that of DAMA and CRCM, and the throughput of EDAMA exceeds that of DAMA by 10% on average. The reason is that EDAMA introduces a cross-layer design method and multiple MAC buffers, thus overcoming the problem of self-induced blocking.

Reference [35] proposed a MAC protocol with directional orientation. All senders and receivers operate only in directional mode, which eliminates the gain asymmetry problem that arises when the omnidirectional and directional modes are used together. In the protocol, the sending node must cache the location information of the neighbor node to select the appropriate sector for information transmission. The paper does not describe how the location information is obtained.

Reference [36] proposed two opportunity enhancement strategies using directional antennas for the 802.11 protocol. One is to enable nodes transmit data more flexibly using the location information of stations. Simulation results showed that this strategy allows more transmission, and the network throughput is significantly improved. The other is to change the access mode of the 802.11 MAC data queue. If the data transmission at the top of the node queue in the transmission direction is blocked, the station detects whether the transmission direction of the next data item in the node queue in the transmission direction interferes with the ongoing transmission. If not, the station can quickly transmit this data item instead of the data at the top of the queue. The first strategy requires the comprehensive modification of the physical and MAC layers. The second strategy must correctly judge the transmission direction of packets in the node queue. The implementation is complex.

6.2.2.4 Multihop RTS

Reference [37] proposed a multihop RTS-packet-based multichannel MAC (MMAC) protocol, which removes the implicit condition that the directional and omnidirectional transmission ranges in the DMAC protocol are the same. By using the high gain of directional antennas, a larger transmission range is obtained while maintaining a certain level of spatial multiplexing. The protocol assumes that the sending node has a multihop path to the destination node to send RTS packets, and CTS, data, and ACK packets are directionally transmitted between the sending and destination nodes by a one-hop route.

Figure 6.6 shows an example that illustrates the general flow of the protocol. If node A wants to send data to T, node A first sends a DRTS in the direction of node F. Then A sends a multihop RTS along the directional/omnidirectional path A–B–C–F, which is called a forwarding RTS. When it is forwarded, it has the highest priority and does not need to back off. When the NAV in the forwarding direction of an intermediate node is not exhausted, the RTS frame is discarded. This RTS does not require an answer. In addition, the receiving direction of A needs to point to F so that it is ready to receive the directional CTS from node F. When node F receives the forwarded RTS frame, it needs to perform physical carrier listening, wait for the

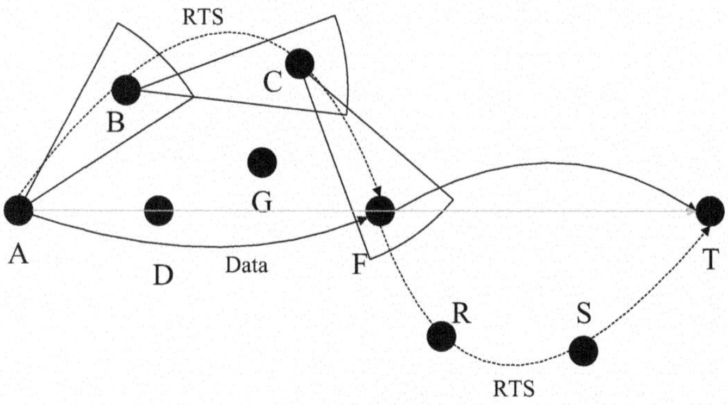

Fig. 6.6 Illustration of multihop RTS protocol for forwarding from node A to node F

SIFS time, and then send a CTS to node A. Because the receiving direction of node A points to F, A sends data frames to F after receiving the CTS. F sends an ACK frame to A after receiving the data. Data transmission from F to T occurs by a similar process.

The MMAC protocol can greatly reduce the number of hops between the source and destination nodes by establishing directional sending and receiving links between nodes. Simulation results [37] showed that the MMAC protocol outperforms IEEE 802.11, but it depends on a specific network topology and traffic types in the network system.

6.2.2.5 MDMAC

The basic DMAC protocol causes hidden terminals and deafness due to omnidirectional and directional mode switching and directional mode transmission. Gain asymmetry causes hidden terminals, and failure to hear RTS/CTS causes hidden terminals and deafness; thus, the channel is not fully utilized. Moreover, the basic DMAC protocol requires listening in omnidirectional mode and switching to directional mode for communication. Reference [38] proposed a bootlace-lens-based DMAC Protocol, the modified DMAC (MDMAC) protocol MDMAC, which operates only in directional mode. Channel listening is performed by a beamforming network consisting of a bootlace antenna and three directional antennas with coverage angles of 120°. Simulation results [38] showed that the DMAC protocol based on a bootlace lens has better throughput performance than the basic DMAC and IEEE 802.11 protocols. It not only avoids the problem of hidden terminals, but also eliminates the problem of deafness to a great extent.

6.2.2.6 TDMA-Based 2P Protocol

Raman and Chebrolu proposed a MAC protocol suitable for long-distance 802.11 mesh networks [39]. Each node in the network is configured with multiple wireless transceivers connected by Ethernet links. Each transceiver corresponds to a wireless link. The directional antennas of two communication nodes must be aligned to form a wireless link in the network. When the system is working, each node sends or receives on all links simultaneously. There are only two states: synchronous sending and synchronous receiving. Adjacent nodes switch alternately and operate in different states. Therefore, the protocol is described as a two-phase (2P) protocol. The RTS/CTS mechanism is not adopted for sending and receiving packets. Data packets are sent directly, and ACK replies are sent from the opposite end. Raman also studied the channel allocation problem of 2P protocols in networks with multiple channels [40] and discussed several channel allocation schemes using the coloring algorithm in graph theory.

6.2.2.7 DMesh Protocol

WMNs are considered to be effective for solving the last-kilometer problem in broadband access. The three main factors affecting the adoption of WMNs are high throughput, cost-effectiveness, and easy deployment. Das et al. proposed a new network architecture, directional mesh (DMesh), to improve the performance of WMNs [41]. In mesh networks, multiple antennas and channels are used to achieve spatial and frequency separation, and higher throughput is obtained while maintaining cost-effectiveness and easy deployment. The nodes in DMesh adopt multiple antennas. *K* directional antennas are added to the original omnidirectional antennas of each mesh router. The omnidirectional interface is used to provide a robust connection and send and receive control information. The directional interface is responsible for connecting wireless links. The OLSR protocol, which is called the directional OLSR (DOLSR) protocol, is extended to the routing layer. The 802.11 protocol is used in the MAC layer. Reference [41] focused on distributed channel management. A heuristic channel allocation scheme was obtained by designing a distributed algorithm. DMesh uses the DOLSR protocol to construct a high-throughput routing tree with the gateway as the root node. All interfaces are initially configured on a default channel. The parent node selects a new channel according to a directional channel allocation scheme and sends an assign message to the child node. Then the parent and child nodes use the iwconfig command to set their directional interfaces to the selected channel. That is, only the channel for the one-hop link is selected, and the parent node determines the channel to be used. Channel allocation is recalculated every 300 s. DMesh determines whether omnidirectional or directional antennas are used and whether channel allocation technology is used. The conservative distributed channel allocation (DCA), aggressive DCA, and measurement-based DCA schemes have been proposed for use with directional antennas and multichannel technology. Simulation results [41] showed that the throughput of DMesh is 231% higher than that of an omnidirectional antenna network with multiple transceivers and channels, and the data packet delay is greatly reduced.

6.2.3 Performance Comparison of DMAC Protocols

Because of the directionality of UV NLOS communication coverage, we compare current MAC protocols based on directional antennas; here we ignore DMAC protocols that require complex hardware support, such as the MDMAC protocol. Table 6.4 compares the performance of DMAC protocols [42]. D and O represent directional and omnidirectional transmission, respectively.

Many early DMAC protocols replace adaptive array antennas with beam-switched antennas. Because the beamforming algorithm of adaptive array antennas is complex, many researchers used adaptive array antennas in wireless ad hoc networks to improve network performance. However, many antenna models proposed in the literature are unsuitable for practical applications. A study [46] showed that the network throughput of an actual antenna model is reduced by 36% compared with that of the

Table 6.4 Comparison of various DMAC protocols

MAC protocol	RTS		CTS		Data		ACK		Hide/expose terminals	Deafness	Neighbor location information
	TX	RX	TX	RX	TX	RX	TX	RX			
Nasipuri [30]	o	o	o	o	D	d	–	–	Not mentioned	Not mentioned	Not required
DMAC-1 [31]	d	o	o	o	D	o	d	o	Not mentioned	Not mentioned	Need
DMAC-2 [31]	o/d	o	o	o	d	o	d	o	Not mentioned	Not mentioned	Need
MMAC [37]	d	o	d	d	d	d	d	d	Not mentioned	Not resolved	Multihop RTS routing
Circular DMAC [32, 33]	d	d	d	d	d	d	d	d	Partially resolved	Partially resolved	Location table
DAMA [34]	d	d	d	d	d	d	d	d	Partially resolved	Partially resolved	Not resolved
Directional virtual carrier sensing [43]	d	d	d	d	d	d	d	d	Not mentioned	Not mentioned	AOA cache
Lal et al. [44]	d	d	d	d	d	d	d	d	Not mentioned	Not mentioned	Need
DBTMA/DA [45]	o	d	d	d	d	d	d	d	Partially resolved	Not mentioned	Need

ideal antenna model. The effects of the sidelobe of directional antennas also degrade the network performance, and this problem also be addressed. Reference [47] studied location-dependent carrier monitoring and the interference caused by the antenna sidelobe and proposed a reservation-based DMAC protocol (RDMAC) for wireless multihop networks. The implementation of this protocol is based on the reservation and transmission periods. The reservation period consists of detection and beam indication periods; it can reduce the interference between neighbor nodes and increase the network throughput. Both theoretical analysis and a simulation [47] showed that the RDMAC protocol can achieve higher throughput and lower latency. The access process of this protocol is ORTS–OCTS–DRTS–DCTS–DDATA–DACK (where O and D represent omnidirectional and directional transmission, respectively). This protocol introduces additional control frame interaction overhead, which wastes a certain amount of bandwidth.

In directional transmission and reception using directional antennas, problems arise because the antenna beams are aligned simultaneously because of the directivity of the beam. Table 6.4 compares the performance of current DMAC protocols. Because the directional antennas used in the literature are relatively ideal, reference [48] adopted a hybrid directional antenna model. Considering the effects of directional antenna sidelobe, the upper limit of the network capacity of a random network was analyzed theoretically using a directional antenna and multichannel technology. It was concluded that the upper limit of the network capacity is related to the ratio of the sidelobe and main lobe radii, where the network capacity decreases as this ratio increases. Therefore, in network simulations, it is necessary to use the hybrid directional antenna model to analyze the network performance to obtain more accurate simulation results.

6.2.4 Problems Caused by DMAC Protocols

6.2.4.1 Directional Hidden Terminals

Hidden terminals occur mainly because it is impossible for a node to listen to interference when switching to directional mode from omnidirectional mode. The communication distance of a node is affected by the omnidirectional and directional gain of the antenna. Networks using directional antennas have three types of neighbor nodes: omnidirectional/omnidirectional, omnidirectional/directional, or directional/directional. The communication distances among the three types neighbors increases in that order. When nodes switch between omnidirectional and directional operation, the communication distance changes, which results in directional hidden terminals.

(1) Hidden terminals caused by gain asymmetry

A neighbor node is in the omnidirectional listening state when it is idle. Therefore, it may not receive RTS or CTS packets sent by a directional/directional neighbor

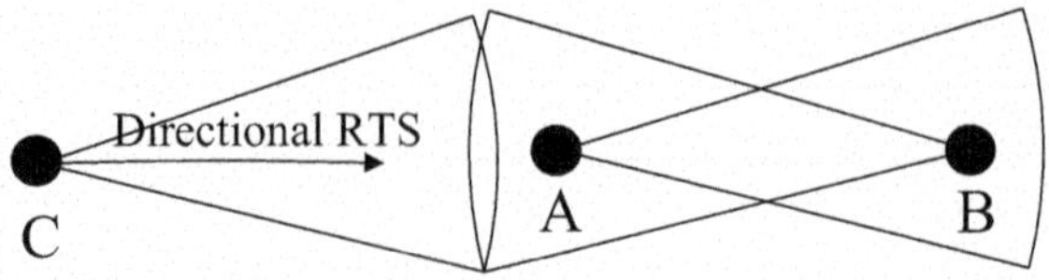

Fig. 6.7 Hidden terminal caused by gain asymmetry

node, and it does not back off. When a sending node forwards received data packets in the direction of the receiving node, it may interfere with the normal reception of the data. Thus, the node can be considered to be a hidden terminal. This hidden terminal problem is caused by gain asymmetry. As shown in Fig. 6.7, node B is in the omnidirectional receiving state when it is idle. Even if node C sends packets to node B directionally, node B cannot listen. When node C is idle, it is also in the omnidirectional state. When nodes A and B communicate with each other, the directional CTS of node B cannot be heard by node C. If node C sends signals to node B while node A is sending data to node B, it interferes with data reception by node B.

This problem is caused by the high gain of directional antennas, which causes the transmission distances of directional/directional and directional/omnidirectional neighbors to vary. In the CRCM protocol, directional RTS/CTS are sent in all directions, and neighboring nodes are notified to set the DNAV for backoff before communication to solve this problem. Current practice is generally to reduce the transmission power of the antenna during directional transmission so that it is consistent with the omnidirectional communication range.

(2) Hidden terminal problems caused by inaudible RTS/CTS

In the scenario shown in Fig. 6.8, because directional antennas are used, communication between nodes A and B and nodes C and D can be conducted simultaneously. Assuming that communication between A and B occurs before that between C and D, node B has not received an RTS/CTS notification, so it does not back off in the direction of node C or D. If node B sends packets to node C or D at this time, the communication between nodes C and D is disrupted. The reason is that neighboring nodes around the two nodes currently communicating do not receive RTS/CTS notifications. In the MAC protocol for omnidirectional antennas, two pairs of nodes close to each other cannot communicate simultaneously, but such simultaneous communication is very likely to occur with directional antennas, resulting in this situation.

6.2.4.2 Deaf Nodes

Deafness is the most common problem in MAC protocols using directional antennas. A node using directional antennas can receive only from the direction with which it is aligned.

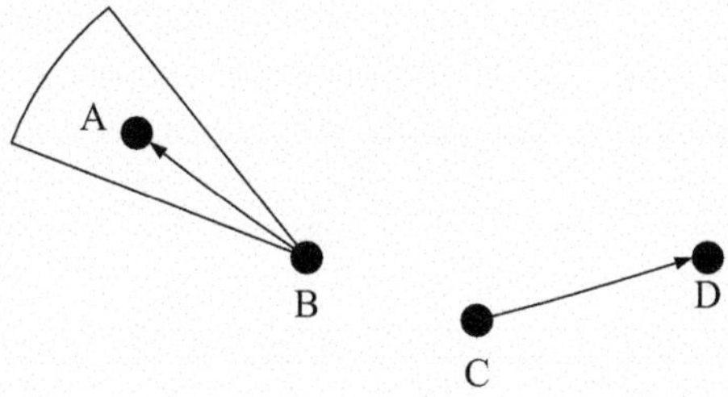

Fig. 6.8 Hidden terminal problems caused by unreceived RTS/CTS

As shown in Fig. 6.9, when nodes A and B use directional antennas to send packets for communication, node A (in either the transmitting or receiving state) cannot hear the RTS sent by node C. This phenomenon is called deafness. After node C sends an RTS to node A for timeout, it adds a backoff window and resends the RTS, resulting in unfairness that puts node C at a disadvantage when competing with other nodes to access A. Finally, after resending the RTS many times, node C is likely to assume that node A is not reachable, and routing updates and oscillations occur. When node C sends an RTS, node D, which is nearby, may listen to this RTS and make an unnecessary backoff. In addition, it is possible that each node is deaf, resulting in deadlock (Fig. 6.10).

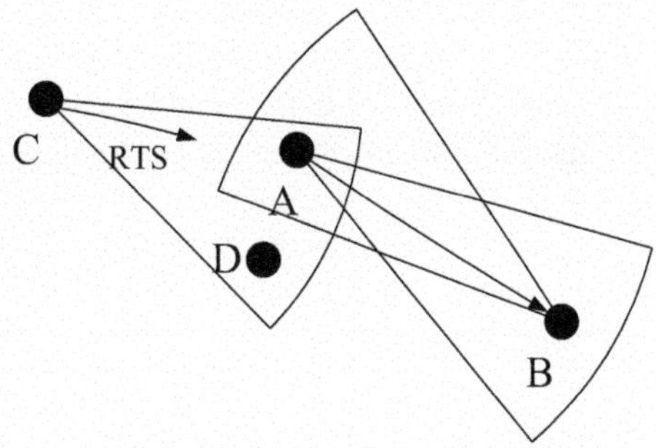

Fig. 6.9 Transmission failure and unfairness caused by deafness

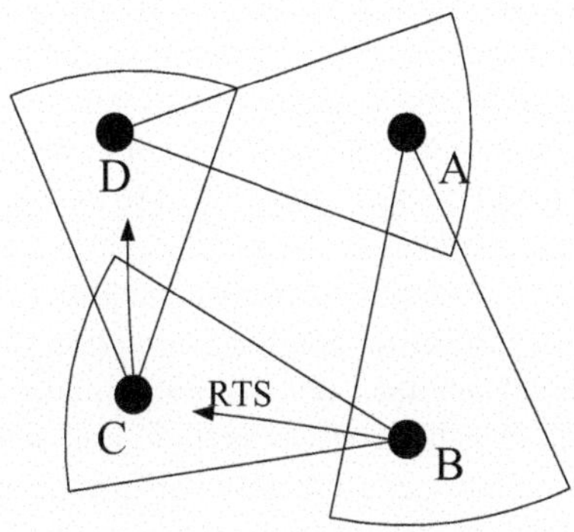

Fig. 6.10 Deadlock caused by deafness

The deafness problem is due mainly to the inability to notify other nodes to back off or set the DNAV before communication after RTS/CTS packets are sent directionally. It not only degrades the performance of the MAC protocol, but also affects the upper layer protocol. When the sender sends an RTS but does not receive a CTS within the scheduled time, it considers that a conflict has occurred and expand its CW to participate in the backoff and channel competition again. After the CW is expanded, the ability to compete for the channel.decreases, which wastes considerable network bandwidth resources and causes serious network unfairness.

6.2.4.3 Determination of Neighbor Location Information

In omnidirectional MAC protocols, node location is not needed, because the communication coverage of nodes is a circular area. If a neighbor node is in the circular area, it can receive information sent by the node. In directional MAC protocols, the transmitting and receiving nodes must be strictly aligned. That is, the main lobe directions of the directional antennas at the transmitting and receiving nodes must be aligned; otherwise, packets cannot be received, even if the directional antenna at the receiving end is aligned with the transmitting end. Of course, the sidelobe of the directional antenna can also transmit information. However, this transmission requires that the sender and receiver have each other's location information.

Many methods have been proposed to determine the location information of neighboring nodes. One is to use GPS to determine the location of each node in the network in advance. Nodes are in the omnidirectional state when idle so that they do not miss information from any direction; an example is the DMAC protocol. However, using GPS may be expensive, and only directional/omnidirectional and omnidirectional neighbors can communicate. Directional/directional neighbors cannot communicate directly, and directional hidden terminal problems can occur. The second method is to use the direction of arrival of a smart antenna to estimate the direction with the maximum signal gain; thus, when a packet is received from a neighbor, the orientation of the neighbor node can be determined. This method exploits the advantages of omnidirectional antennas. The location of the destination node need not be known before communication is initiated. The receiving end can always receive packet information from any direction, but at the cost of large delays.

In networks using directional antennas, the main problem during network initialization is that the nodes do not know how to switch antennas. If the random discovery strategy is used, that is, if the node antenna points in random directions according to probability to discover neighbors, this strategy is disrupted, and the node antenna may not be able to discover neighbors after several attempts. Reference [49] proposed a deterministic neighbor discovery strategy for static wireless networks using directional antennas, which is particularly suitable for mesh backbone networks. In typical military mesh networks for long-distance communication, six fixed-beam directional antennas can efficiently perform directional transmission and reception and realize high network capacity.

6.3 Performance Analysis of UV NLOS Communication Network

UV communication has become a hot topic in research on wireless optical networks because of its advantages of NLOS communication, all-weather and continuous operation, and strong anti-interference ability, and the development of small, inexpensive, lightweight, and highly reliable UV optoelectronic devices. Theoretical and experimental studies have been conducted on the atmospheric channel transmission model; modulation, coding and decoding; communication effects under different conditions; and various factors affecting the UV communication performance. Most experimental studies considered point-to-point communication. However, there is little research on the performance of networks using UV NLOS communication. Here, a UV NLOS communication model is established, and the performance of networks using the three modes of UV NLOS communication is simulated under different network topologies using the ns-2 [50, 51] network simulation software.

6.3.1 Theoretical Analysis of UV NLOS Communication Network Performance

An analysis of the coverage of UV NLOS communication reveals that the coverage of the NLOS(A), NLOS(B), and NLOS(C) modes is similar to that of omnidirectional transmission and reception, directional transmission with omnidirectional reception, and directional transmission and reception, respectively, in networks with directional antennas. Therefore, we used the method in reference [52] to analyze the network performance of the three communication modes.

We used an interference model based on the transmitter, as shown in Fig. 6.11, where the interference area is defined as the area that the sender can cover. That is, transmission interferes with all nodes except the receiver. For simplicity, we used the same transmission distance for each node. The UV NLOS(A) mode, with omnidirectional transmission and reception, is represented by a circle with the node at the center; the effective distance between the transmitter and receiver is the radius. This configuration also represents the directional transmission and reception mode of UV NLOS communication. The center angles of the sector when the node is transmitting and receiving directionally are α and β, respectively. This assumption is reasonable for an analysis of the improvement in the channel spatial multiplexing rate for the three modes of UV NLOS communication.

Suppose that the effective transmission distances of nodes Tx1 and Tx2 during data transmission are r_1, r_2 and $(1+\Delta)(r_1+r_2)$, respectively. Then, to avoid interference, the distance between them should be at least 3, that is,

$$|Tx_1 - Tx_2| \geq (1 + \Delta)(r_1 + r_2) \tag{6.3}$$

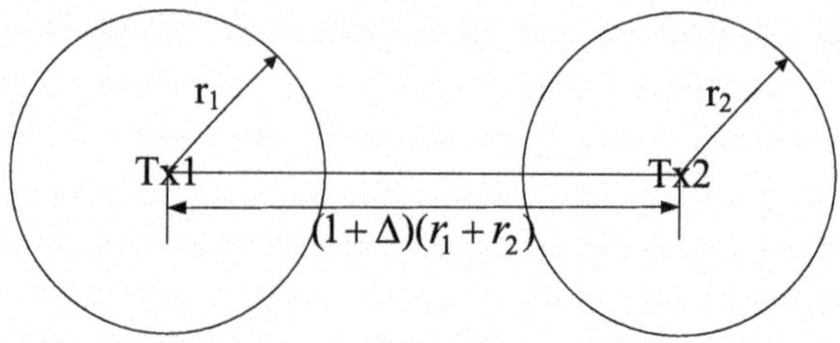

Fig. 6.11 Interference model based on transmitter

where $\Delta > 0$ is a conservative condition in the protocol to avoid the simultaneous transmission of neighbor nodes on the same channel and ensure that no signal conflict occurs. For simplicity, we set each node to have the same transmission distance.

We analyze the upper limit of the transmission capacity of the three modes of UV NLOS communication in a plane network with unit circular area. The conditions of the analysis are as follows [52]. (1) There are n nodes in the network; their positions are fixed and artificially determined (not randomly distributed). (2) In a time of T s, the network transmits a total of λnT bits of data. (3) The average transmission distance from all the source nodes to the destination nodes within T s is $\overline{L}$ m. According to (2), the transmission capacity of this network is $\lambda n\overline{L}$ bit-m/s. (4) Each node can transmit W bits of data per second.

Next, we prove the upper limit of the transmission capacity for three modes of UV NLOS communication according to document [52].

6.3.1.1 NLOS(A): Omnidirectional Transmission/Omnidirectional Reception

Bit $b(1 \leq b \leq \lambda nT)$ is assumed to arrive by hop $h(b)$ at the destination node by two-hop transmission from the source node. The transmission distance of hop h is r_b^h. . According to (3), we have

$$\sum_{b=1}^{\lambda nT}\sum_{h=1}^{h(b)} r_b^h \geq \lambda nT\overline{L} \tag{6.4}$$

At most $n/2$ nodes are transmitting data at any time; therefore,

$$H \approx \sum_{b=1}^{\lambda nT} h(b) \leq \frac{WTn}{2} \tag{6.5}$$

From (6.3):

$$\sum_{b=1}^{\lambda nT}\sum_{h=1}^{h(b)} \pi(1+\Delta)^2(r_b^h)^2 \leq WT \tag{6.6}$$

That is,

$$\sum_{b=1}^{\lambda nT}\sum_{h=1}^{h(b)}\frac{1}{H}(r_b^h)^2 \le \frac{WT}{\pi(1+\Delta)^2 H} \tag{6.7}$$

Notice that the quadratic function is convex. Therefore,

$$(\sum_{b=1}^{\lambda nT}\sum_{h=1}^{h(b)}\frac{1}{H}r_b^h)^2 \le \sum_{b=1}^{\lambda nT}\sum_{h=1}^{h(b)}\frac{1}{H}(r_b^h)^2 \tag{6.8}$$

From (6.7) and (6.8):

$$\sum_{b=1}^{\lambda nT}\sum_{h=1}^{h(b)} r_b^h \le \sqrt{\frac{WTH}{\pi(1+\Delta)^2}} \tag{6.9}$$

Then we substitute (6.4) into (6.9) to obtain

$$\lambda nT\overline{L} \le \sqrt{\frac{WTH}{\pi(1+\Delta)^2}} \tag{6.10}$$

Finally, (6.11) is obtained by substituting (6.5) into (6.10). The upper limit of the transmission capacity for omnidirectional transmission and reception of UV NLOS(A) communication, $\lambda n\overline{L}$, is given by [52]

$$\lambda n\overline{L} \le \frac{1}{\sqrt{2\pi}}\frac{1}{(1+\Delta)}W\sqrt{n} \quad \text{(bit-m/s)} \tag{6.11}$$

6.3.1.2 NLOS(B): Directional Transmission, Omnidirectional Reception

For NLOS(B) communication, according to the interference model of the transmitter shown in Fig. 6.12, the interference area becomes $\frac{\alpha}{2\pi}\pi(1+\Delta)^2 r^2$.

Through a similar derivation process, the transmission capacity of UV NLOS(B) directional transmission and omnidirectional reception can be obtained [52]:

$$\lambda n\overline{L} \le \frac{1}{\sqrt{\alpha}}\frac{1}{(1+\Delta)}W\sqrt{n} \quad \text{(bit-meters per second)} \tag{6.12}$$

Fig. 6.12 Interference model based on transmitter during directional transmission

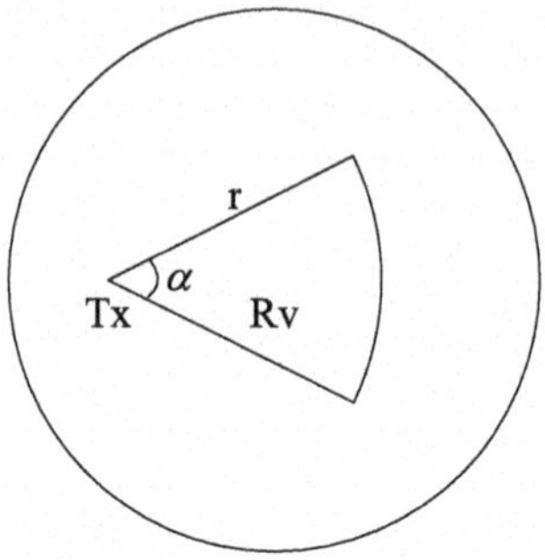

6.3.1.3 NLOS(C): Directional Transmission, Omnidirectional Reception

Figure 6.13 shows the interference model for directional transmission and reception.

In NLOS(C) communication, data transmission and reception are both directional. We assume that all receivers have the same directional coverage, and the probability of this coverage in each direction is fixed. The probability that the receiver covers the transmitter is $\beta/2\pi$. Therefore, the conditional probability of interference at the receiver is $\beta/2\pi$; that is, $\beta/2\pi$ of the receivers in the transmission range experience interference. Therefore, the conditional interference area is

$$\frac{\beta}{2\pi}\left[\pi(1+\Delta)^2(r_b^h)^2\frac{\alpha}{2\pi}\right]=\frac{\alpha\beta(1+\Delta)^2(r_b^h)^2}{4\pi} \tag{6.13}$$

The transmission capacity of UV NLOS(C) directional transmission and reception can be obtained by the derivation shown in (6.6)–(6.11) [52]:

$$\lambda n\overline{L}\leq\sqrt{\frac{2\pi}{\alpha\beta}\frac{1}{(1+\Delta)}}W\sqrt{n}\quad(\text{bit}-\text{meters per second}). \tag{6.14}$$

Fig. 6.13 Model of directional transmission and reception **a** with and **b** without interference

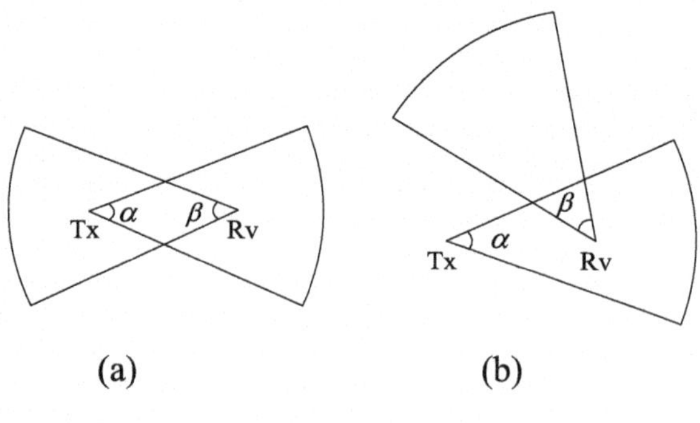

A comparison of (6.11), (6.12), and (6.13) reveals that when the UV NLOS(B) mode is adopted, the transmission capacity has a gain of $\sqrt{2\pi/\alpha}$. When the UV NLOS(C) mode is adopted, the transmission capacity has a gain of $2\pi/\sqrt{\alpha\beta}$. Note that $\overline{L}$ is the average transmission distance from the internal node to the destination node in unit area, which can be regarded as a constant. Therefore, the throughput is proportional to the transmission capacity and also has the same gain.

6.3.2 *UV NLOS Communication Model*

This section describes a UV NLOS communication model, which includes models of the UV multi-transceiver node, UV directional static routing (UV-DSR) protocol supporting UV NLOS communication, and transmission in UV NLOS communication.

6.3.2.1 Node Model in UV NLOS Communication

As described in previous sections, the UV NLOS(A) mode uses omnidirectional transmission and reception, and the transmitting and receiving devices of nodes all face the sky, so it is relatively simple to realize networked communication. The NLOS(B) and NLOS(C) modes use directional transmission with omnidirectional reception and directional transmission and reception, respectively. For these modes, directionality must be fully considered to realize networked communication, because these modes do not have circular coverage areas. When these two modes are adopted, the elevation angle, divergence, and field of view of each node are fixed after network initialization. When the distance between nodes increases, data transmission cannot be completed by point-to-point UV communication, and intermediate nodes are required for forwarding. Directional forwarding by nodes in UV NLOS communication is realized by driving a single UV light source to rotate 360° using a micro-electromechanical system or by using a UV multi-transceiver node structure. Here, we used the latter. Each UV node has one or more interfaces, each with a set of UV transceivers, and each set of UV transceivers has a certain directional coverage during communication.

Figure 6.14 illustrates a configuration of UV multi-transceiver nodes. Node A has two UV interfaces, which are used to communicate with nodes B and E. Node B has three UV interfaces, which are used to communicate with nodes C, D, and A. When the network topology is determined, the number of interfaces required by each node in the network is fixed. When nodes A and B communicate, the UV transceiver of the corresponding UV interface is used. Communication between other pairs of nodes in the network also uses the corresponding UV interface. This node forwarding model fully reflects the characteristics of directional UV NLOS communication. When the UV NLOS(B) mode is adopted, the transmission gain is directional, and the reception gain is 1 (that is, reception is omnidirectional). When the NLOS(C) mode is adopted,

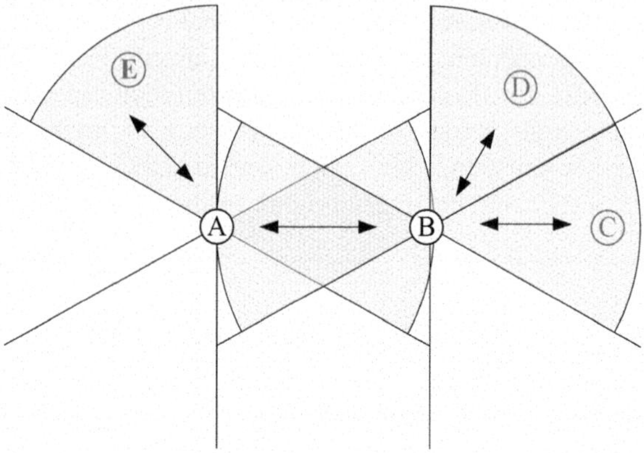

Fig. 6.14 Schematic diagram of UV multi-transceiver node communication

its transmission and reception gains are directional. The implementation code of the receive gain is as follows:

```
double DirAntenna::getRxGain(double x, double y, double z, double lambda) {

double angle;
double dist_ = sqrt(x*x + y*y + z*z);
double gain;
if(dist_ = = 0.0).
{

    return 1.0;

}
angle = atan2(y,x);
angle = angle*180/M_PI;
if(angle < 0) angle + = 360;

    gain = (gainVals[((int)angle)%360]);
    return gain;

}
```

6.3.2.2 UV Multi-transceiver Node Model

To support the simulation of the UV multi-transceiver node structure using ns-2, we propose a UV multi-transceiver node model based on the multi-interface model [53]. Unlike that in reference [53], the model in reference [54] improves the network performance not by using one interface per channel, but by using multiple interfaces in a single channel. In addition, the coverage of each interface of the node is different so as to improve the spatial multiplexing capability of the channel. Figure 6.15 shows the original node model used in ns-2 [55], and Fig. 6.16 shows the UV multi-transceiver node model proposed in reference [54]. A comparison reveals that the UV multi-transceiver node model replicates the components under the routing agent, except for the channel model and address resolution protocol (ARP) module, including logical link control (LLC), interface queue (IFQ), MAC, and network interface (NetIF) modules. Each interface is configured with a set of UV transceivers. When data packets are transmitted to the channel, the channel sends them to different interfaces for reception according to the physical addresses of the packets. We modified the link layer file to include the interface number used to transmit data packets The ARP module, which is connected to the link layer, can parse the interface number used by the data packets, and the data packets are transferred to the appropriate queue according to the interface number and then to the corresponding MAC layer. Finally, they are transmitted to the channel through the network interface. After receiving the data packets, the channel selects the appropriate interface to upload them. To support network simulation research on UV NLOS(B) and NLOS(C) directional communication, we added the UV-DSR protocol, which specifies the interface number used by nodes in communication. Interface numbers can be configured for nodes as required.

The UV multi-transceiver node model is implemented as follows. (1) In the ns-lib.tcl file the simulator procedure for the node interface number (numif) is added to define the interface numbers of the node; then the numifs variable is added in the node configuration process. (2) In the ns-mobilenode.tcl file, the add interface process is called only once; therefore, a for loop should be added to this process. According to the number of node interfaces, the appropriate number of link layers, queues, MAC layers, network interfaces, and antennas should be created. Finally, multiple network interfaces should be associated to the same channel. (3) The ARP module in ns-2 resolves the physical (MAC) address of the next-hop target node of the packet, and the packet is put into the interface queue. Because the ARP depends on the address of the node, it is not applicable to nodes with multiple interfaces. We are at link layer 11 The add ARP entry command is added to the.cc file so that we can easily manually set the ARP table in the script. The corresponding ARP.cc and ARP.h files are also modified to add the declaration and implementation, respectively, of the ARPadd function. The physical (MAC) address of the next hop of the node is specified in the added ARP table. The MAC address depends on the network interface and UV light source.

6.3.2.3 Directional Static Routing Protocol

The UV-DSR protocol is based mainly on the AODV routing protocol. The interface number variable is added by changing the routing table structure. In this way, the routing table records which interface the next-hop node selects to receive packets when they arrive at the next-hop node. The routing table entries include the source node, hop count, next-hop node, destination node, and interface number. By adding the command command, we can easily add the DSR table in the script. The specific implementation is as follows. (1) Because the routing table in our new routing protocol is written in the TCL script, it must be written in the routing protocol

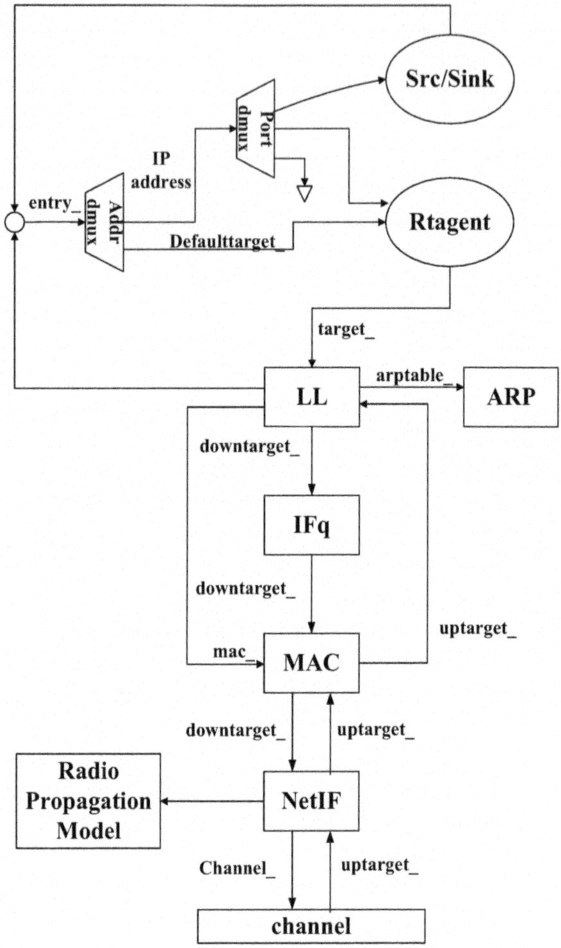

Fig. 6.15 Wireless node model

WLStatic. To add and update the routing table, addstaticroute is added to the command function of the.cc file. (2) In the wlstatic_rtable.h file, the wlstatic_rt_entry class adds the interface number variable to the rt_update process of the routing table. The update function also adds an interface number. In this way, the routing table records which interface the next-hop node selects to receive a packet when it arrives at the next-hop node. When the channel receives the packet, it sends it to the corresponding network interface to continue uploading. (3) The wlstatic.cc file adds support for multiple interfaces everywhere. For example, in the forwarding function, packets should be sent to the corresponding link layer according to the interface number used by the route, which is found in the routing table. (4) The process of creating a new routing agent module must be defined in the ns-lib.tcl file as create–wlstatic–agent. In addition, support for new routing protocols must be added to

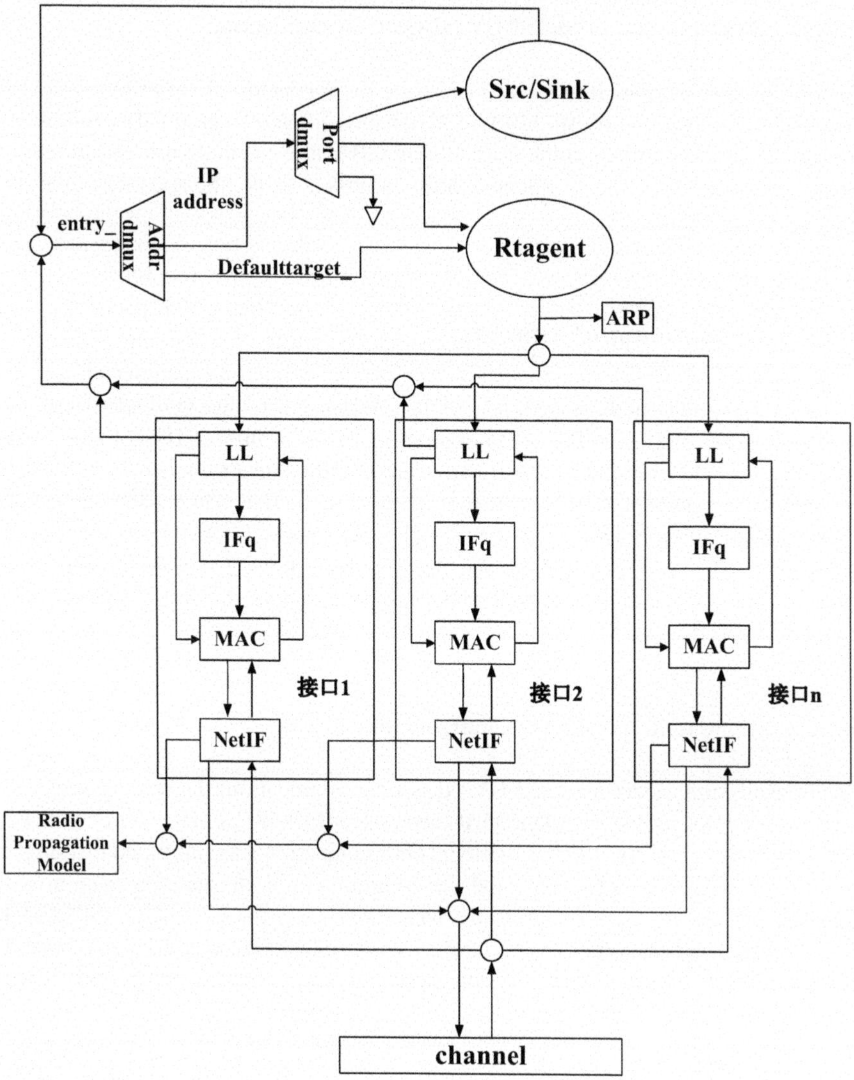

Fig. 6.16 UV multi-transceiver node model

the process of creating a wireless node. (5) In the ns-mobilenode.tcl file, the add-target and add-target-rtagent TCL commands add support for the WLStatic protocol, following the AODV protocol. (6) To add a new routing protocol, the files Common/packet.h, Trace/cmu-trace.h, Trace/cmu-trace.cc, and Makefile are modified.

6.3.2.4 Transmission Model of UV NLOS Communication

According to (6.2), which is used to calculate the received optical power of a UV NLOS communication link, the UV NLOS transmission model is added by following the two-ray ground model in the ns-2 software. Because the node has various additional angle attributes, the mobilenode.h file, mobilenode.cc file, and other relevant files should be updated.

6.3.3 Simulation and Analysis

To verify the validity of the node model of the UV multiplexer, we simulated the three communication modes of UV NLOS communication in different topologies. The appropriate modules in the ns-2 software were modified according to Sects. 6.4.2.1 and 6.4.2.3 to build a UV NLOS communication model. Then the model was compiled, and the test script was written. The GDB and tcl-debug tools were used to test the.cc,.h, and.tcl files. After successful debugging, the running results were obtained, and the results were analyzed and plotted using the Awk language and gnuplot, respectively.

6.3.3.1 Simulation Parameter Configuration

The transmission model assumed UV NLOS communication; the routing protocol used the UV-DSR protocol, and the MAC layer used IEEE 802.11 DCF. The coverage of UV NLOS(B) and NLOS(C) directional communication was simplified to a single sector. The angle was set to 45°, and the communication distance between nodes was 200 m. The transceiver elevations were (90°, 90°) for NLOS(A), (90°, 30°) for NLOS(B), and (20°, 60°) for NLOS(C). We simulated the network performance of the three UV NLOS communication modes in cross, parallel, chain, and lattice topologies.

6.3.3.2 Simulation Results and Analysis

(1) Cross topology

The cross topology is shown in Fig. 6.17. Data are transmitted between node 0 and nodes 1, 2, and 3. Figure 6.18 shows the network throughput of this topology at different constant bit rates (CBRs). (See Appendices D-1 and D-2 for the simulation script and throughput analysis procedure, respectively.)

The UV NLOS(C) mode has the largest throughput, followed by NLOS(A), and NLOS(B). The reason is that UV NLOS(C) is used for directional transmission and reception, and data can be transmitted between two node pairs simultaneously.

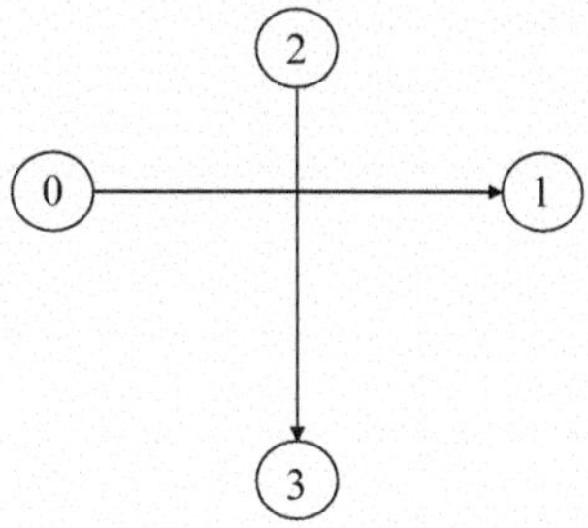

Fig. 6.17 Cross topology

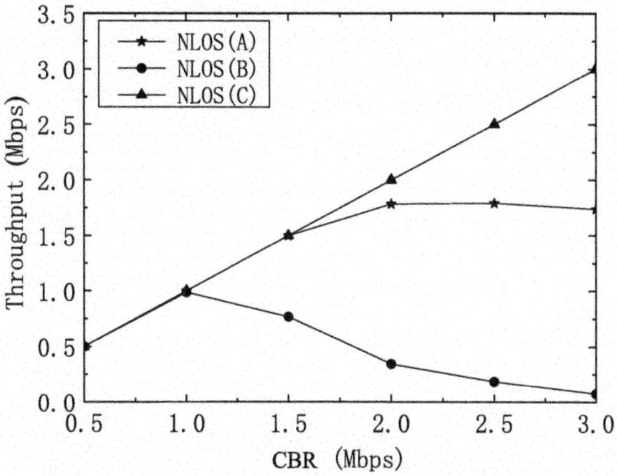

Fig. 6.18 Throughput of cross topology

When NLOS(A) is adopted, only one pair of nodes can communicate at a time. When NLOS(B) is adopted, node 2 sends RTS frames to node 3 because directional transmission and omnidirectional reception are used. If node 0 cannot hear the transmission of node 2, the channel is considered to be idle. Thus, the RTS frame is sent to node 1, resulting in a receiving conflict between the CTS frame returned by node 3 to node 2 and the RTS frame sent by node 0 to node 2. Node 0 becomes a directional hidden terminal to node 2. The presence of the directional hidden terminal reduces the network performance, and thus the throughput of this communication method in this topology is the lowest.

(2) Parallel topology

The parallel topology is shown in Fig. 6.19. Data are transmitted between nodes 0 and 1 and between nodes 2 and 3. Figure 6.20 shows the network throughput of this topology at different CBRs. The UV NLOS(C) and NLOS(B) modes have the same throughput, which is larger than that of the NLOS(A) mode. The reason is that when the UV NLOS(C) and NLOS(B) modes are adopted, two pairs of nodes can communicate simultaneously, which greatly improves the spatial multiplexing rate of the channel. In the NLOS(A) mode, because omnidirectional and omnidirectional

Fig. 6.19 Parallel topology

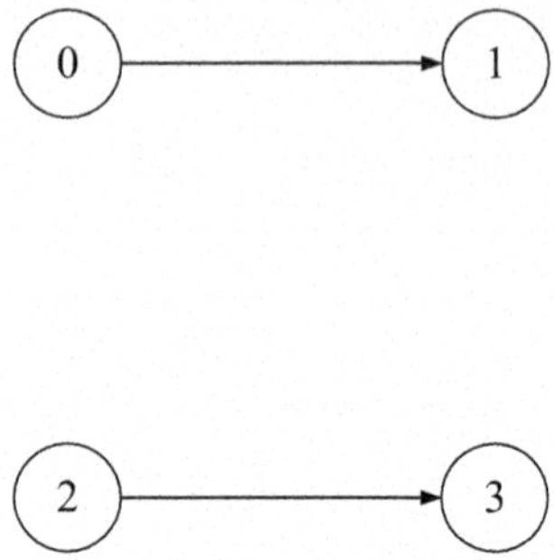

reception are used, when nodes 0 and 1 communicate, node 2 hears the transmission of node 0 and determines that the channel is busy. After communication between nodes 0 and 1 ends, nodes compete for the channel to send data to node 3.

Fig. 6.20 Throughput of parallel topology

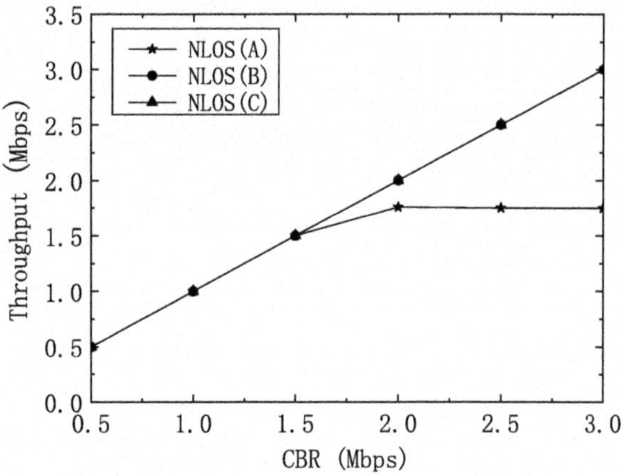

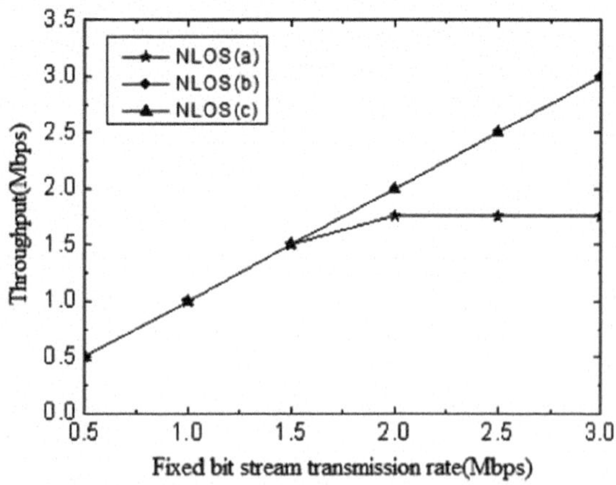

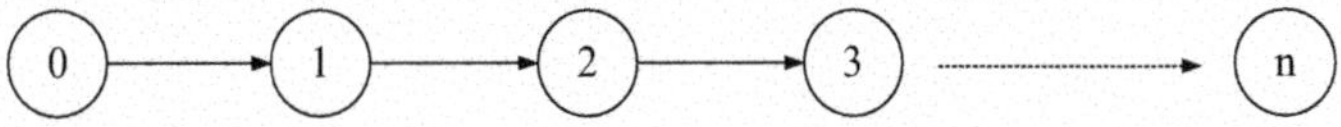

Fig. 6.21 Chain topology

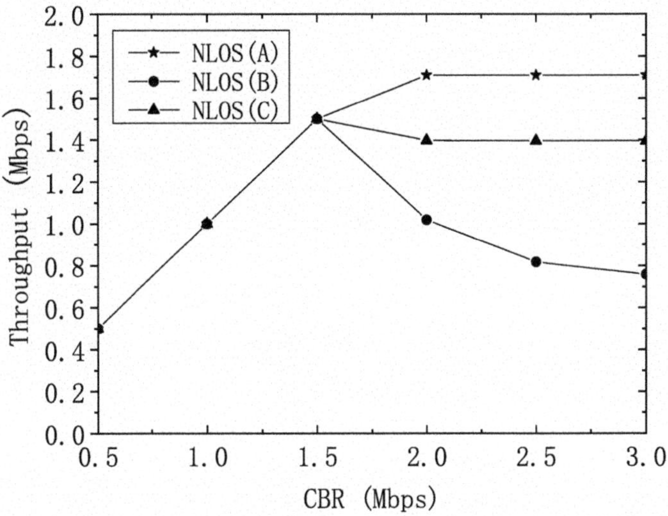

Fig. 6.22 Throughput of chain topology with $n = 3$

(3) Chain topology

A network with the n-node chain topology in Fig. 6.21 was simulated. Node 0 sends a stream to node n at various CBRs ranging from 0.5 to 3 Mbps. The throughput of networks with 3, 4, 5, and 6 nodes in the chain topology is shown in Figs. 6.22, 6.23, 6.24, and 6.25, respectively.

The simulation results are summarized as follows.

(a) For a chain topology with $n = 3$, 4, 5, or 6 nodes, the throughput of the UV NLOS(C) mode is always higher than that of the UV NLOS(B) mode. The reason is that when the UV NLOS(B) mode is adopted, directional hidden terminals are a serious problem because of omnidirectional reception by nodes. For example, when $n = 4$, node 2 sends RTS packets to node 3. Node 1 cannot receive the RTS sent by node 2. Node 1 determines that the channel is idle and sends RTS packets to node 2. Because node 2 receives from all directions, the CTS packets sent by node 3 to node 2 conflict with the RTS packets sent by node 1 to node 2. Node 1 becomes a directional hidden terminal of node 2, which affects communication between nodes 2 and 3. The case is similar for other values of n. When the UV NLOS(C) mode is adopted, node 2 does not receive RTS packets sent by node 1, and node 1 cannot affect communication between nodes 2 and 3.

(b) With increasing number of nodes, the throughput of the UV NLOS(C) mode gradually increases and exceeds that of the UV NLOS(A) mode when $n = 6$.

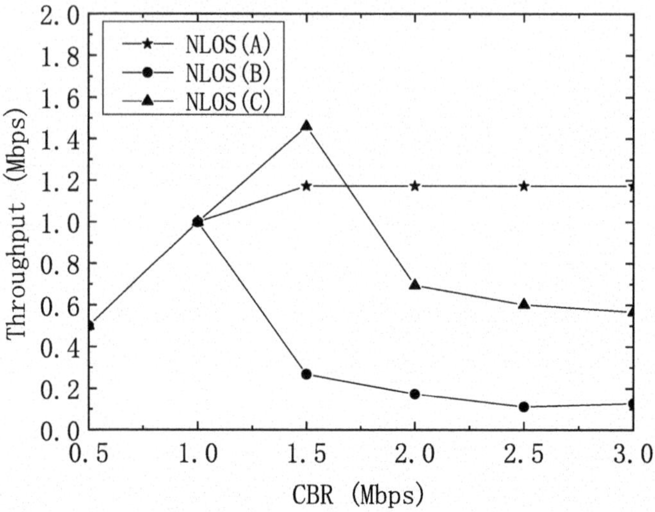

Fig. 6.23 Throughput of chain topology with $n = 4$

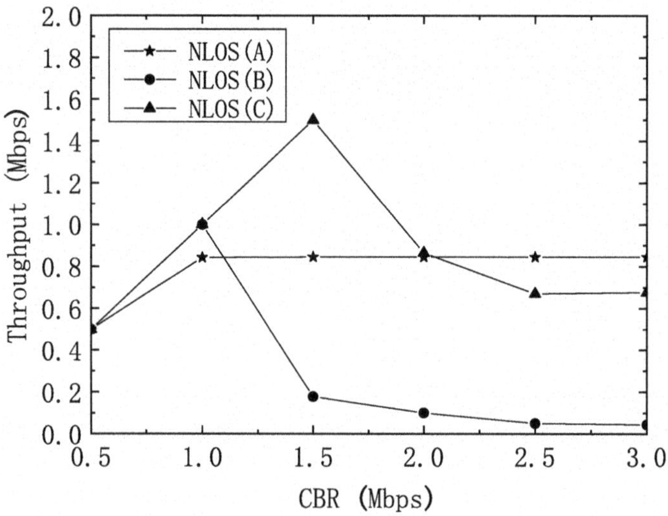

Fig. 6.24 Throughput of chain topology with $n = 5$

When n is relatively small, the use of UV NLOS(C) communication causes deafness. For example, when $n = 3$, nodes 1 and 2 conduct directional communication. Node 0 considers the channel to be idle because it cannot receive RTS packets sent by node 1; therefore, it sends an RTS packet to node 1. Because nodes 1 and 2 communicate directionally, node 1 cannot receive the RTS packet from node 0 until communication with node 2 ends. If node 0 does not receive a

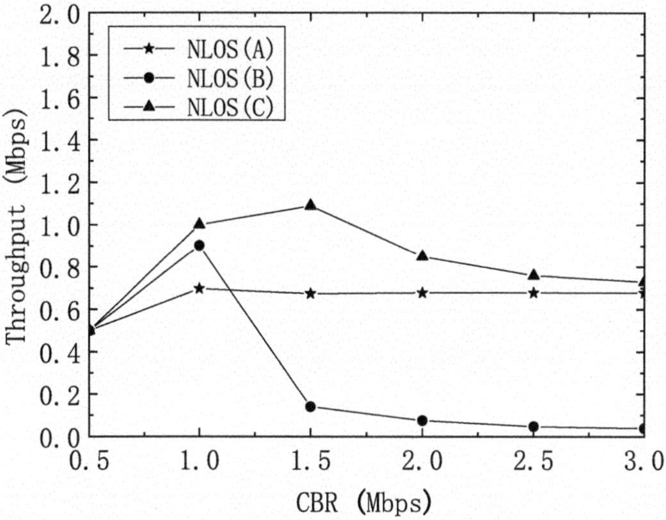

Fig. 6.25 Throughput of chain topology with $n = 6$

CTS packet from node 1 in reply, it resends the RTS packet until the maximum number of retransmissions is reached. Node 0 becomes deaf to node 1 during directional communication between nodes 1 and 2. With increasing number of nodes, when the UV NLOS(C) mode is adopted, the number of node pairs communicating simultaneously increases. For example, when $n = 4$, nodes 0 and 1 and nodes 2 and 3 can communicate simultaneously without affecting each other. However, when the UV NLOS(A) mode is adopted, only one of these node pairs can communicate. The increase in the number of nodes decreases the effect of deafness and increases the spatial multiplexing rate of the channel. Therefore, when $n = 6$, the network throughput of the UV NLOS(C) mode exceeds that of the UV NLOS(A) mode.

(c) For the chain topology, the UV NLOS(B) and NLOS(C) modes have the problems of directional hidden terminals and deafness, respectively. The problem of directional hidden terminals is more serious. The throughput of the UV NLOS(B) mode is always the lowest. With increasing number of nodes and increasing channel spatial multiplexing rate, the effect of deafness on the network is reduced, and the performance of the UV NLOS(C) mode is optimal.

(4) Lattice topology

The lattice topology used in the simulation is shown in Fig. 6.26. From 0 to 100 s, six CBR data streams send data simultaneously. Figure 6.27 shows the network throughput for CBRs ranging from 160 to 960 kbps. The throughput is largest for the UV NLOS(C) mode, because directional communication improves the spatial multiplexing rate of the channel. The higher spatial multiplexing rate greatly reduces the effect of deafness on the network. When the communication rate is less than 500 kbps in the UV NLOS(B) mode, the higher spatial multiplexing rate of the channel can reduce the effect of directional hidden terminals. However, with increasing communication rate, the problem of directional hidden terminal becomes more serious. As a result, the throughput of the network is lower than that in the UV NLOS(A) mode. The presence of directional hidden terminals greatly offsets the effect of improved channel spatial multiplexing rate on the network performance.

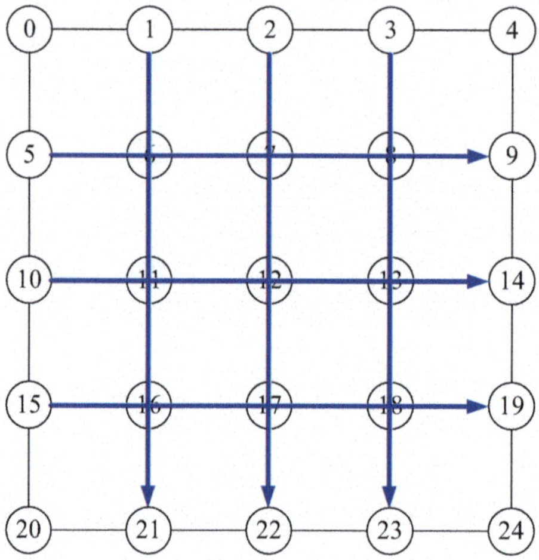

Fig. 6.26 5 × 5 Lattice topology

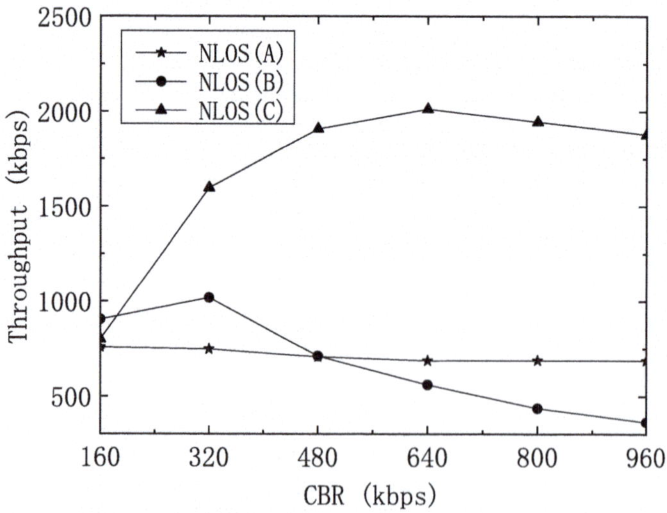

Fig. 6.27 Throughput of 5 × 5 lattice topology

6.4 MAC Protocol for UV Directional Access

6.4.1 *Analysis of MAC Protocols for UV NLOS Communication*

An analysis of the MAC protocol reveals that in the UV NLOS(A) mode, nodes send and receive omnidirectionally and cover a circular area. Conventional omnidirectional MAC protocols such as IEEE 802.11 DCF can be used. In the UV NLOS(B) and NLOS(C) modes, the nodes use omnidirectional and directional reception, respectively, and the coverage is not a circular area. Consequently, a node can communicate with only some neighbor nodes; that is, neighbor nodes can communicate with a sending node only if they are within its coverage.

Current DMAC protocols use directional antenna technology, which is realized by combining the directional and omnidirectional modes of the antenna. The data and ACK frames are transmitted and received directionally, but the RTS and CTS frames for channel reservation are transmitted and received by various combinations of omnidirectional and directional protocols, for example, the ORTS/OCTS, DRTS/OCTS, and multihop RTS protocols described in Sect. 6.2.2. This communication implemented by directional and omnidirectional cooperation is obviously not suitable for our UV NLOS(B) and NLOS(C) communication methods because the UV transmitting and receiving nodes are initially set at various angles, including the transmitter and receiver elevation angles, divergence angle, and field of view. In current DMAC protocols, it is impossible to continuously adjust these four angles during packet transmission. In fact, data packets are sent very quickly. If the sending node does not successfully receive a response from the receiving node within a fixed time, the data are retransmitted. The adjustment of various angles of the UV transceiver may not be completed during this period. Therefore, after we determine which UV NLOS communication mode is used between the transmitting and receiving nodes, either NLOS(B) (directional transmission and omnidirectional reception) or NLOS(C) (directional transmission and reception) is chosen as the transmission and reception modes of all frames between them. In this section, we propose a UV DMAC (UVDMAC) based on perceived angles.

6.4.2 *UVDMAC Protocol*

6.4.2.1 Direction Angle of UV NLOS Communication Coverage

In Chap. 2, we analyzed the coverage of various modes of UV NLOS communication. We derive the direction angle of coverage according to reference [56]. As shown in Fig. 6.28, the divergence angle of the transmitting node is denoted by ϕ_t, the field of view of the receiving node is θ_t, the receiving elevation is ϕ_r, the distance between the transmitting and receiving nodes is θ_r, and the power at the transmitting end is

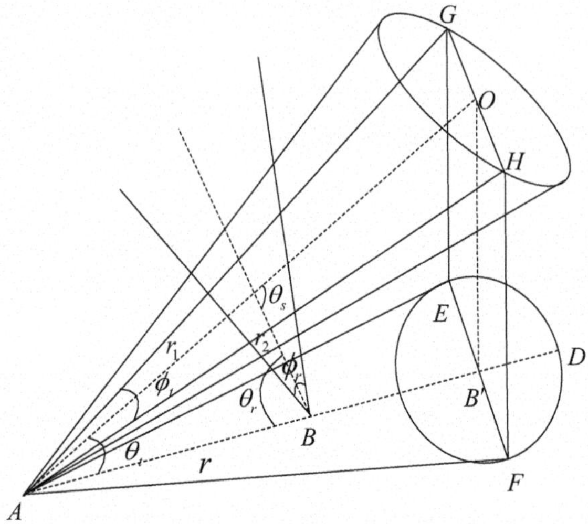

Fig. 6.28 Projection stereogram of UV NLOS communication

limited. The farthest transmission distance is r. As shown in the figure, a cone with a height of r_1 and diameter of GH is projected onto the ground. The projection area is EAFD, and the direction angle of visible coverage is $\angle EAF$.

From Fig. 6.28, $B'E = OG$, $\tan(\frac{\phi_t}{2}) = \frac{OG}{OA}$, $\cos\theta_t = \frac{AB'}{OA}$, and $\tan(\frac{\angle EAF}{2}) = \frac{B'E}{AB'} = \frac{OG}{AB'} = \frac{OA \cdot \tan(\frac{\phi_t}{2})}{OA \cdot \cos(\theta_t)} = \frac{\tan(\frac{\phi_t}{2})}{\cos(\theta_t)}$. Therefore, the coverage angle is

$$\angle EAF = 2 * a\tan(\frac{\tan(\frac{\phi_t}{2})}{\cos(\theta_t)}) \tag{6.15}$$

The direction angle of the UV NLOS communication coverage depends on the divergence angle ϕ_t and emission elevation angle θ_t.

6.4.2.2 Network Models and Assumptions

We assume the following network environment. (1) Nodes know the location information of neighboring nodes in advance. This assumption is true when the network topology is fixed. (2) Each node is equipped with multiple UV transceivers, and each UV transceiver covers a certain angle range. Multiple transceivers are used to communicate with neighboring nodes. (3) When a node transmits data, the upper layer can determine which direction of the UV transceiver to use for data transmission.

6.4.2.3 Protocol Description

(1) Channel reservation

The channel reservation mechanism of UVDMAC before communication is similar to those of the DMAC and basic MAC protocols. Suppose the upper layer of the network knows the direction of the receiver. The sender and receiver are initially in the idle (channel scanning) mode, and the UV transceivers in all directions from each node are operating. When node A has data to send to node B, it selects the UV transceiver in the direction of node B to send an RTS request to node B, and the UV transceivers in other directions from node A stop working. Then node A waits for the receiving end to reply with a CTS within a fixed time interval (the SIFS). If a CTS is not received within the SIFS time, the backoff window of node A is doubled, and a backoff value between 0 and the backoff window value is randomly selected. The RTS request is sent until channel contention again succeeds.

(2) Receiving RTS and sending CTS

After receiving the RTS request from the node A, node B replies with a CTS to node A. At this time, the activity of UV transceivers in other directions from node B is suspended. The direction of the CTS packet is consistent with the direction of the RTS; that is, the UV transceiver in the same direction is selected to receive the RTS and send the CTS.

(3) Receiving CTS and exchanging data/ACK

When node A sends the RTS packet to node B, it uses the UV transceiver in the same direction to wait for the CTS packet. When node A receives the CTS packet from node B, it indicates that channel reservation was successful. Node A selects the UV transceiver in the RTS direction to send the data. When node B successfully receives the data, it uses the UV transceiver in the CTS direction to reply to node A with an ACK packet.

(4) DNAV update

Each node has multiple transceivers corresponding to multiple DNAV values. After receiving the RTS request, which is not sent to neighbor nodes around node A, the DNAV in the corresponding direction is updated according to the Duration/ID field value in the RTS request. That is, the DNAV of the UV transceiver direction required for communication between itself and node A is updated. Similarly, the neighbor nodes around node B update the DNAV value in the appropriate direction according to the received CTS. When a node judges whether the channel is idle and performs carrier listening, it needs to perform physical and virtual carrier listening. The virtual carrier-sense function determines only whether the DNAV corresponding to the communication direction of the UV transceiver is 0. If the DNAV in this direction is 0, the virtual carrier-sense function determines that the channel is idle.

Otherwise, it determines that the channel is busy. It is not necessary to know whether the DNAV in other directions is 0.

(5) Angle perception

During network initialization, the node needs to identify neighbor nodes at different elevations. Therefore, multiple interfaces are required to broadcast at different elevations and record the neighbor nodes when each interface of the node sends to different elevations. Each node maintains an interface elevation neighbor table. The records in the table include the interface number, sending elevation, and neighbor nodes. Before each communication, the node looks for an appropriate neighbor node in this table. If it succeeds, one-hop communication can be performed. The node selects the appropriate interface according to the table and sends the elevation for information transmission. Otherwise, for multihop communication, the node selects the transmission elevation that can reach the farthest communication distance, which can reduce the number of hops required for information transmission. Because each type of interface has a different communication coverage, the use of the interface elevation neighbor table fully exploits the advantages of channel spatial multiplexing.

6.4.3 Simulation and Analysis

(1) Chain topology

For the chain topology with four nodes shown in Fig. 6.29, node 0 sends to node 3 at different CBRs. The data packet length is 1000 bytes, and the simulation time is 100 s.

The simulation results in Figs. 6.30 and 6.31 show that the improved directional access protocol results in higher throughput and lower delay than the original UV directional access protocol. The reason is that the angle cannot be adjusted by the original directional transmission and reception protocol based on UV NLOS communication; therefore, nodes 0 and 3 communicate using three hops. However, our modified directional access protocol based on angle perception can adjust the transmission elevation of node 0 at the transmission end; therefore, the direction angle of the coverage of the transmission end decreases, and the communication distance of node 0 increases. This increase enables node 0 to communicate directly with node 2 without forwarding through node 1, and then node 2 forwards the data to node 3. The node reduces the number of forwarded packets by adjusting the sending elevation, which increases the throughput and reduces the average delay of packet transmission.

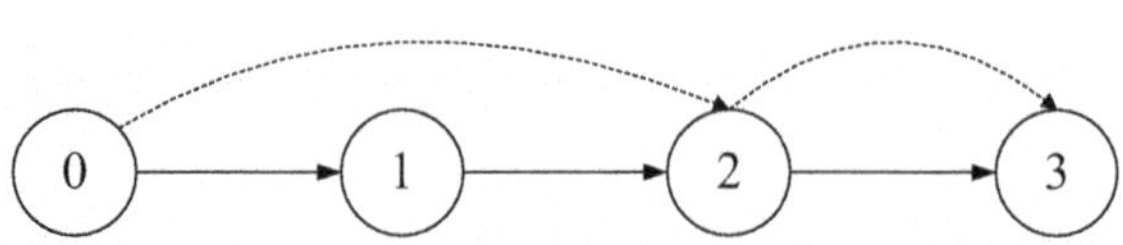

Fig. 6.29 Chain topology

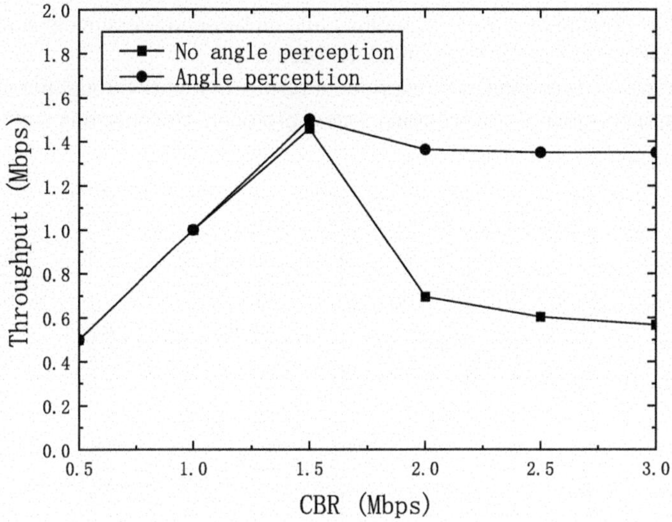

Fig. 6.30 Throughput of chain topology

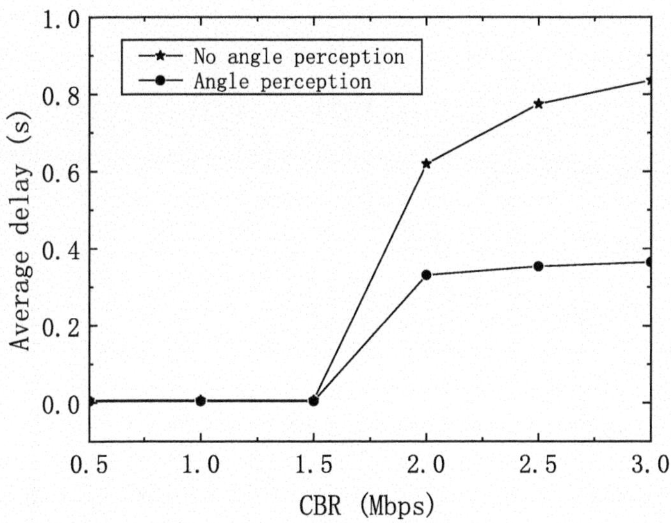

Fig. 6.31 Average delay of chain topology

(2) Lattice topology

The 5 × 5 lattice topology shown in Fig. 6.32 was simulated, where six pairs of nodes transmit data simultaneously: nodes 1 → 21, 2 → 22, 3 → 23, 5 → 9, 10 → 14, and 15 → 19. The data packet length is 1000 bytes, and the simulation time is 100 s. The obtained network throughput and end-to-end average delay at various CBRs are shown in Figs. 6.33 and 6.34, respectively.

The improved directional access protocol results in higher throughput and lower delay than the original UV directional access protocol. The reason is that the angle

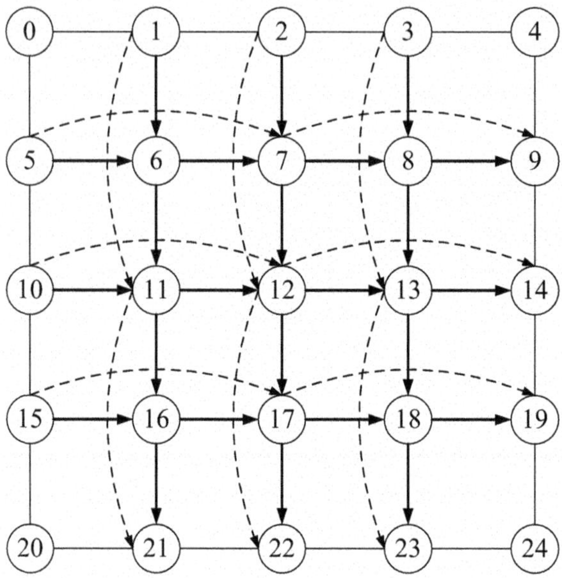

Fig. 6.32 5 × 5 Lattice topology

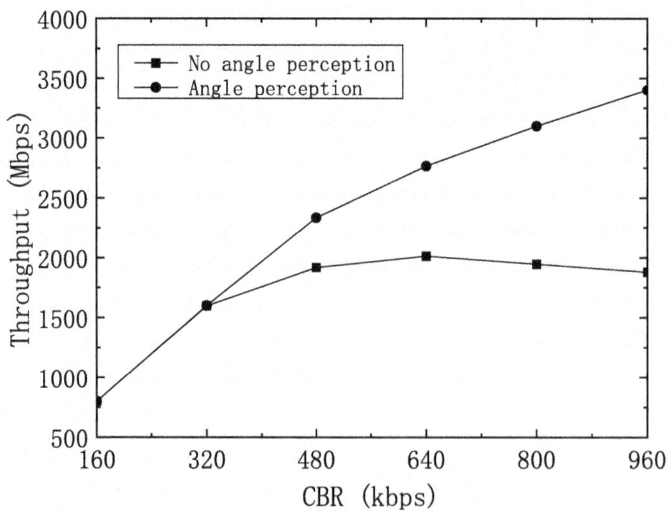

Fig. 6.33 Throughput of network with 5 × 5 lattice topology

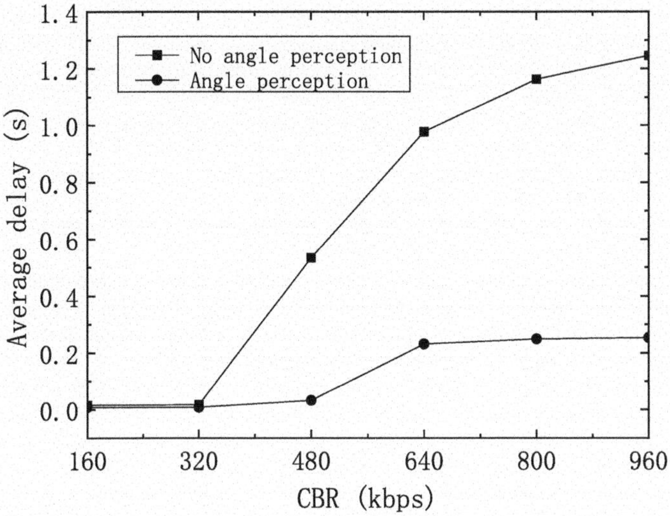

Fig. 6.34 Average delay of network with 5 × 5 lattice topology

cannot be adjusted by the original directional transmission and reception protocols based on UV communication; therefore, nodes 1 and 21 can communicate only using four hops. However, our modified directional access protocol based on angle perception adjusts the transmission elevation of node 1 at the transmitting end. Consequently, the direction angle of the coverage at the transmitting end decreases, and the communication distance of node 1 increases, This increase enables node 1 to communicate directly with node 11 without forwarding through node 6. Similarly, node 11 can communicate directly with node 21 because of the angle adjustment. Therefore, data transmission from node 1 to node 21 requires forwarding only once, by intermediate node 11. When the original protocol is adopted, communication between nodes 1 and 21 is relayed through nodes 6, 11, and 16 before reaching node 21. The node reduces the number of forwarded packets by adjusting the sending elevation, and the time required to transmit a packet is also reduced. More packets can be transmitted per unit time, and thus the directional access protocol based on angle perception can improve the throughput. In addition, the decrease in packet forwarding time can reduce the end-to-end transmission delay of packets.

References

1. Zhang Y, Guo D (2007) Principle and technology of wireless mesh network. Electronic Industry Press, Beijing
2. Translated by Guoda, Zhang Y, pengxiaochuan (2008) Wireless mesh networks: architecture, protocols and standards. Electronic Industry Press, Beijing

3. Perkins CE, Royer EM (1999) Ad-hoc on-demand distance vertor routing. In: Proceeding of the 2nd IEEE workshop on mobile computing systems and applications, New Orleans, pp 99–100
4. Zhong YJ, Yuan DF (2003) Dynamic source routing protocol for wireless ad hoc networks in special scenario using location information. Int Conf Commun Technol 2:1287–1290
5. Perkins CE, Bhagwat P (1994) Highly destination sequenced distance vertor routing (DSDV) for mobile computers. ACM SIGCOMM Comput Commun Rev 24(4):234–244
6. Jacquet P, Muhlethaler P, Clausen T et al (2001) Optimized link state routing protocol for ad hoc networks. IEEE INMIC, Le Chesnay, pp 62–68
7. Haas ZJ, Pearlman MR (2001) The performance of query control schemes for the zone routing protocol. IEEE/ACM Trans Netw (TON) 9(4):427–438
8. Chen G, Xu ZY, Ding HP et al (2009) Path loss modeling and performance trade-off study for short-range non-line-of-sight ultraviolet communication. Opt Exp 17(5):3929–3940
9. Ding HP, Xu ZY, Sadler BM (2010) A path loss model for non-line-of-sight ultraviolet multiple scattering channels. EURASIP J Wirel Commun Netw 1–12
10. He QF, Sadler BM, Xu ZY (2009) Modulation and coding tradeoffs for non-line-of-sight ultraviolet communications. In: Proceedings of SPIE, vol 7464, no 74640H, San Diego, pp 1–12
11. Wang P, Gao J (2008) Research on performance of marine UV communication. Int Conf Commun Technol 11:371–374
12. Xu ZY (2007) Approximate performance analysis of wireless ultraviolet links. ICASSP 3:577–580 (Honolulu)
13. Ding HP, Chen G, Majumdar AK et al (2009) Modeling of non-line-of-sight ultraviolet scattering channels for communication. IEEE J Sel Areas Commun 27(9):1535–1544
14. Gary AS, Melissa N, Mrinal I et al (2000) NLOS UV communication for distributed sensor systems. Proc of SPIE, vol 4126, San Diego, pp 83–96
15. Wang JL, Luo T, Dai M et al (2009) UV NLOS communications atmospheric channel model and its performance analysis. World Congr Comput Sci Inf Eng 1:85–88
16. Tan P, Chan MC (2006) AMCM: adaptive multi-channel MAC protocol for IEEE 802.11 wireless networks. In: International conference on broadband communications networks and systems, San Jose, pp 1–10
17. Ramanathan R, Redi J, Santivanez C et al (2006) Ad hoc networking with directional antennas: a complete system solution. IEEE J Sel Areas Commun 23(3):496–506
18. Haas ZJ, Deng J (2002) Dual busy tone multiple access-a multiple access control scheme for ad hoc networks. IEEE Trans Commun 50(6):975–985
19. Dai HN, Ng KW, Wong CW et al (2008) On the capacity of multi-channel wireless networks using directional antennas. IEEE INFOCOM 112:628–636
20. Abramson N (1970) The aloha system-another alternative for computer communication. In: AFIPS fall joint computer conference, New York, pp 281–285
21. Kleinrock L, Tobagi F (1975) Packet switching in radio channels: part I-carrier sense multiple-access modes and their throughput-delay characteristics. IEEE Trans Commun 23(12):1400–1416
22. IEEE Std.802.11 (1997) Wireless LAN media access control (MAC) and physical layer (PHY) specifications, pp 71–87
23. IEEE 802.11e/D13.0 (2005) Medium access control (MAC) quality of service (QoS) enhancements, pp 48–108
24. Fan WF, Gao DY, Tsang DHK et al (2004) Admission control for variable bit rate traffic in IEEE 802.11e WLANs. APCC/MDMC'04, vol 1, Beijing, pp 272–277
25. Gao D, Cai J, Chen CW (2008) Admission control based on rate-variance envelop for VBR traffic over IEEE 802.11e HCCA WLANs. IEEE Trans Veh Technol 57(3):1778–1788
26. Chen K (2007) Design and research of directional MAC layer protocol in ad hoc networks. Master Thesis of University of Science and Technology of China, Hefei, pp 27–33
27. Zander J (1990) Slotted aloha multihop packet radio networks with directional antennas. Electron Lett 26(25):2098–2100

28. Chu W (2007) Research on MAC mechanism of wireless mesh network based on directional antenna. Master's thesis of Xi'an University of Electronic Science and Technology, Xi'an, pp 18–24
29. Shan Z, Li L (2010) MAC protocol based on directional antenna in ad hoc networks. Comput Eng 2(36):21–24
30. Nasipuri A, Ye S, You J et al (2000) A MAC protocol for mobile ad hoc networks using directional antennas. In: Proceedings of IEEE wireless communications and networking conference, vol 3, Chicago, pp 1214–1219
31. Ko YB, Shankarkumar V, Vaidya NH (2000) Medium access control protocols using directional antennas in ad hoc networks. In: Proceedings of IEEE conference on computer communications, vol 1, no 3, Tel Aviv, pp 13–21
32. Korakis T, Jakllari G, Tassiulas L (2003) A MAC protocol for full exploitation of directional antennas in ad-hoc wireless networks. In: Proceeding of the 4th ACM international symposium on mobile ad hoc networking and computing, Annapolis, pp 98–107
33. Jakllari G, Broustis J, Korakis T et al (2005) Handling asymmetry in gain in directional antenna equipped ad hoc networks. In: IEEE 16th international symposium on PIMRC, vol 2, Berlin, pp 1284–1288
34. Gossain H, Cordeiro C, Joshi T et al (2006) Cross-layer directional antenna MAC protocol for wireless ad hoc networks. Wirel Commun & Mob Comput 6(2):171–182
35. Shihab E, Cai L, Pan JP (2009) A distributed asynchronous directional-to-directional MAC protocol for wireless ad hoc networks. IEEE Trans Veh Technol 58(9):5124–5134
36. Nadeem T (2009) Enhancements for IEEE 802.11 networks with directional antennas. In: IEEE 34th conference on local computer networks, Zurich, pp 205–208
37. Choudhury RR, Yang X, Ramanathan R et al (2002) Using directional antennas for medium access control in ad hoc networks. In: Mobile computing and networking, New York, pp 59–70
38. Tomar GS, Verma S, Rai M (2006) Modified directional medium access protocol using bootlace lens. In: IEEE region 10 annual international conference (TEN-CON' 06), Hong Kong, pp 1–4
39. Raman B, Chebrolu K (2004) Revisiting MAC design for an 802.1-based mesh network. In: Third workshop on hot topics in networks (HotNets-III), San Diego
40. Raman B (2006) Channel allocation in 802.11-based mesh networks. In: The 25th annual conference on computer communications, Barcelona, pp 1–10
41. Das SM, Pucha H, Koutsonikolas D et al (2006) DMesh: incorporating practical directional antennas in multi-channel wireless mesh networks. IEEE J Sel Areas Commun Spec Issue Multi-Hop Wirel Mesh Netw 24(11):2028–2039
42. Dai H, Ng KW, Wu MY (2006) An over view of MAC protocols with directional antennas in wireless ad hoc networks. In: International conference on wireless and mobile communications, Bucharest, p 84
43. Ye F, Sikdar B (2004) Distance-aware virtual carrier sensing for improved spatial reuse in wireless networks. In: Global telecommunication conference, vol 6. Dallas, pp 3793–3797
44. Lal D, Toshniwal R, Radhakrishnan R et al (2002) A novel MAC layer protocol for space division multiple access in wireless ad hoc networks. In: International conference on computer communications and networks. Miami, pp 614–619
45. Huang ZC, Shen CC, Srisathapornphat C et al (2002) A busy-tone based directional MAC protocol for ad hoc networks. MilY Commun Conf Anaheim 2:1233–1238
46. Takai M, Martin J, Bagrodia R et al (2002) Directional virtual carrier sensing for directional antennas in mobile ad hoc networks. In: Proceedings of the 3rd ACM international symposium on mobile ad hoc network & computing, Lausanne, pp 183–193
47. Chang JJ, Liao WJ, Hou TC (2009) Reservation-based directional medium access control (RDMAC) protocol for multi-hop wireless networks with directional antennas. In: IEEE international conference on communications, Dresden, pp 1–5
48. Wang J, Kong L, Wu MY (2010) Capacity of wireless ad hoc networks using practical directional antennas, Sydney, pp 1–6
49. Kumar SH, Seah W (2008) Efficient neighbour discovery algorithm for maritime mesh networks with directional antennas. In: International conference on ITS telecommunications, Phuket, pp 6–11

50. Xu L, Pang Bo, Zhao Y (2003) NS and network simulation. People's Posts and Telecommunications Press, Beijing
51. Ke Z, Cheng R, Deng D (2009) NS2 simulation experiment—multimedia and wireless network communication. Electronic Industry Press, Beijing
52. Yi S, Pei Y, Kalyanaraman S (2003) On the capacity improvement of ad hoc wireless networks using directional antennas. In: Proceedings of the 4th ACM international symposium on mobile Ad Hoc networking & computing, Annapolis, pp 108–116
53. Calvo RA, Campo JP (2009) Adding multiple interface support in NS-2[User Guide]. http://www.google.com.hk/search?q=Adding+multiple+interface+support+in+NS-2&client=aff-360homepage&hl=zh-CN&ie=gb2312
54. Hou Z (2011) Research on access protocol of ultraviolet wireless optical mesh network, Xi'an: Master's thesis of Xi'an University of Technology
55. Fang L, Liu S (2008) NS-2 network simulation foundation and application. National Defense Industry Press, Beijing
56. Taifei Z, Yanling F, Xizheng K, Hua H (2010) Research on node coverage in "solar blind" ultraviolet communication network. J Opt 30(8):2229–2235

Chapter 7
Routing Protocol for Ultraviolet Self-organizing Networks Based on Node Location and Speed Information

The main requirements for wireless ad hoc network routing protocols are rapid convergence, loop-free routing, the avoidance of infinite computing problems, low control and management overhead, and the support of one-way channels [1]. Mobile ad hoc networks (MANETs) are temporary multihop networks composed of a group of mobile nodes and a wireless transceiver; they do not need a central control node and are easy to deploy. They are a hot topic of research on network technology. Conventional MANETs operate in the radio-frequency band. With the rapid introduction of new services, the limited bandwidth of wireless networks has become a great challenge. This chapter describes the use of ultraviolet (UV) communication with wireless ad hoc networks. On the basis of an in-depth analysis of routing protocols for MANETs and a UV propagation model, a routing protocol suitable for UV ad hoc networks is proposed.

7.1 Routing Techniques for Mobile Self-organizing Networks

Most protocols used in fixed wired networks rely on the distance vector algorithm [2] or link-state algorithm [3], both of which assume that the network topology is relatively stable. Neither of these algorithms is suitable for use in MANETs, where the topology can change at any time. The routing of conventional wired networks also depends on routers or distributed routing databases, and it is impossible for nodes in MANETs to permanently maintain the information in routing tables. For these reasons, the MANET working group of the Internet Engineering Task Force has studied and consolidated several draft protocols for MANETs and classified these protocols into three categories: flat routing protocols, cluster routing protocols, and routing protocols based on geographic location information. Figure 7.1 shows the classification of routing protocols for mobile self-organized networks.

X. Ke, *UV Light Self Organizing Network Theory*, Optical Wireless Communication Theory and Technology, https://doi.org/10.1007/978-981-97-7854-6_7

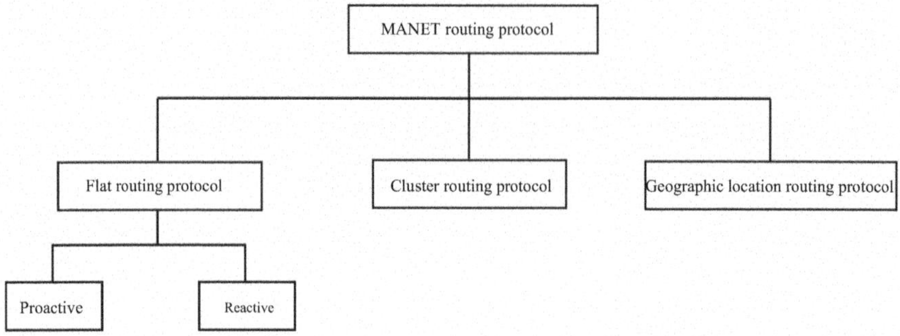

Fig. 7.1 MANET routing protocol classification

7.1.1 Flat Routing Protocols

All nodes in a flat routing protocol are on the same level, and there is no functional or structural difference between nodes; all nodes are equal. Flat routing protocols can be further divided into proactive and reactive protocols.

(1) A priori routing protocols

A priori routing protocols, which are also known as proactive or table-driven routing protocols, are similar to link-state protocols in fixed wireline networks and were proposed early in MANET research. In an a priori routing protocol, each node maintains at least one route to other nodes. When the network topology changes or the route is blocked, the node publishes its information to the network through a new path, and other nodes in the network update their routing tables in a timely manner to maintain uniform, accurate routing information. The routing table makes it possible for a source node to send packets directly to a destination node using the information in the table. The delay in sending packets is small, but the overhead of maintaining the routing table is large; therefore, this type routing protocol is inefficient when the network topology changes frequently. Classical a priori routing protocols include the destination-sequenced distance vector (DSDV) protocol [4] and the system- and traffic-dependent adaptive routing algorithm (STARA) [5]; the performance of DSDV is relatively good.

The DSDV routing protocol is an a priori protocol based on the Bellman–Ford algorithm. In the DSDV protocol, each node maintains a routing table of next-hop nodes. When a node sends a packet, it adds a sequential number as an identifier to that packet. This serial number is broadcast by all nodes that receive the packet and stored in the next-hop node routing table of those nodes. DSDV uses both event-triggered and time-triggered methods to update network links. To reduce the length of routing packets, it uses two methods to update packets: full and incremental updating. The main advantages of the DSDV protocol are that it eliminates routing loops and uses a shortest-circuit-first algorithm, which greatly reduces the overhead of routing control messages. The disadvantage is that it is difficult to adapt to MANETs, which change rapidly [6].

The routing metric of the STARA protocol uses the average delay instead of the more commonly used routing hops to identify suitable routes. Consequently, it takes into account factors such as the wireless channel capacity and queueing delay when routing packets. Each node i uses an improved end-to-end acknowledgement protocol to calculate the average delay A for each pair of source and destination nodes (i, d), as follows:

$$D_{ik}^{d}(t) = \frac{1}{1-\lambda}\sum_{l=0}^{\infty}\lambda^{l}D_{ik}^{d}(t-l) \tag{7.1}$$

where the forgetting factor $\lambda \in [0, 1]$ is used to adjust the proportional relationship between the historical and current delay values; $k \in N$, where N denotes the set of all neighboring nodes that can be reached by node i in one hop.

$$P_{ik}^{d}(t) = P_{ik}^{d}(t-1) + a(t)\cdot(D_{i}^{d}(t) - D_{ik}^{d}(t)) \tag{7.2}$$

According to (7.2), data packet traffic is assigned to different neighboring nodes so that all available paths have the same delay. Note that this path-averaging estimation mechanism does not require the support of bidirectional channels or clock synchronization between nodes. The STARA algorithm uses an estimate based on the expected delay as a distance parameter. This mechanism allows the algorithm to adapt not only to changes in the network topology but also to changes in the network service; it can be used in unidirectional MANET links.

(2) Reactive routing protocols

Reactive routing protocols, also commonly known as on-demand routing, begin to search for routes when a source node is ready to send a data packet to a destination node. In reactive routing protocols, each node does not need to maintain a routing table for all nodes in the network. The route lookup process is initiated only when no route is found to be available when the packet is ready to be sent. Reactive routing has less routing overhead than a priori routing, but the route-finding process is relatively slow in the absence of available routes. Reactive routing is usually divided into two parts: route discovery and route maintenance. When a source node wants to send a data packet to a destination node but finds no available route, the source node initiates a route discovery process. The simplest route discovery process is flood routing, which is illustrated in Fig. 7.2.

Source node S wants to send a data packet to destination node D. S broadcasts the route request (RREQ). Neighboring nodes M and N continue to broadcast the RREQ to their neighboring nodes and load their own information into the RREQ packet until the RREQ packet reaches D. Node D receives one or more paths back to S. It selects an optimal path according to arrival time or path length and generates a route reply (RREP) group back to S. Node S receives a route to D after receiving the RREP group and starts sending data packets. When the network topology changes and the existing

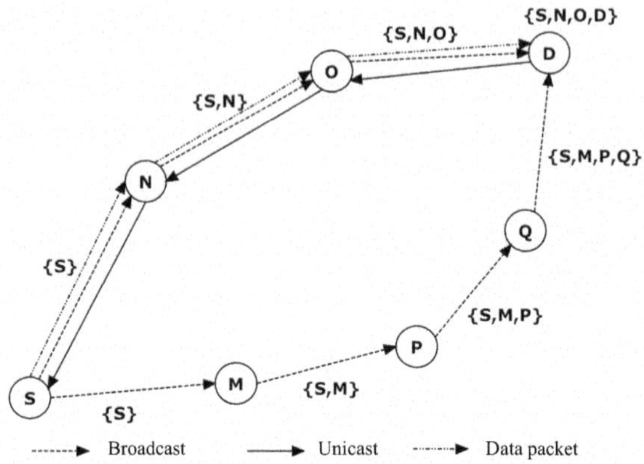

Fig. 7.2 Schematic diagram of flood routing process

route fails, S first removes the route to D and re-initiates the RREQ. The advantage of the flooding algorithm is that the RREQ packet is almost certain to reach the destination node; the disadvantage is that the routing overhead is very large. Classical reactive routing protocols include ad hoc on-demand distance vector (AODV), dynamic source routing (DSR), and the temporally ordered routing algorithm (TORA).

The AODV protocol is a reactive routing protocol based on DSDV; it is designed for MANETs with tens to thousands of mobile nodes. In the AODV protocol, when an intermediate node receives a RREQ packet, it can reply to the RREQ using this path because it can implicitly obtain the path from itself to the source node when the destination node finally receives the RREQ packet. The protocol is characterized by its hop-by-hop packet forwarding approach and the inclusion of multicast routing protocol extensions. Its main drawback is that it does not support unidirectional channels.

DSR was the first reactive routing protocol. It uses a source routing algorithm instead of a hop-by-hop routing method, and it also consists of two processes: route discovery and route maintenance. Each node in the network must maintain a routing table, which is updated when a new route is discovered. The header of each data packet contains information about each hop node from the source node to the destination node; therefore, this protocol is called a source routing algorithm. The advantage of DSR is that nodes need to maintain only routes to the nodes with which they communicate, and the protocol overhead is small; DSR also uses route caching techniques to reduce the route discovery overhead. In addition, one route discovery process may generate multiple routes to the destination, which also facilitates route selection. Because each data packet needs to carry routing information in its header, the disadvantage of DSR is that the additional overhead of data packets is large. In addition, RREQ messages are flooded, making it possible for neighboring nodes to propagate conflicting RREQ messages and possibly generate duplicate broadcasts; cached and expired routes affect the accuracy of route selection.

The TORA protocol is a reactive routing protocol based on the directed acyclic graph (DAG) algorithm [7]; it is divided into three processes: route discovery, route maintenance, and RREQ. Route discovery in TORA is similar to that in other reactive routing protocols in that it first broadcasts the RREQ group to the network. However, the DAG algorithm is used in the RREP. The main process is as follows. Each node in the network is assigned a height value relative to the source node. The source and destination nodes have the maximum and minimum height values, respectively. On this basis, the heights of neighboring nodes are compared, and the node with the larger height points to the node with a smaller height, thus forming one or more directed paths. The TORA protocol ensures that there are no loop routes in the network; however, its routing overhead is too large for use as the primary routing protocol for MANETs.

7.1.2 Cluster Routing Protocols

The use of flat routing protocols in MANETs does not give good results; in fact, the appropriate routing algorithm for MANETs must be chosen according to the network topology characteristics. Flat routing protocols are more suitable for MANETs with tens to thousands of nodes where the distance between nodes does not differ greatly. Moreover, the use of flat routing protocols generally combines a priori and reactive routing protocols. If only a priori routing protocols are used, each node needs to maintain a routing table to other nodes when the network is very large, which uses considerable system resources. Moreover, changes in the network topology can result in route failures, and the overhead of route maintenance is quite high. If only reactive routing protocols are used, the route lookup time is very slow when nodes in the network are separated by very large distances, which is undesirable for services with high real-time requirements such as voice and video conferencing. Therefore, if a mixture of a priori and reactive routing protocols can be used, the results may be better. In practice, many researchers have proposed similar schemes. Because hybrid routing often uses a clustered structure, it is also known as a cluster routing protocol. Within a certain small area, unclustered routing is performed by an a priori routing protocol. A certain amount of routing information is maintained at the local scale so that data packets can be exchanged quickly. When two or more communication nodes are far apart, reactive routing protocols are used to reduce the routing overhead.

In cluster routing protocols, the network is divided into several clusters. Nodes are divided into two types: cluster head nodes and ordinary nodes. Nodes in the same cluster (including cluster head and ordinary nodes) jointly maintain the routing information in the cluster. The cluster head node records and manages the topology information of the cluster and exchanges its own topology information with other cluster head nodes. The diagram in Fig. 7.3 illustrates the cluster protocol.

The cluster routing protocol has two main purposes. The first is to reduce the routing message overhead required for exchanging routing information and the system overhead required for maintaining routing tables by reducing the number of routing switching nodes, which is consistent with the concept of layering in fixed wired networks. The second purpose is to stabilize the routes in each cluster after

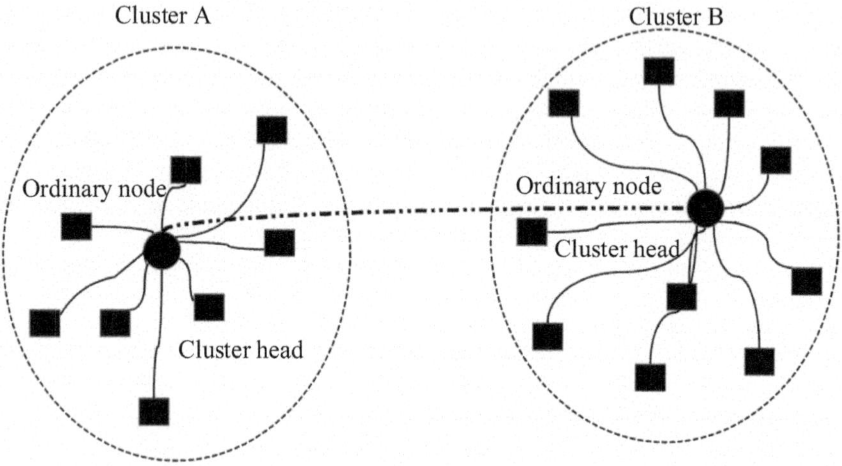

Fig. 7.3 Schematic diagram of cluster protocol

the network is divided into several subnets, which can greatly reduce the impact of topology changes on routes. The characteristics of cluster routing make it suitable for large-scale and decentralized MANETs, and it has good scalability. The disadvantage of cluster routing protocols is that nodes in clusters are highly dependent on the cluster head node. The reliability and stability of the cluster head node strongly affect the performance of the entire network, and the topology management required to support the movement of ordinary nodes to different clusters adds to the system overhead. Although MANETs are used mainly to supplement backbone communication networks, their scale is not very large, and the role of cluster routing protocols is not obvious. Most proposed or applied MANET routing protocols are based on flat routing. However, with the gradual popularization of MANET applications, especially the expansion of large-scale networks, cluster ad hoc routing protocols are expected to become a focus of research. Typical cluster routing protocols include core-extraction distributed ad hoc routing [8], the zone routing protocol, and cluster-head gateway switch routing [9].

7.1.3 Geolocation-Based Routing Protocols

Mobile ad hoc routing protocols using geolocation information have become a hot topic of research, and the use of GPSs can further improve the routing efficiency. Classical routing protocols based on geolocation information include the distance routing effect algorithm for mobility (DREAM) [10], greedy perimeter stateless routing (GPSR) [11], and location-aided routing (LAR) [12].

The DREAM protocol is an a priori routing protocol based on node location information. It provides distributed, loop-free, multipath routing and can be applied in MANETs with high-speed mobile nodes. DREAM uses the distance factor (distance effect) and mobility rate, which are related to the routing update frequency and packet message survival time, to minimize the routing packet overhead. The distance factor is calculated assuming that the further apart two nodes are, the more slowly they appear to move relative to each other. The DREAM protocol requires the support of a GPS module. The node can use information from the GPS module to obtain these two parameters.

In the DREAM protocol, each node maintains a location table, which is used to record the location information of all other nodes in the network, and each node periodically broadcasts a hello message to inform its neighbors of its location. In addition to the neighboring nodes, the closer nodes receive periodic control packets more frequently. The main process of the DREAM protocol is as follows. First, the source node calculates a direction vector from itself to the destination node according to the location information in the table and then selects a set of neighboring nodes in the direction vector. If the set is empty, the data packet is broadcast to every node in the network by flooding. Otherwise, the set is encapsulated into the header of the data packet and transmitted with the data packet, and only the nodes included in the set should process the data packet. After the nodes in the set receive the data packet, they select their neighbor node sets again as described above, update the header of the data packet, and forward the data packet. If the selected set is empty, the node discards the data packet. When the destination node receives the data packet, it returns an acknowledgement (ACK) packet in the same way. If the source node does not receive an ACK of the array packet header from a node in the set of neighbors within a certain time period, it retransmits the packet again by flooding.

The GPSR protocol is also based on location information, but it uses location information only when a source node sends or intermediate nodes forward data packets. It requires little or no node location and routing; thus, it is a state-independent (stateless) routing protocol. It has low routing complexity and is more suitable for MANET environments in which nodes are relatively densely concentrated. In the GPSR protocol, each node periodically broadcasts a hello message (a control group) to give its location information to neighboring nodes. To further reduce the routing overhead, the location information is encapsulated in the data packet sent by the node instead of sent in a separate routing control packet. GPSR assumes that a source node can obtain the location of a destination node using GPS or other methods. It encapsulates this location information in the data packet header, and the intermediate node makes forwarding decisions based on the locations of the destination node and its neighboring nodes.

The main purpose of the LAR protocol is to reduce the routing overhead significantly by using the location information of nodes in the network to narrow down routing queries. The routing overhead is reduced using area and distance policies. Like most routing protocols based on geolocation information, LAR also requires GPS to obtain the location information of nodes. The source node encapsulates its

Fig. 7.4 Schematic diagram of regional policy in LAR

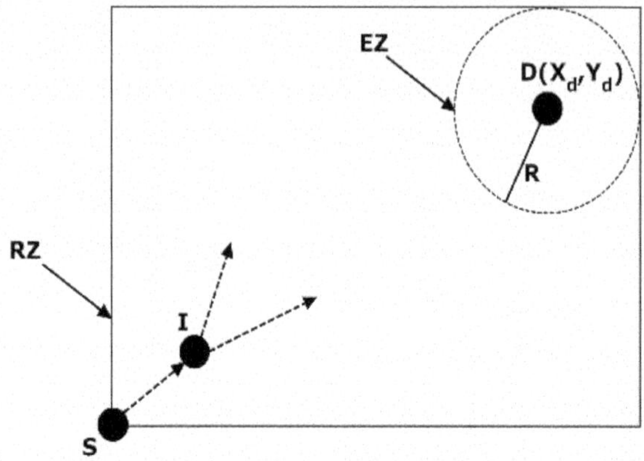

current location and time in the RREQ group when it initiates a RREQ, and the destination node also encapsulates its current location and time in the RREP group. Then the intermediate node can obtain the location information of the source or destination node. To ensure that the routing query range is as accurate as possible, the LAR protocol assumes that each node determines the average speed of other nodes using GPS or other means.

The two policies used by the LAR protocol are different. In the zone policy, the protocol determines the propagation range of the RREQ by calculating the expected zone (EZ) and RREQ zone (RZ) of the destination node; only intermediate nodes in the RZ participate in the route-finding process. As shown in Fig. 7.4, node S wants to send a data packet to node D at time t_1, but S does not have a route to D; thus, it wants to find such a route. Assume that S knows that at t ($t \leq t_1$) the position of D is (X_D, Y_D), and its maximum instantaneous movement speed is v. Then, at t_1, D should be in the circle EZ with (X_D, Y_D) as the center and $R = v(t_1 - t_0)$ as the radius. Let RZ be the smallest rectangular region containing the source node S and EZ. S writes the boundary coordinates of RZ into the RREQ group and broadcasts it; if the node receiving the RREQ is the destination node, it returns a RREP to S. Otherwise the node checks whether it is within the RZ region; if it is, it continues to broadcast the RREQ group.

As shown in Fig. 7.5, the distance policy is used to determine whether the intermediate node forwards RREQ packets according to the distance between itself and the destination node. Suppose that S initiates a routing process to D at time t_1, and S knows the location of D at time t (X_D, Y_D). At t_1 S calculates the distance $DIST_S$ from itself to D and encapsulates $DIST_S$ and (X_D, Y_D) in the RREQ packet. When an intermediate node receives a RREQ packet from S, it first calculates the distance $DIST_i$ from itself to the destination node D. If $DIST_i \leq DIST_s + \delta$ (where δ is a given constant), the node processes the RREQ packet, replaces $DIST_s$ with $DIST_i$, and forwards it. Otherwise, it discards the RREQ. Subsequent nodes use the same method to decide whether to forward RREQ packets. If node i is the destination node, a routing response packet is generated and returned to S.

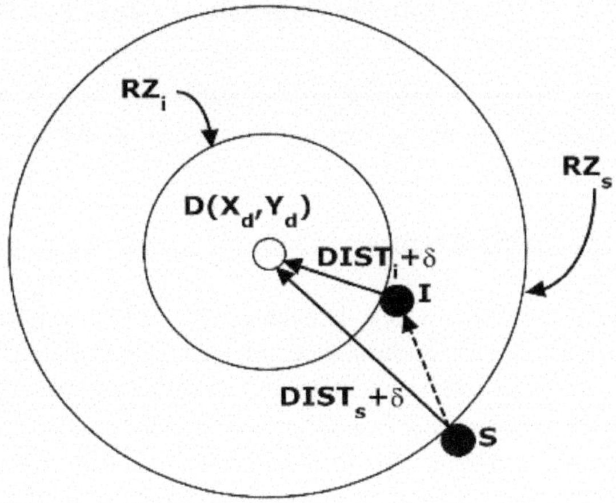

Fig. 7.5 Schematic diagram of distance policy in LAR

LAR limits the RREQs to the RZ area, which can effectively solve the broadcast storm problem and decrease the time required for route lookup. Simply speaking, a broadcast storm occurs when broadcast data fill the network and cannot be processed; the data use a large amount of network bandwidth, resulting in route failure or even complete paralysis. A data frame or packet is transmitted to each node on the local network segment (defined by the broadcast domain), which is broadcast. Because of design and connection problems with the network topology, or for other reasons, many broadcasts are copied in the network segment, spreading data frames and resulting in network performance degradation or even network paralysis [13]. However, LAR requires GPS support. If the destination node moves very fast or faster than the speed recorded in the source node's routing table, the search is likely to fail.

7.2 Performance Comparison and Analysis of Classical Routing Protocols

Reference [14] analyzed the physical layer and link layer models of UV communication systems in detail. From the models, we can conclude that in the non-line-of-sight A [NLOS(A)] mode, ultraviolet (UV) communication is similar to radio communication, which is essentially omnidirectional. In the NLOS(C) mode, UV communication is similar to the directional antenna mode in radio communication. According to the analysis in the previous chapter, if a UV communication system forms an ad hoc network, it must have the support of its own routing protocol to communicate. Here, we analyze and compare the routing protocols of classic MANETs and propose a routing protocol suitable for UV ad hoc networks.

7.2.1 Algorithm Mechanism of Classical Protocols

Many mature protocols are available for use in MANETs. We compare the performance of two classical protocols, DSR and TORA.

7.2.1.1 DSR Protocol

The DSR protocol is a demand-based routing algorithm [15]. Its most important feature is that it uses the source route, which is initiated by the source node to establish a route to the destination node, and sends the source node of the packet until it reaches the destination. That is, a sequence table lists the node addresses through which the path passes. These paths are stored in the route buffer, and the packet header of the data packet carries the source route. This source routing method avoids the need for data packets to constantly update the route through intermediate nodes; thus, it allows nodes to store the latest routing information in their routing buffer for future use when forwarding or inadvertently receiving data packets. Because all operations of the protocol are based on demand, data packets can respond dynamically to changes in the current path as needed. The DSR protocol uses two important mechanisms: route search and route maintenance.

The route search mechanism of the DSR protocol is as follows. When a source node in the ad hoc network wants to send a data packet to a destination node, the route search program broadcasts a RREQ packet to the network (the packet records the passing node address in a sequence table), and each node that receives the RREQ packet broadcasts it again, but discards duplicate route search packets it receives. The RREQ grouping format is shown in Table 7.1.

When the destination node or an intermediate node with a path to the destination in the routing cache receives the RREQ, it sends an acknowledgement (ACK) packet in response to send the route in the RREQ packet back to the source node. Because wireless links are asymmetric, ACK packets cannot simply be returned to the source node according to the path by which the RREQ arrived. If there is a route back to the source node in the route buffer of the node, the RREQ can return to the source node through this path. Otherwise, the route search program is started. To ensure that the two nodes do not search for each other and cause a route search cycle, the route search message must be accompanied by the route response packet (ACK) to be sent to the source node. After receiving the ACK packet, the source node adds this path to its routing buffer. The ACK format is shown in Table 7.2.

Table 7.1 RREQ data grouping format

Grouping type	Subgroup ID	Additional control information	Source address	Destination address	Information on the list of passed nodes

Table 7.2 ACK data grouping format

Grouping type	Subgroup ID	Additional control information	Source address	Destination address	Path node list

The route maintenance mechanism is as follows. A route is maintained only when it is in use. When a node on the path finds that a data packet cannot be sent to the next-hop node, it finds the route from its own route cache and sends a route error packet to the source node to delete the path from its route cache. If there are still data packets to be sent, but no valid routing information for the destination node is found, the source node restarts the routing search program to obtain a new path. To prevent large-scale channel contention conflicts caused by sending the route response packet (ACK) while a node searches for corresponding routes, it is necessary to delay before sending the ACK:

$$Delay = H \times (h - 1 + r) \tag{7.3}$$

where h is the distance from the node to the destination node, that is, the hop distance of the ACK; r is a random floating-point number; H is the smaller delay constant. To reduce the control overhead, a time to live (TTL) is set for the RREQ packet to limit the search range. The node first searches within the range of a single hop and then increases the TTL by a factor of 2 after failure. The following example illustrates the route search and maintenance mechanisms of the DSR protocol.

When the source node S sends data to the destination node D, it first checks for an unexpired route to the destination node in the route buffer. If it finds one, it uses the available route. Otherwise, it starts the route discovery process. It uses flooding to send a RREQ packet to all nodes within communication range; the packet includes the addresses of the source and destination nodes and the unique flag request number. The intermediate node that receives the RREQ message forwards the packet and attaches its own node ID. When the packet arrives at the destination node D or any intermediate node that has a route to D, the node ID passed from S to D or the intermediate node is recorded in the packet, and the path from S to D is established. Figure 7.6 shows a flow chart of the DSR protocol. The specific methods are as follows.

(1) If the source node and request number of the requested packet are on the list, the node has processed the same packet recently, and it discards the packet without further processing.
(2) If the address of the node appears in the routing record of the request packet, the node discards the packet without processing it.
(3) If the destination address of the request packet is consistent with the node address (that is, the node is D), the routing information in the routing record of the RREQ packet is the route from the source node S to the destination node D, and the

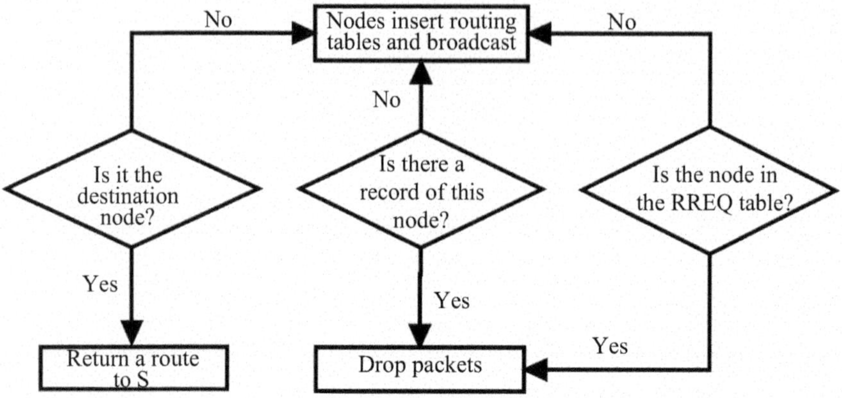

Fig. 7.6 Flow chart of DSR protocol

node generates a routing response packet to the source node on the basis of the packet information.

(4) If the above requirements are not met, the node address is added to the routing record table of the requested packet, and the packet is broadcast to surrounding nodes.

However, this process establishes only the path from S to D. Communication paths in the network should be bidirectional. If two nodes communicate with each other, the wireless coverage of one node, A, may be larger than that of the other node, B. A can send information to B, but B cannot send to A. Here, the path from D to S has not been completely determined. Next, node D or an intermediate node sends a RREP to the source node S, which contains the route information from S to D and reverses the route from S to D to supply the reply message. This process is similar to a reverse route discovery with D as the source node and S as the destination node. After these processes are completed, information exchange can begin. Each node maintains a request list, which records the RREQ packets processed by the node recently. This list can prevent mobile nodes in an ad hoc network from repeatedly processing the same route discovery packets.

7.2.1.2 TORA Protocol

The TORA protocol is an adaptive distributed routing algorithm based on the link reversal method; it is used mainly for high-speed dynamic multihop wireless networks. As a source-initiated on-demand routing protocol, it finds multiple routes from the source node to the destination node. The main feature of TORA is that when the network topology changes, control messages are propagated only within the local area where the topology has changed. Therefore, nodes need to maintain routing information only for neighboring nodes. The disadvantages of TORA are high overhead for route establishment and maintenance and route oscillation when multiple nodes simultaneously select and delete routes. Because the algorithm is based on synchronous clocks, time differences can result in route failure, and the algorithm has the potential to oscillate, which affects the route establishment time.

The TORA protocol consists of route generation, maintenance, and deletion. The height of the destination node (i.e., the propagation sequence number) is initially set to 0. Then the source broadcasts a packet containing the ID and QRY of the destination node, and a node with nonzero height responds with a user datagram protocol (UDP) packet. The height of the node that receives the UDP packet is larger by 1 than that of the node that generated the UDP packet, and the node with the greater height is specified as the upstream node. In this way it is possible to create a DAG from the source node to the destination node. When a node moves, the route must be rebuilt. TORA runs on top of the Internet MANET encapsulation protocol (IMEP). The TORA protocol, like other on-demand routing protocols, first sends a RREQ packet on the network; however, it uses the DAG algorithm in the reply part of the route so that each node in the network maintains a potential energy with respect to a given target node. The potential energy can be obtained by a reverse broadcast from the target node. Nodes that are farther from the target node have higher potential energy, and the target node has lower potential energy. During data propagation, data packets are forwarded from nodes with high potential energy to nodes with low potential energy and eventually flow to the target node. When a local link changes, only the local potential energy must be adjusted, and such changes generally do not affect the network as a whole. The control message is sent only to a small portion of nodes closest to the change in topology so that nodes retain only the routing information of neighboring nodes. Routing is not always optimal in this algorithm, and suboptimal routes are often used to reduce the overhead of route discovery.

7.2.2 *DSR and TORA Performance Comparison*

To better illustrate the differences between these two protocols, we simulated them using OPNET simulation software.

7.2.2.1 Mobile Workstation Trajectory Configuration

OPNET supports the simulation of mobile nodes and controls the movement of mobile nodes by configuring the movement trajectory. A trajectory is defined in the properties of the mobile workstation so that the workstation moves according to this trajectory during the simulation. The steps for defining the trajectory are as follows [16].

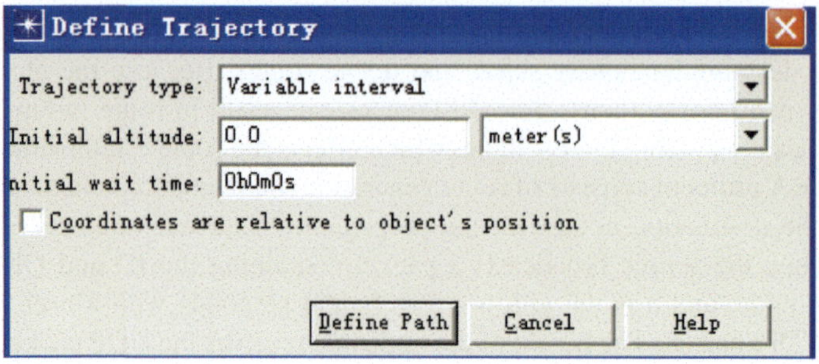

Fig. 7.7 Define trajectory menu

(1) In the Topology menu, Define Trajectory is selected, as shown in Fig. 7.7.

The type of trajectory, fixed interval or variable interval, can be selected as needed; the variable type is generally selected. The trajectory is defined by the user. The initial wait time is the time the mobile workstation waits in the initial position at the beginning of the simulation. After these parameters are set, the Define path button is clicked; the mouse tip produces a short line for entering the definition of the trajectory segments.

(2) The icon for the mobile workstation whose trajectory must be defined is left-clicked using the mouse, and the mouse is moved to the place where the mobile workstation will stop in the next step. The screen shown in Fig. 7.8 appears when the workstation is left-clicked.

The "Traverse segment" field is used to define how long the workstation takes to move during this phase. The "End at an altitude of" field specifies the height of the workstation; it can be used to change not only the position but also the height of the mobile workstation as it moves. The "Wait at this point for" field is used to indicate how long the mobile workstation should stay at this point. To define an irregular trajectory, the line is generally divided into segments with different lengths as needed to approximate the actual terrain as closely as possible. The Continue button is clicked to continue, and these steps are repeated until the desired trajectory

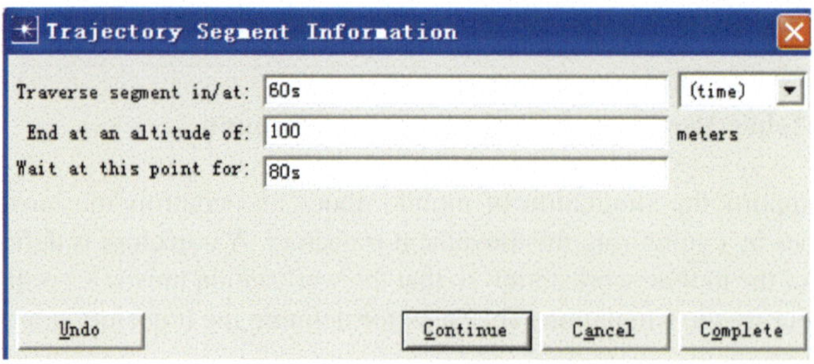

Fig. 7.8 Trajectory segment information menu

has been defined The trajectory is then saved; as many different trajectories as needed can be saved for later use. We defined the following four types of trajectories.

(1) The top two workstations move first to the upper left, then to the right, and finally to the lower left. This is done as follows. Workstation mw_1 is set to wait at the starting point (3750, 2250) for 100 s; move to the upper left to (3000,500), which takes 200 s, and wait there for 100 s; move to the right to (4500,500), which takes 200 s, and wait there for 100 s; move to the lower left to (4000,2000), which takes 200 s, and wait at this point.
(2) The two workstations on the left move first to the lower left for a specified distance, then up for a specified distance, and then to the lower right to complete the trajectory, similar to step 1.
(3) The bottom two workstations move first to the lower left for a specified distance, then to the right for a specified distance, and then to the upper left to complete the trajectory, similar to step 1.
(4) The two workstations on the right move first to the lower right for a specified distance, then up for a specified distance, and then to the lower left.

7.2.2.2 Simulation Analysis of DSR Protocol

The parameters of the simulation are as follows.

Network topology: The network topology consists of eight mobile nodes and one receiver in a 6000 $\times$ 6000 m^2 area. All nodes in the network are configured to run the DSR protocol and perform multi-FTP services. All nodes are mobile and must be configured for mobile tracks.

Emulated service: FTP
Routing protocol: DSR
Devices used: receiver group configuration (Rx Group Config); workstations (Wlan_wkstn)

The model configuration is as follows.

Rx Group Config: The Rx Group Config node is used to accelerate the emulation and eliminate interference caused by receivers within 1000 m of the network. Because the network is mobile, the receiver group needs to be refreshed; the refresh interval is set to 10 s. The receiver_group_config/Receiver Selection Parameters/ Selection Parameters/Distance Threshold(meters) is configured to 1500, and the receiver_group_config/ Duration/Refresh Interval(Seconds) is set to 10.

Workstation Wlan_wkstn: The Ad Hoc Routing Parameter is set to DSR, and the MANET Traffic Generation Parameters/row0/Start Time(Seconds) is set to 100, indicating that the workstation will begin operation 100 s after the simulation begins. The reason is that many parameters must be initialized when the simulation begins, and the 100 s waiting period is used for this purpose. The MANET Traffic Generation Parameters/row0/Packet Inter-Arrival Time(Seconds) is configured as exponential(1), indicating that the packet arrival interval is exponential(1). The MANET Traffic Generation Parameters/row0/Packet Size(bits) is configured as exponential(1024), indicating that the packet size is exponential(1024). The Wireless LAN/Wireless LAN

Parameters/Roaming Capability is set to Enabled, indicating that this workstation is mobile and has mobile capability. In addition, its tracks are configured.

Statistics collected: DSR/Route Discovery Time
DSR/Number of Hops per Route
DSR/Total Traffic Sent
Wireless LAN/Delay(seconds)
Wireless LAN/Through(bits/sec)

The network topology is shown in Fig. 7.9. The simulation time is set to 1 h; the actual simulation time is 12 s on a computer with a 2.21 GHz CPU and 1 GB of memory using the optimized simulation method.

The simulation results and analysis are as follows.

(1) The route discovery time is shown in Fig. 7.10. The average route discovery time is slightly longer, 5 to 8 s, for 10 to 20 min, and decreases to 1 to 2 s after 30 min. The reason for this result may be that the number of workstations in the network is too small to maintain communication among all workstations.
(2) The number of hops per route is shown in Fig. 7.11. It is between 1.25 and 2 in the first 15 min (i.e., when the workstations are moving). After 30 min, that is, after the network topology becomes stable, the number of hops per route is roughly 1.

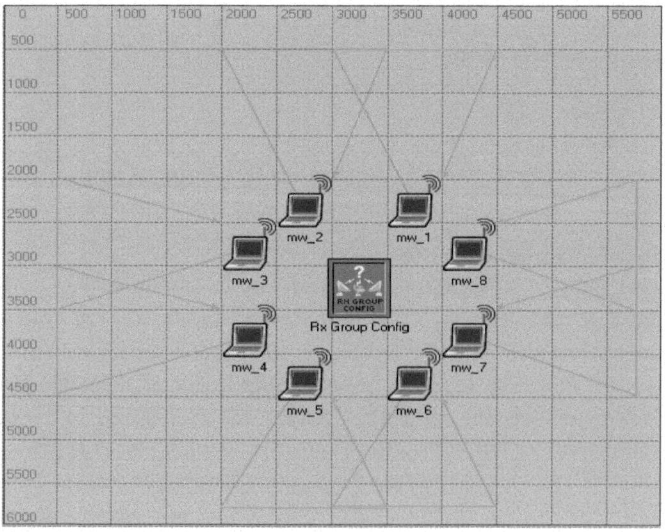

Fig. 7.9 Schematic diagram of mobile DSR network topology

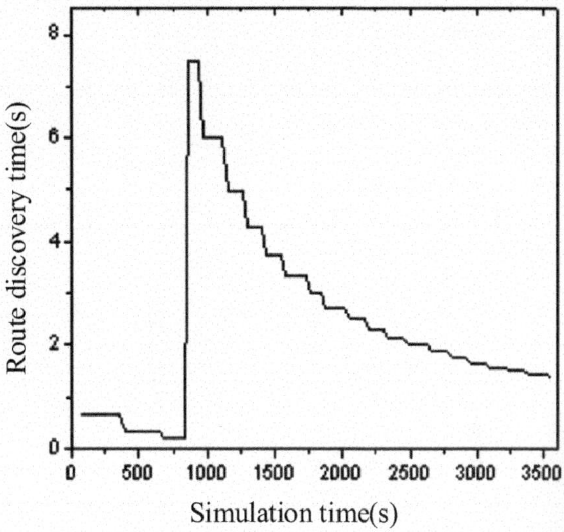

Fig. 7.10 Route discovery time (mobile DSR)

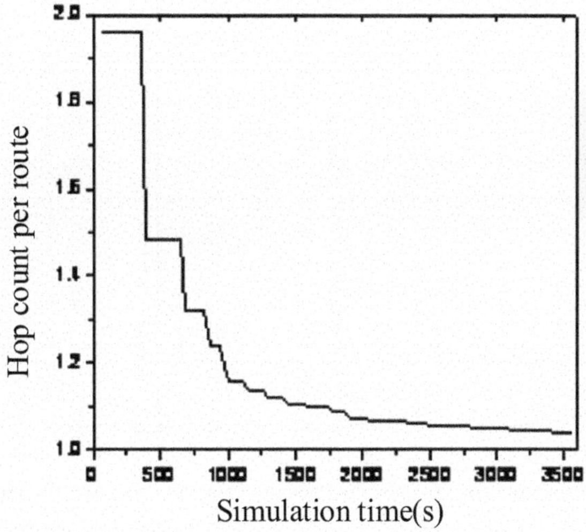

Fig. 7.11 Hops per route (mobile DSR)

(3) The total service delivery is shown in Fig. 7.12. When the network topology becomes stable, the total service delivery is roughly 6000 bps.
(4) The local area network (LAN) time delay is shown in Fig. 7.13. According to the LAN latency, the network latency is roughly 0.001 s, but the network latency is larger (0.005 s) between 13 min and 36 s and 15 min, because the movement of workstations increases the network latency.

Fig. 7.12 Total traffic volume (mobile DSR)

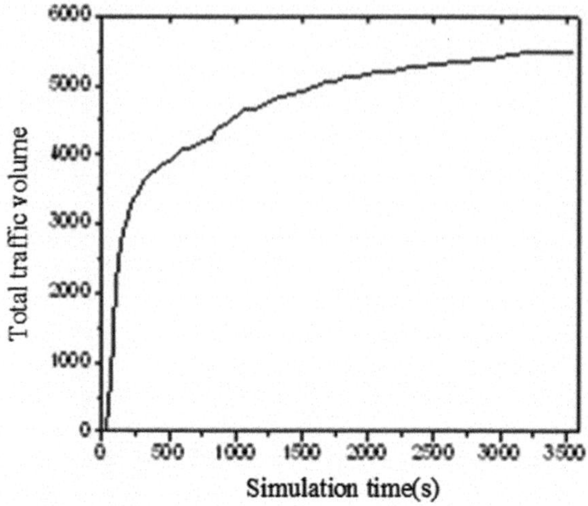

Fig. 7.13 LAN delay (mobile DSR)

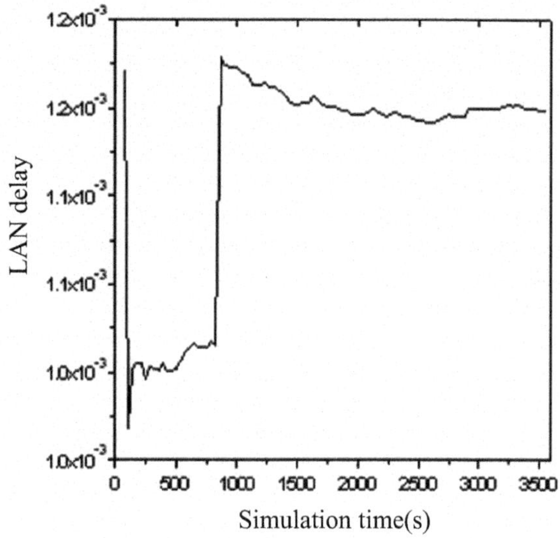

(5) The LAN throughput is shown in Fig. 7.14. The LAN throughput is less than 4000 bps while the workstations are moving, that is, while the network topology changes. When the network topology is fixed, the network throughput exceeds 5000 bps. This result indicates that the throughput is smaller when the network topology is changing and higher when the network topology is fixed.

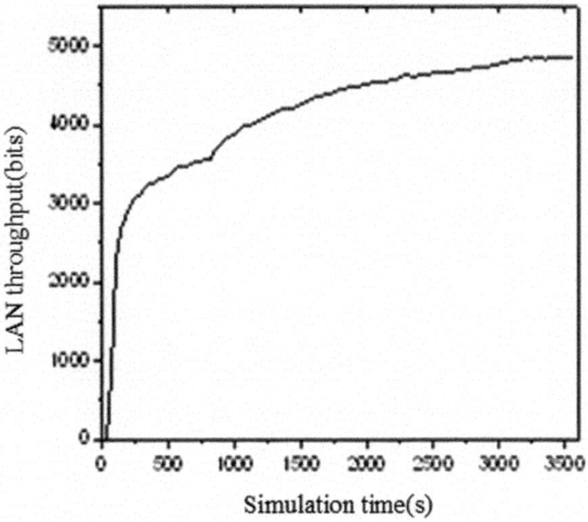

Fig. 7.14 LAN throughput (mobile DSR)

7.2.2.3 Network Simulation Using TORA Protocol

The network topology is shown in Fig. 7.15, which is identical in appearance to Fig. 7.11, but all the workstations in the network use the TORA protocol. The coordinates in the figure show the size of the network in meters.

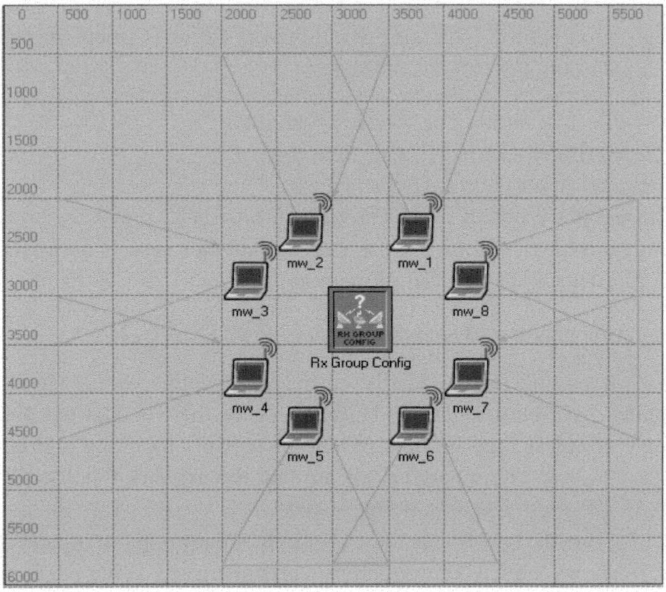

Fig. 7.15 Schematic diagram of mobile TORA network topology

Network topology: The network topology consists of eight mobile nodes and one receiver in a 6000 × 6000 m^2 area. All nodes in the network are configured to run the TORA protocol and perform multi-FTP services. All nodes are mobile and must be configured with mobile tracks.

Emulated service: FTP
Routing protocol: TORA
Devices used: receiver group configuration (Rx Group Config); workstations (Wlan_wkstn)

The model configuration is as follows.

Rx Group Config: The Rx Group Config node is used to accelerate the emulation and to eliminate interference caused by receivers within 1000 m of the network. Because the network is mobile, the receiver group needs to be refreshed; the refresh interval is set to 10 s. The receiver_group_config/Receiver Selection Parameters/Selection Parameters/Distance Threshold(meters) is set to 1500, and the receiver_group_config/ Duration/Refresh Interval(Seconds) is set to 10.

Workstation Wlan_wkstn: The Ad Hoc Routing Parameter is set to TORA, and MANET Traffic Generation Parameters/row0/Start Time(Seconds) is set to 100, indicating that the workstation will begin operation 100 s after the simulation begins. This time is used to initialize parameters. MANET Traffic Generation Parameters/row0/Packet Inter-Arrival Time(Seconds) is configured as exponential (1), indicating that the packet arrival interval is exponential(1). MANET Traffic Generation Parameters/row0/Packet Size(bits) is configured as exponential(1024), indicating that the packet size is exponential(1024). The Wireless LAN/Wireless LAN Parameters/Roaming Capability is set to Enabled, indicating that this workstation is mobile and has mobile capability. In addition, its track is configured.

Statistics collected: number of services generated
TORA_IMEP/IMEP Control Traffic Sent (bits/sec)
LAN Time Delay Wireless LAN/Delay(sec)
LAN Throughput Wireless LAN/Throughput(bits/sec)
The simulation time is set to 1 h; the actual simulation time is 5 s on a computer with a 2.21 GHz CPU and 1 GB of memory.

The simulation results and analysis are as follows.

(1) Fig. 7.16 shows the TORA IMEP control service delivery rate, which is high at the beginning of the simulation. However, when the workstations are moving, the rate is low, less than 100 bps. When the network topology is fixed, the rate exceeds 100 bps. These results indicate that the network topology affects the TORA IMEP control service delivery rate.
(2) Fig. 7.17 shows the LAN delay rate. The LAN latency is approximately 0.0015 s.
(3) Fig. 7.18 shows the LAN throughput, which increases at the beginning of the simulation. However, it is relatively low, less than 3000 bps, when the workstations are moving and increases to nearly 4000 bps when the network topology

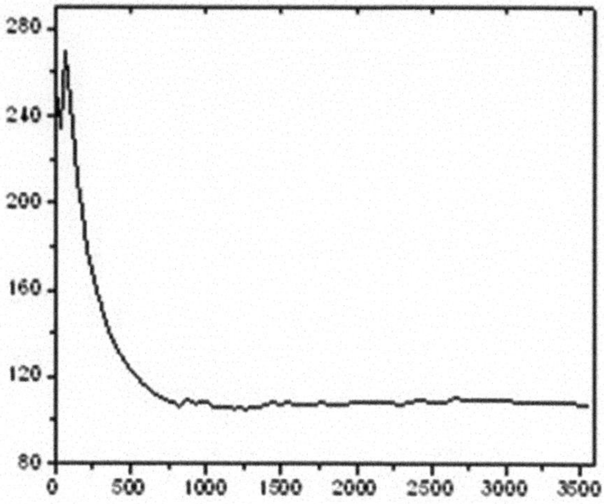

Fig. 7.16 TORA IMEP control service sending rate

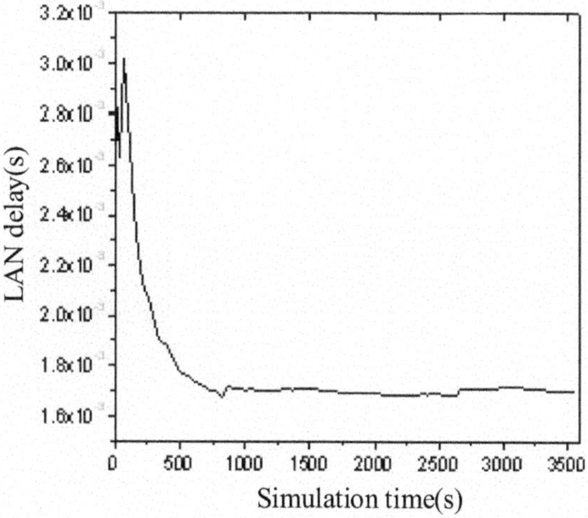

Fig. 7.17 LAN delay (mobile TORA)

is fixed, indicating that the throughput is low when the network topology is variable and high when the network topology is fixed.

Next, the results of the DSR and TORA protocol simulations for MANETs are compared.

Figure 7.19 shows the LAN delay. The LAN latency using the TORA protocol is larger than that using the DSR protocol. The reason is that the network changes rapidly and frequently as time increases. TORA needs to send many IMEP control messages; therefore, the network becomes congested, and the latency is higher.

Figure 7.20 shows the LAN throughput. The throughput of the LAN using the DSR protocol is slightly higher than that of the LAN using the TORA protocol. The reason is that the network topology changes, and to determine the routes between

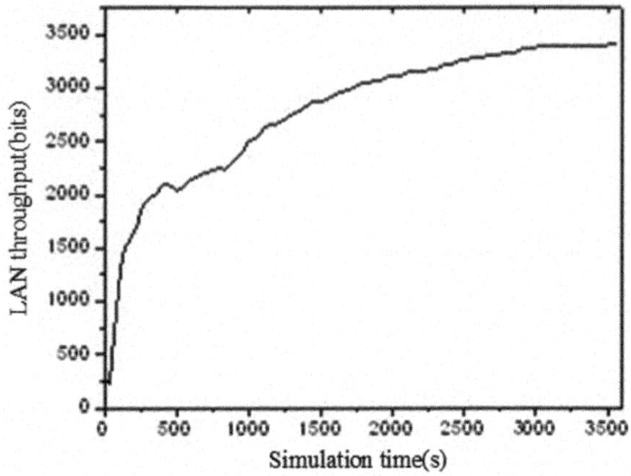

Fig. 7.18 LAN throughput (mobile TORA)

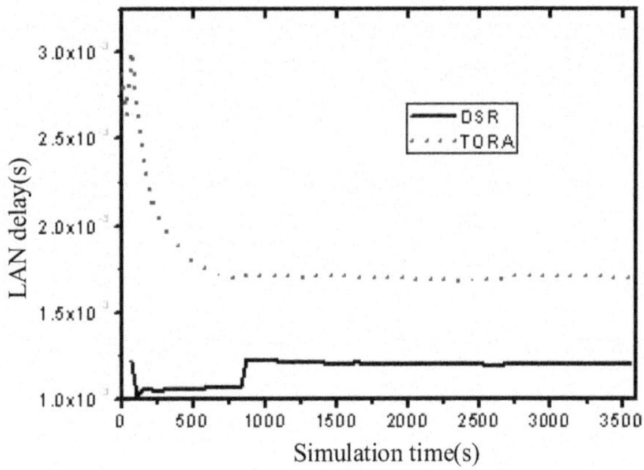

Fig. 7.19 Delay of mobile ad hoc LAN

neighboring nodes, TORA must send many control packets and therefore requires more channel resources.

This analysis indicates that the DSR protocol is more suitable than the TORA protocol when the network topology changes and the emulated service is multi-FTP.

Many classical protocols are used in mobile self-organizing networks; some are a priori, and others are reactive. Because of the variability of the MANET topology, reactive routing protocols have relatively good performance in these networks. We simulated and compared two classical reactive protocols, DSR and TORA, for mobile self-organizing networks using the OPNET Modeler software. The experimental results showed that the parameters of the TORA protocol, such as throughput, routing overhead, and delay, are inferior to those of the DSR protocol when the network topology changes more rapidly, because the TORA protocol must use many IMEP control packets to inform nodes of changes in the network topology. When the network topology changes slowly, TORA performs slightly better than DSR because of the relatively small amount of information in the routing control packets.

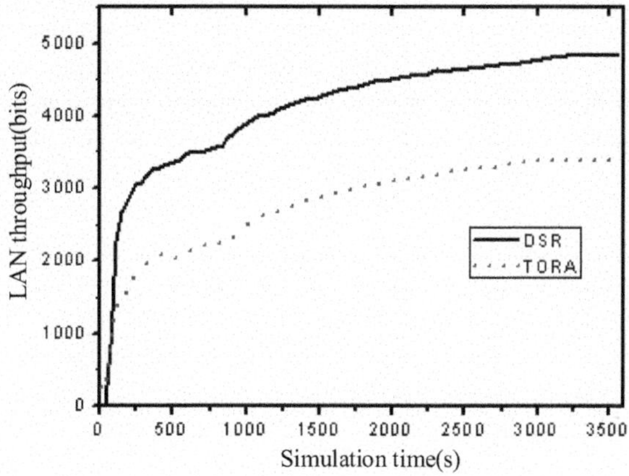

Fig. 7.20 Throughput of mobile ad hoc LAN

7.3 Performance Analysis of UV Self-organized Network Routing

The MANET is an emerging type of wireless network with many unique features. For example, the network topology is variable, and there is no base station or central node; all nodes in the network act as both communication terminals and routers simultaneously. These features make MANETs highly flexible and efficient. Conventional MANETs are used mainly for radio-frequency communication, but the bandwidth in this band has been severely challenged by network services including voice and video. UV communication is a highly desirable complement to conventional MANETs because of its small transmitters and receivers, low cost, MLOS communication capability, high security, very low background noise, and potential high rate communication. The performance, bit error rate (BER), and signal-to-noise ratio (SNR) of the UV MANET routing protocol are analyzed below.

7.3.1 Routing Protocols for UV Communication

According to the scattering model of UV communication [14], we designed a routing protocol suitable for UV self-organized communication. In a simulation, we selected the classical DSR protocol as the routing protocol for the UV self-organized network.

The DSR protocol is a typical reactive routing protocol. As shown in Fig. 7.21, it has two main mechanisms: route discovery and route maintenance. The route discovery mechanism is as follows. If a source node S wants to communicate with a destination node D, first S must broadcast a RREQ packet to its neighbor nodes. This RREQ packet contains useful information such as the addresses of nodes S and D, the sequence number (SN) of this RREQ (which is determined by the source node and used to label this RREQ packet), and a route record area. If a node receives the RREQ packet and is not the destination node, does not have a route to the destination node, or is receiving the RREQ packet for the first time, it broadcasts the RREQ packet. If the node has a route to the destination node in its own route store, it immediately sends a RREP back to the source node S. This RREP contains a complete route from S to D. It also contains additional information such as the destination node (source node of the RREQ), the next-hop node of the RREP, and the same SN as the RREQ. Of course, it is likely that none of the nodes receiving the RREQ have a complete route from themselves to D until the RREQ reaches the destination node D. Thus, when node D receives the RREQ, it generates a RREP itself and passes it back to the source node S with the contents of the above RREP. Next, the destination node discards all RREQ packets with the same SN as this RREP, because they were sent later from other nodes. The route maintenance mechanism is used to maintain existing routes and to ensure that the routes stored in nodes are correct and available. If the data link layer of a node detects a transmission error (that is, a data packet sent from this node that does not reach the next-hop node, where the equal-stop or default time mechanism is used), the node generates a routing error group and sends it to the source node that sent the data packet. When the source node receives a routing error group, it first deletes the routing table entries associated with the link and all subsequent entries. The source node then initiates a new route discovery process to discover a new available route. The DSR protocol is applicable not only to typical MANET networks, but also to our UV self-organizing networks. The communication distance of UV self-organizing networks is very short, so multihop routing and route cache tables are essential.

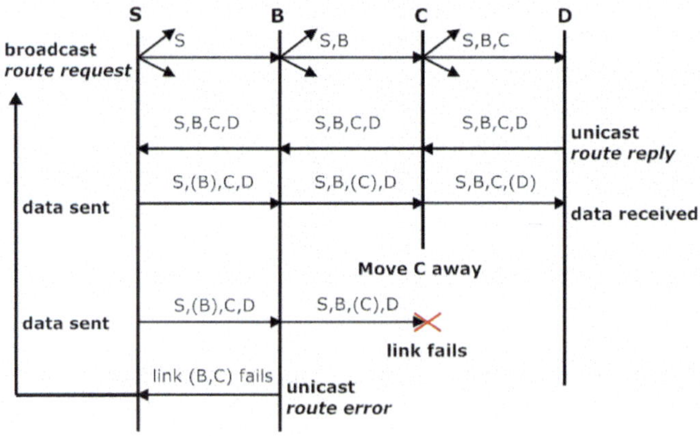

Fig. 7.21 Example of operation flow of DSR protocol

7.3.2 Simulation Model and Parameters

We chose OPNET Modeler [17] as our simulation environment and used OPNET's built-in user interface for geographic settings to set up the geographic environment in our simulation. The most important parameters are as follows.

Total transmitted and received data volumes, end-to-end delay, BER, and SNR.

We assume that this network provides three main services: data packets (FTP), voice service, and video conferencing. The acceptable BER and latency depend on the service type. The maximum BER is 10^{-9} for FTP service and 10^{-6} for voice services, and a BER on the order of 10^{-3} is acceptable for video conferencing. We used the statistical average over several links to calculate the SNR. The parameters used in the simulation are as follows.

- Scene area: $1000 \times 1000\ m^2$
- Maximum transmission distance: 200 m
- Data rate: 2 Mbps
- Mobility: Random movement
- Terrain: Indianapolis
- Simulation time: 1 h
- Modulation method: binary phase shift keying
- Data packet size: 256 bits/packet.

7.3.3 Simulation Results

We obtained the following results using OPNET simulation software. As shown in Fig. 7.22, the curves for total packets sent and total packets received for FTP are very similar, indicating that most data packets can reach the receiving end. In addition, the data transmission rate is approximately 4 KB/s. However, only 15% of the total data volume is transmitted by FTP service, which indicates that most of the data in the network represent routed information, and the network utilization is still relatively low.

Figure 7.23 shows the simulation results for voice services. The received data packets decrease rapidly when the simulation time exceeds 1200 s. This result indicates that many data packets cannot reach the destination node after 1200 s; in addition, the routing overhead increases sharply, indicating that most of the network overhead is routing overhead, which affects the network bandwidth and eventually data transmission. The data transmission rate is approximately 4–6 KB/s. Figure 7.24 shows that for video conferencing, the performance is similar to that of FTP (Fig. 7.23); however, data loss occurs after 2500 s, indicating that routing overhead takes up most of the bandwidth after 2500 s.

Fig. 7.22 Transmission and reception of FTP service

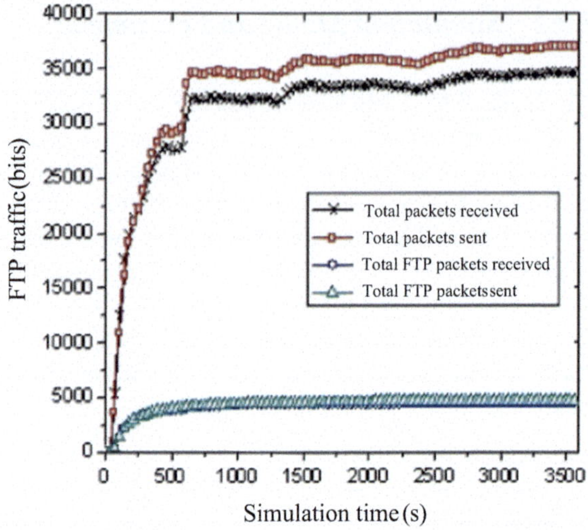

Fig. 7.23 Transmission and reception of voice services

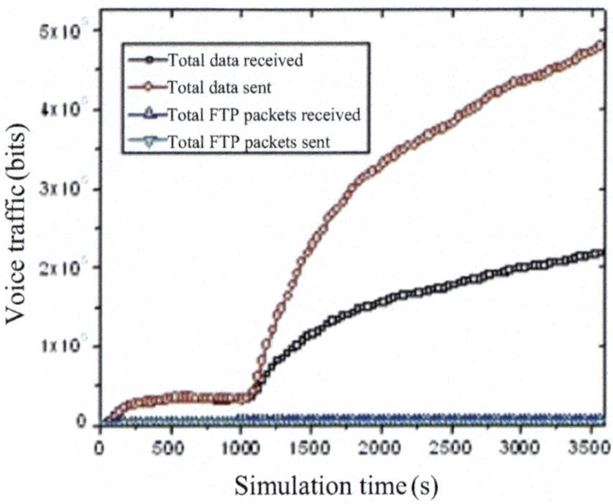

Figure 7.25 shows the BER curves of each service. The BER exceeds the thresholds for FTP and voice transmission; by contrast, this network meets the requirements for video conferencing.

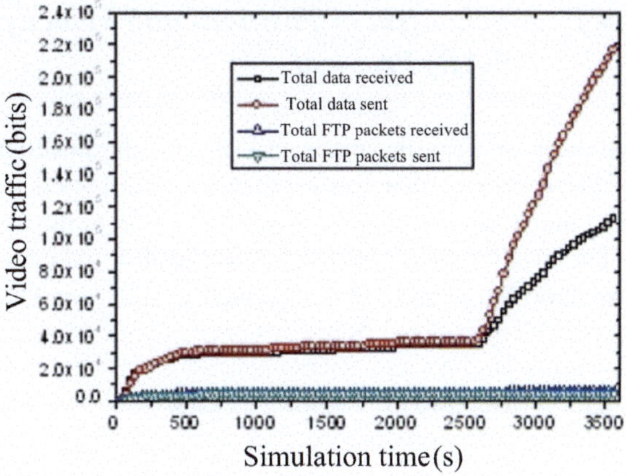

Fig. 7.24 Transmission and reception of video services

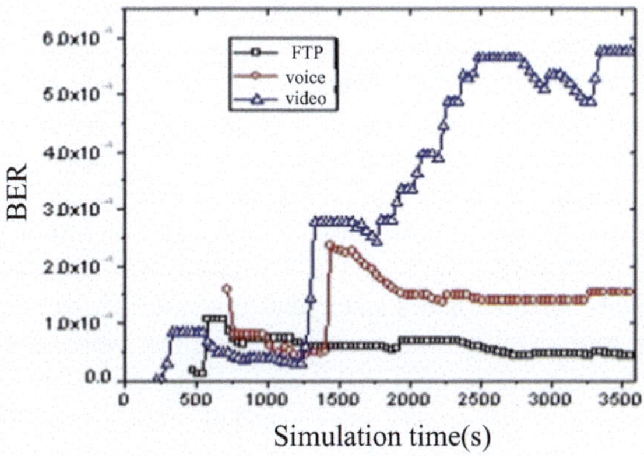

Fig. 7.25 BER curves of each service

7.4 Routing Protocol Based on Node Location and Speed Information

Many routing protocols for MANETs have been proposed, but these classical routing protocols do not take into account the fact that routing failure can be very frequent if network nodes are moving at very high speed. Route failure occurs when a node moves at high speed because the node easily leaves the broadcast range of another node, and if this node is a relay node on another node's route, this route fails, resulting in communication failure.

7.4.1 Mechanism of Position- and Velocity-Aided Routing (PVAR) Protocol

We propose a routing protocol [18, 19], the position- and velocity-aided routing (PVAR) protocol, that uses geolocation information and information on the speed of node movement to maintain the effectiveness of the route for as long as possible and to effectively reduce the routing overhead of the network. The simulation script of this algorithm is given in Appendices E-1, E-2, and E-3.

7.4.1.1 Selection of Neighbor Nodes

In the PVAR protocol, we assume that nodes or communication terminals move at a very high speed (at least 10 m/s) in the network. We also assume that nodes move in a straight line toward their destination, and the direction does not change until they reach their destination.

In MANETs, the neighbor nodes of a source node (central node) are the nodes within the one-hop broadcast range of the source node, and the source node can communicate with other nodes only by broadcasting; that is, the source node can communicate only with its neighbor nodes. To communicate with non-neighboring nodes, the source node must communicate hop by hop through its neighboring nodes to reach the destination node. Therefore, the source node must maintain information on its neighbor nodes and update this information periodically. However, the nodes in a MANET can be moving at any time, whereas their broadcast range is fixed. Therefore, if network nodes are moving at a relatively high speed, their neighbor nodes can change very quickly, and the network topology can become quite unstable as a result. These rapid changes can result in communication failure because the neighbor node of a sending node is likely to be an intermediate node used to reach another node, and if the neighbor node is no longer in the broadcast range of the source node, the route from the source node to the destination node fails. Because the route discovery mechanism in the PVAR protocol we designed is based on the broadcast mechanism, the selection of neighbor nodes is highly important for this protocol. If a neighbor node moves roughly parallel to the source node at a comparable speed, it will be a neighbor node of the source node for a long time, and we call it a good candidate neighbor. If the neighbor node of the source node moves in the opposite direction, the two nodes move away from each other and eventually leave each other's broadcast range. If the neighbor node is a relay node from the source node to another node, all nodes on the route are affected. In addition, the failure of a route increases the routing overhead in the network, as nodes on this route all initiate the route discovery process again to repair the failed link. As the number of such neighbor nodes increases, the route becomes more likely to fail, and the routing overhead is higher. We refer to this class of neighbor nodes as bad candidate neighbors. Of course, in the actual simulation, certain criteria must be set to distinguish between good and bad candidate neighbors. Because the nodes in our simulation are moving

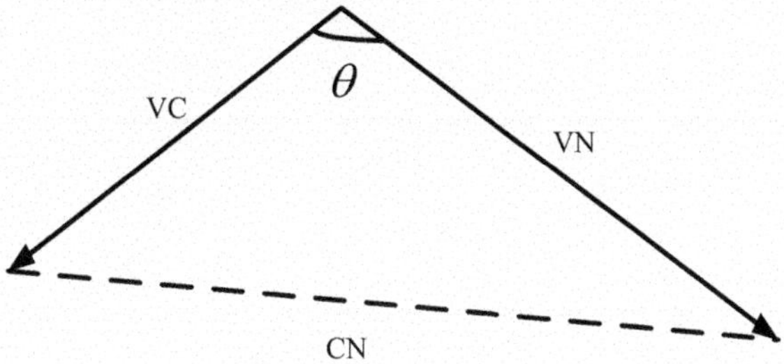

$$\theta = \arccos \frac{\left|\overrightarrow{VC}\right|^2 + \left|\overrightarrow{VN}\right|^2 - \left|\overrightarrow{CN}\right|^2}{2 \bullet \left|\overrightarrow{VC}\right| \bullet \left|\overrightarrow{VN}\right|}$$

Fig. 7.26 Calculation of angle between VC and VN using cosine definition

at relatively high speeds, we use the node position and direction vector of node motion to determine which nodes should eventually become relay nodes for routing. Suppose the unit direction vector of the central node in the direction of motion is VC, and the unit direction vector of motion of one of its neighboring nodes is VN. If we know the precise values of VC and VN, we can calculate the angle between VC and VN using the cosine theorem. If we use a threshold value to calculate the unit vectors in the direction of motion of the neighbor nodes and the unit vector in the direction of motion of the central node individually, we can determine whether the neighbor node is a good or bad candidate. Figure 7.26 illustrates the calculation of the angle between VC and VN using the cosine definition.

The unit direction vector of velocity alone is not sufficient to determine whether a neighbor node can be a relay node from the central node to its destination node. We must also introduce the concept of node location into the PVAR protocol. If the distance between a neighbor node and the central node exceeds a predefined threshold, the neighbor node is judged to be a bad candidate neighbor; otherwise, it is a good candidate neighbor. A neighbor node can be considered a real neighbor by the central node, and thus a relay node on the route, only if it is found to be a good candidate under both criteria. In our simulations, the threshold value of the angle between VC and VN is 135°. The threshold value of the distance between the central node and neighboring nodes is 220 m (the default communication distance is 250 m). Figure 7.27 shows a schematic diagram of decision-making about neighbor nodes.

7.4.1.2 Route Discovery Process

In developing the PVAR protocol, we were concerned primarily with establishing an efficient route discovery process. Our goal was to obtain a more reliable route discovery process with less routing overhead.

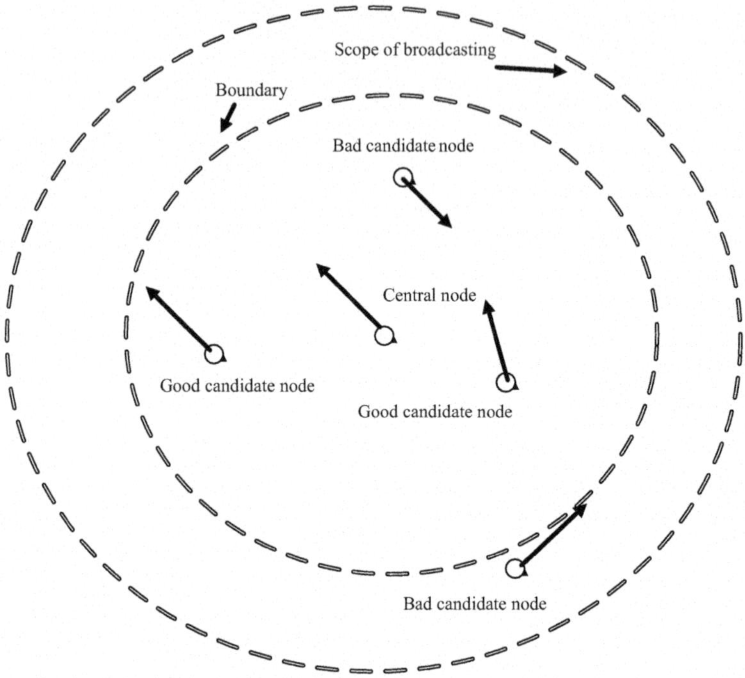

Fig. 7.27 Schematic diagram of neighbor node judgment

(1) Route discovery by flooding

The route discovery mechanism of the PVAR protocol is implemented by flooding, much like that in the DSR or AODV protocol. When node S wants to communicate with node D, which is not its neighbor node, S must initiate a route discovery process. In the previous discussion, we learned that S generates an SN and sends it to its neighbor nodes in a RREQ group. In this case, the neighbor nodes are the real neighbor nodes that were identified using the method described above. In fact, each node receives this RREQ packet and processes it differently. Here, however, we describe the basics of the route discovery mechanism. When a neighboring node receives this RREQ group, it first checks whether it has a sufficiently new route to the destination node D. In this context, sufficiently new means that the SN of the route to D stored by the neighboring node in its route cache area is greater than or equal to the SN in the received RREQ group. If this neighbor node has such a route, it generates a RREP and returns it to S. After receiving this RREP, S discards all RREPs with SNs less than or equal to its SN. S can then deliver the data packet to D through this route, and the route discovery process is completed. If the neighboring node does not have an existing route to D, it continues to broadcast the RREQ packet to its neighboring nodes. Each node forwards the RREQ packet according to this mechanism until it reaches the destination node D. After receiving the RREQ packet, D generates a RREP and returns it to the source node S along the route of the RREQ packet. Whenever S receives a RREP, it updates the relevant route in its route store area. In the PVAR protocol, each node maintains an SN, which is loaded in each route control packet it sends. When a node receives a route control packet, it first determines whether the SN in the packet is greater than the SN maintained by the node. If it is, the node processes the route control packet; otherwise, the node discards it.

(2) Route discovery process in PVAR

The route discovery process of the PVAR protocol is based on the flooding mechanism. However, using information such as the geographic location and node speed, we changed the route discovery mechanism based on flooding to increase its reliability and reduce the routing overhead. These changes are as follows. If a source node S wants to communicate with another node D, and S does not have a route to D, S generates a RREQ group and initiates the route discovery process. In addition to the source node address, destination node address, and SN we mentioned earlier, S encapsulates its speed and location information in the RREQ group. If S knows the location and speed of D, it calculates the distance between itself and D and encapsulates it in the RREQ group, where it is denoted by DIST. S also encapsulates the coordinate and speed information of D in the RREQ group. If S does not have this information, it sets the DIST, coordinates, and speed of D to an agnostic flag bit to notify other nodes; we set this flag to -1 in our simulation. When an intermediate node M receives the RREQ packet, it first determines whether it is the destination node; if it is, it generates a RREP, which includes the location and speed information of D. Then D updates its SN to match the SN in the RREQ group. Here we do not pass the route answer group back to S along the route from D to S, but rather take a flooding approach so that more nodes receive the location and speed information of D. As a result, the information in the network is more up-to-date, and the selected route is less likely to fail. In addition, this enables as many nodes as possible to learn about a reliable route to D. If this route does not fail in a short time, these nodes do not need to initiate a route discovery process to find a reliable route to D. This process greatly reduces the routing overhead. When the source node S receives this RREP, the route discovery process is complete, and S can send upper layer data based on this route. If M is not the destination node D itself, it first checks whether it has a sufficiently new route to D (that is, a route with an SN greater than the SN of the RREQ group). If M has such a route, it encapsulates its stored information about the location and speed of D in the RREP group and returns it to the source node S. If M is not the destination node and does not have a sufficiently new route to D, it acts as a relay node. First, M checks the data values of the DIST area and the coordinates and velocity of D in the RREQ packet. If all of these values are -1, S does not have any information about D. M then checks whether it has any information about D. If M does not have any information about D, it leaves the part of the RREQ packet with a value of -1 as it is. Then it broadcasts the RREQ group to its neighboring nodes. If M has information about D, it calculates the distance from itself to D. If the distance from M to D is less than the DIST value in the RREQ packet, the distance from M to D is less than the distance from S to D. M replaces the DIST value in the

Type(8)	Dst(32)	Src(32)	XD(32)	YD(32)	VD(32)	DIST(32)	SN(32)

Type: Packet Type Dst: Destination Address Src: Source Address
XD: X coordinate of destination YD: Y coordinate of destination
VD: Velocity of destination DIST: Distance between Dst and Src
SN: Sequence Number

Fig. 7.28 RREQ packet header

RREQ packet with the distance from itself to D and broadcasts the RREQ packet to its neighboring nodes. Each node that receives a RREQ packet forwards it in this way until it reaches D. This situation may not be common in simulations. However, if the route from S to D fails, S must initiate a new route discovery process to determine a route to D. All nodes on this route may be affected. It is possible, however, that the route failure is not caused by the movement of D, in which case the location and speed information of D are still useful. Figure 7.28 shows the format of the RREQ packet header used in the simulation. Figure 7.29 shows a flow chart of the PVAR algorithm.

7.4.2 Simulation Model and Results

We simulated the PVAR protocol in the ns-2 environment [20]; the simulation script is given in Appendix E-4. We compared the PVAR protocol with the standard AODV protocol that comes with ns-2. The reason is that AODV is a classical protocol, and the standard model in ns-2 can be used as a reference. We simulated networks with 50 and 100 nodes. The nodes are restricted to a square area of $670 \times 670\ m^2$. The nodes have velocities of 1–10, 10–20, or 20–30 m/s, and the velocities follow a Gaussian distribution. At the beginning of the simulation, all nodes are randomly distributed in the plane of the square. As it moves toward its destination, each node stops moving for a time period of 0 s (no pause) or 10, 20, or 30 s. The total simulation time is 100 s.

We set up 20 constant bit rate (CBR) flows as data packet sources throughout the network; they are generated by 20 different nodes. Each CBR flow sends data packets with a size of 128 bytes at 10 packets per second. When a CBR service is generated, its destination node is randomly generated. Each node can communicate only with its neighboring nodes by broadcasting; if a data packet cannot be sent to the destination node because of a failed route, the source node simply discards the packet. The other simulation parameters are as follows.

- MAC layer protocol: 802.11
- Queue length at the data link layer: 100 packets
- Propagation model: Two-ray ground
- Antenna mode: Omnidirectional.

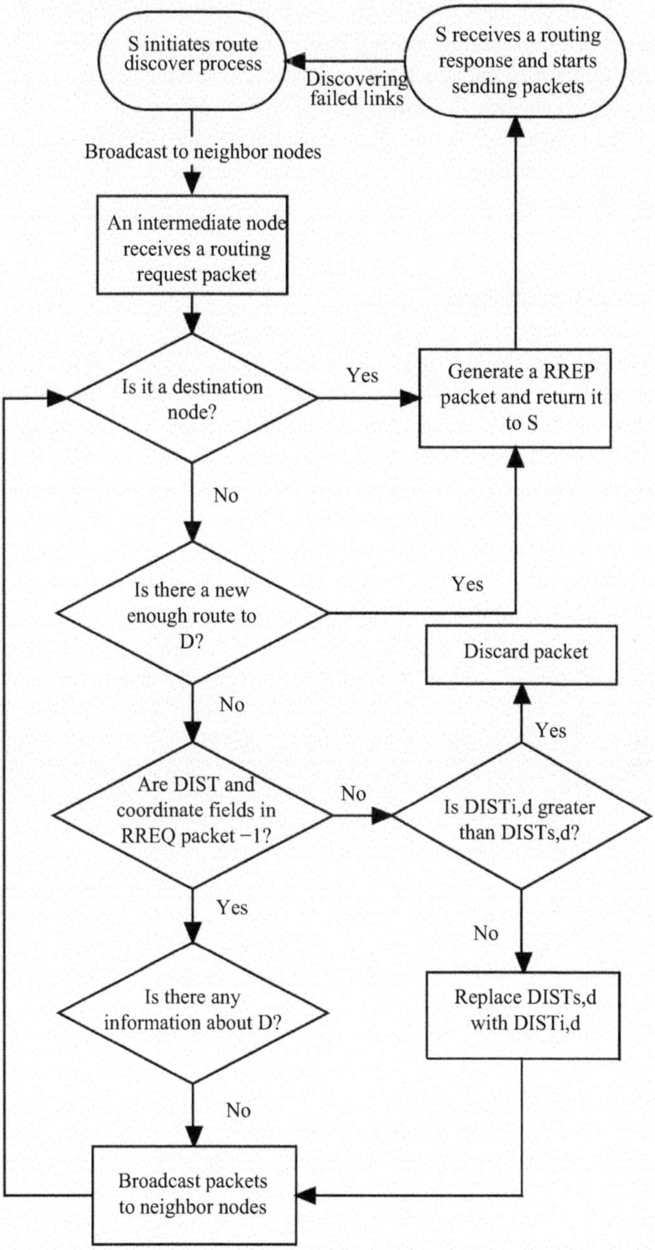

Fig. 7.29 Flow chart of PVAR algorithm

To evaluate the performance of PVAR, two parameters were selected for comparison: the packet arrival rate and routing packet overhead. The packet arrival rate is the ratio of the number of packets that successfully reach the destination node to the total number of packets sent in CBR traffic. The routing packet overhead is defined as the ratio of the number of routing control packets to that of CBR data packets. The simulation scripts for testing the arrival rate and overhead are given in Appendices E-5 and E-6, respectively.

7.4.2.1 Group Arrival Rate

Figure 7.30 shows the group arrival rate performance of a 50-node network using the PVAR or AODV protocol. The lowest packet arrival rate of the PVAR protocol is 0.988, and its packet arrival rate is more stable when the node speed is 10 m/s. Of course, AODV also performs well. The performance of both the PVAR and AODV protocols decreases when the node speed is between 10 and 20 m/s. The reason is most likely that the possibility of route failure increases greatly with increasing node speed. However, when the pause time is 30 s, the packet arrival rate of the AODV protocol suddenly decreases, which indicates that it is less stable than the PVAR protocol. When the speed of the nodes is between 20 and 30 m/s, the packet arrival rate of AODV exhibits considerable jitter, which also indicates the relative instability of the AODV protocol. However, PVAR shows very good stability. We conclude that the PVAR protocol shows better packet arrival rate performance than the AODV protocol, as it has less jitter and a higher packet arrival rate.

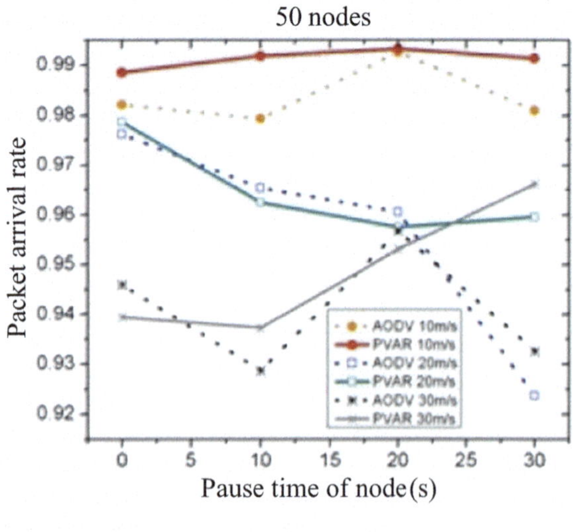

Fig. 7.30 Packet arrival rate in 50-node network [19, 20]

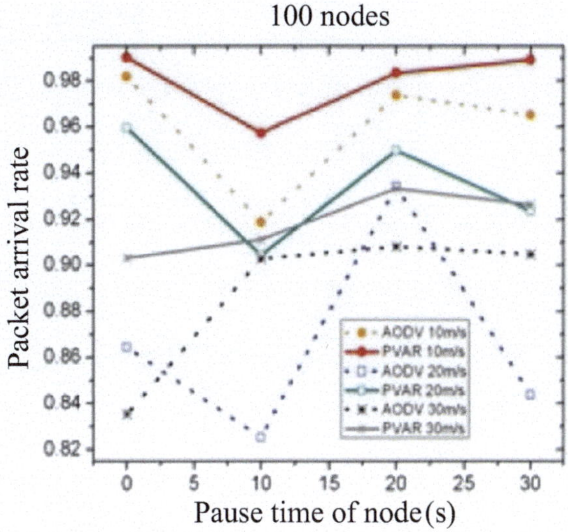

Fig. 7.31 Packet arrival rate in 100-node network [19, 20]

Figure 7.31 shows the results for a 100-node network. The jitter of the packet arrival rate is higher for both protocols compared to that for the 50-node network for each pause time, and the absolute value of the packet arrival rate is also lower than that for the smaller network. However, the PVAR protocol performs significantly better than the AODV protocol in this simulation. The packet arrival rate of PVAR protocol is at least approximately 90% at any node speed, whereas the packet arrival rate of the AODV protocol is below 90% at node speeds of 20–30 m/s. PVAR performs better when the network topology is more complex and more variable. It can also be said that AODV is not suitable for use in this network environment.

7.4.2.2 Routing Overhead

Figure 7.32 shows the routing overhead performance of a 50-node network using the PVAR or AODV protocol. As the node speed increases, the routing overhead also increases. The reason is most likely that routing failure becomes more likely with increasing node speed. In addition, despite the jitter, the overall performance of PVAR is better than that of AODV.

Figure 7.33 shows the results for a 100-node network. When the node speed is 10 m/s, PVAR outperforms AODV and maintains a very low routing overhead. However, when the node speed is 20–30 m/s, the performance of both PVAR and AODV degrades rapidly, and the jitter is also high. The reason is that the nodes must keep repairing routes that have failed when they are moving at high speed. In addition, the overall routing overhead of PVAR is less than that of AODV; there are two main reasons. (1) The neighbor nodes are selected neighbor nodes, which more reliable as relay nodes for routes. (2) When a node receives a RREQ group,

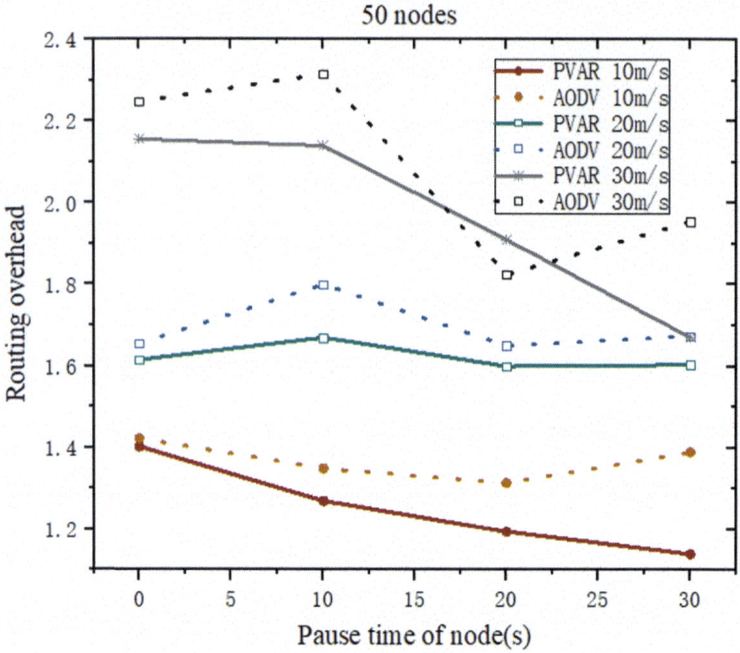

Fig. 7.32 Routing overhead of 50-node network [19, 20]

it first determines whether it is closer to the destination node. If it is further from the destination node than the source node, it discards the RREQ packet; the routing overhead is greatly reduced, but the performance is not affected. The experimental results show that PVAR outperforms AODV, and it is more suitable as a routing protocol for MANETs when the nodes are moving very fast (20–30 m/s).

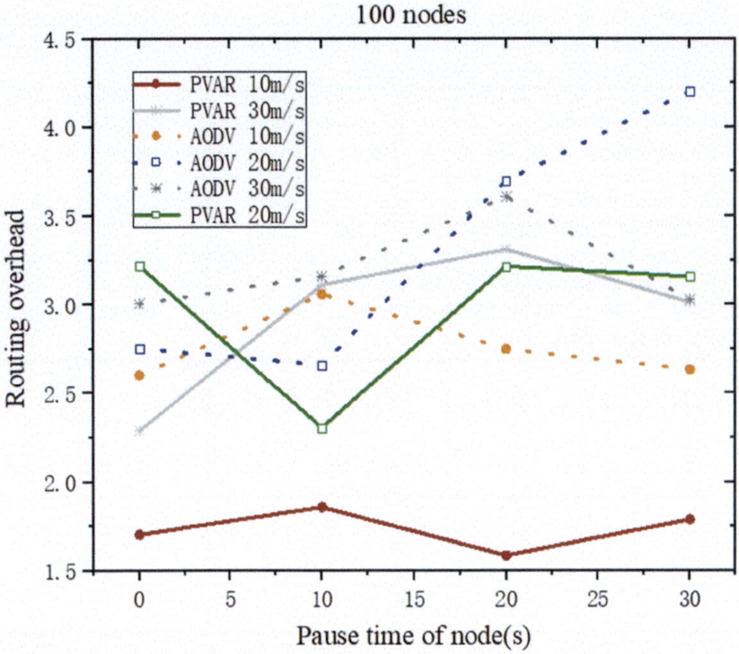

Fig. 7.33 Routing overhead of 100-node network [19, 20]

References

1. Hean KO, Hean LO, Elok RT et al (2006) Scalability study of ad-hoc wireless mobile network routing protocol in sparse and dense networks. In: The 2nd international conference on distributed frameworks for multimedia applications, pp 1–8
2. Ford FD (1962) Flows in networks. Princeton University Press, Princeton
3. Mcquillan J, Richer I, Rosen E (1980) The new routing algorithm for the ARPANET. IEEE Trans. Commun 28(5):711–719
4. Perkins CE, Bhagwat P (1994) Highly dynamic destination-sequenced distance-vector routing (DSDV) for mobile computers. In: ACM SIGCOMM'94 conference on communications architectures, protocols and applications, vol 4, no 2, pp 234–244
5. Gupta P, Kumar P (1997) A system and traffic dependent adaptive routing algorithm for ad hoc networks. In: The 36th conference on decision and control, vol 1, no 1, pp 2375–2380
6. Chen NS (2007) Research on QoS routing protocols in wireless mobile self-organized networks. Wuhan University of Technology, Wuhan, pp 1–20
7. Gafni E, Bertsekas D (1981) Distributed algoritluns for generating loop-free routes in networks with frequently changing topology. IEEE Trans Commun 29(1):11–18
8. Sivakumar R, Sinha P, Bharghavan V (1999) CEDAR: core extraction distributed ad hoc routing. IEEE J Sel Areas Commun 17(8):1454–1465
9. Chiang C-C, Wu HK, Liu W et al (1997) Routing in clustered multihop mobile wireless networks with fading channel. Proc IEEE 1(4):197–211
10. Basagni S, Chlamtac I, Syrotiuk VR et al (1998) A distance routing effect algorithm for mobility. ACM/IEEE MobiCom 1(1):1
11. Karp B, Kung HT (2000) GPSR: greedy Perimeter stateless routing for wireless networks. Proc MOBICOM 1(1):243–254

12. Koy VN (1998) Location aided routing in mobile ad hoc networks. In: The fourth annual ACM/IEEE international conference on mobile computing and networking, vol 1, no 1, pp 66–75
13. http://baike.baidu.com/view/197497.htm (2010)
14. Feng Y, Zhao T, Ke X et al (2010) Research on the coverage area of communication in the solar-blind UV communication network. Acta Optica Sinica 30(8):2229–2235
15. Zheng ChuangMing, Zhang ShengHua (2005) Analytical comparison of TORA and DSR routing protocols for Ad hoc networks. Data Commun 3(1):17–18
16. Long H (2006) OPNET modeler and computer network simulation. Xi'an University of Electronic Science and Technology Press, Xi'an
17. Zhang M, Dou HL, Chang HT (2007) OPNET modeler and network simulation. People's Post and Telecommunications Publishing House, Beijing
18. Yang P, Ke X, Zhao T (2009) Study of ultraviolet mobile ad hoc network. Symp Photonics Optoelectron 1(1):352
19. Yang, P-L (2010) Research on routing protocols for ultraviolet self-organizing networks. Master's thesis, Xi'an University of Technology, Xi'an
20. Luping F, Shihua L (2008) NS-2 network simulation fundamentals and applications. National Defense Industry Press, Beijing

Chapter 8
Routing Protocol Based on Ant Colony Algorithm for Ultraviolet Communication Networks

Wireless mesh networks (WMNs) are high-capacity, high-speed distributed wireless networks. WMNs have typically been based on radio-frequency communication, but with increasing demand for various services, wireless networks are facing enormous challenges in terms of bandwidth and transmission rate. Solar-blind ultraviolet (UV) communication is an ideal transmission medium for WMNs because of its capacity for non-line-of-sight (NLOS) communication, low eavesdropping rate, 24-h operation, and high-speed transmission. Existing WMN routing algorithms do not have uniform evaluation criteria because of differences in network size and use. However, the ant colony algorithm has potential application prospects in research on WMN routing algorithms because of its features such as distributed computing, positive feedback, greedy heuristic search, and strong robustness. This chapter introduces a routing protocol based on the ant colony algorithm for UV communication systems on the basis of an analysis of WMN routing protocols and a UV communication propagation model.

8.1 Wireless Mesh Networks

8.1.1 Wireless Mesh Network Concepts

To realize ubiquitous wireless communication, WMN technology emerged with the development of mobile ad hoc networks (MANETs) in the mid-1990s.

WMNs can be classified as infrastructure/backbone WMNs and hybrid WMNs, as illustrated in Figs. 8.1 and 8.2, respectively. These network architectures can be constructed using various radio technologies, in which a mesh router connects to the Internet and forms self-configuring, self-repairing connections. Mesh routers typically use two types of radio technology, one for communication with the backbone network and the other for communication with users. This network architecture can

X. Ke, *UV Light Self Organizing Network Theory*, Optical Wireless Communication Theory and Technology, https://doi.org/10.1007/978-981-97-7854-6_8

be applied to small local area networks. As shown in Fig. 8.2, a hybrid WMN is composed of an infrastructure/backbone WMN and a client WMN. The mesh client can access the network through the mesh router or mesh client. The infrastructure/backbone WMN provides connections to other networks, such as the Internet, Wi-Fi, Worldwide Interoperability for Microwave Access (WiMAX), mobile communication, and sensor networks. Client-provided bypass capability extends the WMN coverage.

Although WMNs differ from MANETs, they use some wireless local area network (WLAN) technologies. Strictly speaking, however, the WMN is a new broadband wireless network structure, a high-capacity, high-speed distributed network, which

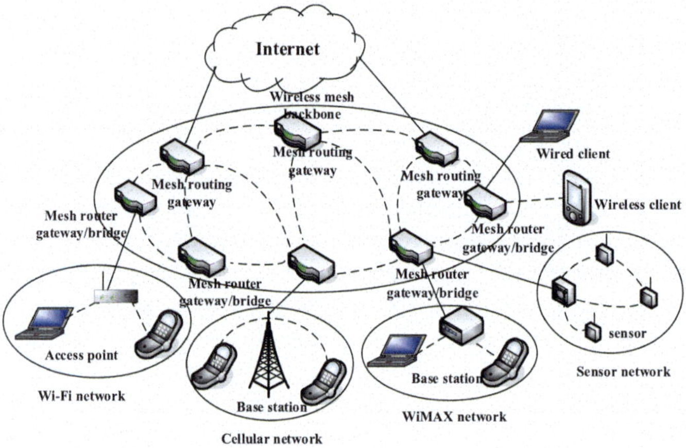

Fig. 8.1 Infrastructure/backbone WMN [1]

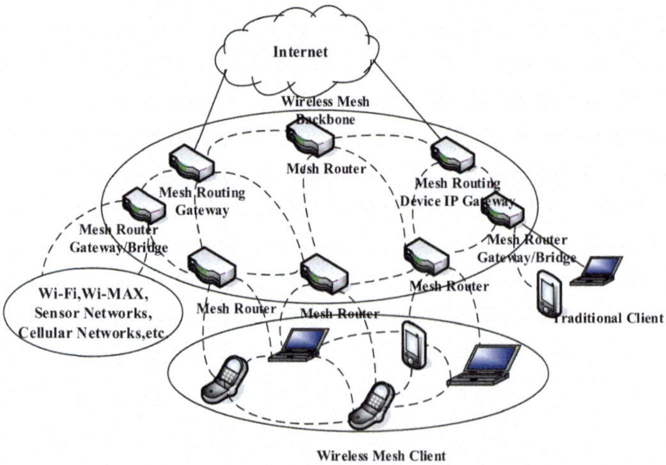

Fig. 8.2 Hybrid WMN

differs from conventional wireless networks. The network topology of WMNs is similar to that of MANETs, but most nodes in the network are essentially static and do not use batteries to supply power, and the topology changes little. For single-hop access, WMNs can be viewed as a special type of WLAN, and WMNs generally do not exist as a separate type of network, but as a wireless extension of the Internet. Gateway nodes, also known as neighborhood access points (APs), provide high-speed Internet connections, and users in homes or offices connect to gateway nodes through their own wireless APs. For areas beyond the signal coverage of gateway nodes, user nodes are responsible for relaying or forwarding business, thus realizing inexpensive and fast signal transmission over a wide area. This type of networking eliminates expensive infrastructure investment at the beginning of network construction and has many valuable advantages over conventional point-to-multipoint wireless access, such as self-organization, self-regulation, self-healing, scalability, and support for ad hoc networks. Therefore, WMNs can form a highly reliable network architecture.

Because of their high reliability, good scalability, and low investment cost, WMNs are being considered as a new scheme to solve the "last kilometer" problem in wireless access. They have been written into the IEEE 802.16 Wireless Metropolitan Area Network (WMAN) or Wireless Municipal Area Network standard (commonly known as WiMAX) and the series of standards in IEEE 802.15; they are currently being included in the formulation of the IEEE 802.11 s standard. In addition, previous studies have found that WMNs are among the best technologies for building future WMAN core networks. They are very likely to challenge 3G technology; they are a potential technology for building B3G/4G, and they are also a feasible technology for building large-scale MANETs [2–6].

8.1.2 Routing Technology for Wireless Mesh Networks

Like wireless multihop ad hoc networks, WMNs provide routing as their core function. Routing protocols provide the necessary routing capabilities to WMN networks. The routing process must be reliable, fast, and cost-effective. It has the following features. (1) The routing protocol in a WMN can not only select routes according to the minimum number of hops, but also selects routes after comprehensive evaluation of various performance metrics. (2) To provide fault tolerance and robustness, routing protocols can quickly select alternative links to avoid service interruption when wireless links fail. (3) Routing protocols should be able to use traffic engineering technology to balance loads among multiple routes and maximize the use of system resources. (4) Routing protocols must support both routers and user terminals. Therefore, the main task of the routing protocol is to find routes between source and destination nodes. This functionality allows communication between nodes over a better or optimal route through wireless multihop transmission.

Because WMNs vary in size and have a wide range of uses, there is no uniform standard for evaluating their routing algorithms. Consequently, the term "optimal

Table 8.1 Classification of common routing protocols used in WMNs

Classification	Agreement	Metric
Ad hoc routing protocols	LQSR	ETX
	SrcRR	ETX
	MR-LQSR	WCETT
Flood-controlled routing protocols	LOLS	ETX或ETT
	MMRP	未指定
	OLSR	Hop, ETX, ML或ETT
Communication-aware routing protocols	AODV-ST	ETX或ETT
	Raniwala and Chiueh [8]	Hops or load-balancing measurement
Opportunistic routing protocols	ExOR	Unidirectional ETX
	ROMER	Number of hops or delay

routing algorithms" is seldom used. WMN routing protocols were classified in reference [7]. As shown in Table 8.1, they can be divided into four types: ad hoc, flood control, communication-aware or tree-based, and opportunistic routing protocols.

8.1.2.1 Ad Hoc Routing Protocols

Routing protocols based on ad hoc networks are a new type of protocol based on the characteristics of WMNs; they include link-quality source routing (LQSR), high-throughput routing (SrcRR), a high-throughput routing protocol for 802.131 mesh networks (DRAFT) [9], and multifrequency link-quality source routing (MR-LQSR, also multi-radio link-quality source routing) [10].

The LQSR protocol is a source link-state routing protocol based on dynamic source routing (DSR). It has four parts: (1) neighbor node discovery, (2) weight assignment for links from the node to neighbor nodes, (3) information transfer between nodes, and (4) finding the optimal path to the destination node using link weight information. The LQSR protocol modifies the route discovery and maintenance mechanisms and adds a new connection maintenance., which supports the collection of link information. When a node receives a route request, it attaches information about its address and upstream link to the route request packet. When a route replies, the route reply transmits the path and route information to the source node. After the path is stored in the link cache, the node must maintain the link information in the cache to keep it up-to-date and accurate. Moreover, LQSR uses two mechanisms, reactive and special, to maintain link information. LQSR is implemented as follows. (1) The send buffer is used to store packets during the route discovery phase; (2) the maintenance buffer is used to store data packets during route maintenance; (3) the route request table is used so that the same route request is not processed multiple times. Like other routing protocols, LQSR incurs additional overhead, as follows.

(1) Because of its routing updates and other operations, it has more operational overhead. (2) Source routing and other fields in each packet increase the volume of the packet. (3) During data transmission, all nodes along the route use HMAC-SHA1 to generate new HASH values to reflect changes in LQSR packet headers, which also increases network overhead. (4) Because it is more secure than wired equivalent privacy, that is, because the end node uses AES-128 to encrypt and decrypt valid data, it greatly increases the computing load of nodes.

The SrcRR protocol was proposed to address the drawbacks of the conventional DSR protocol and combine it with the expected transmission count (ETX) criterion. It is also a reactive routing protocol and is very similar to the DSR protocol in that it uses a link cache. In SrcRR, all nodes maintain a link cache to store the ETX information they obtain by listening. When the ETX information of a link changes, the node reruns the Dijkstra minimum route weight algorithm to calculate the optimal path to other nodes. To ensure that the routing information is up-to-date, if a message has not been refreshed for 30 s, it is deleted as obsolete information. When a node needs to send data to a node in the network and there is no path to that node in the cache, the node initiates a route request. An intermediate node receives the route request, appends its node number and the ETX value of the current link to the route request, and forwards it. A node unconditionally forwards the first route request it receives. If subsequent requests for the same route carry a better path than that previously forwarded, the node forwards the route request again. This process ensures that the destination node receives the optimal path. When the destination node receives a route request, it sends a route reply to the source node. When the source node adds the path to the link cache, all packets waiting to be sent are transmitted through the route with the minimum ETX. When an intermediate node forwards data, it updates the ETX of the link so that the link cache between the source and destination nodes is up-to-date. When the quality of the path used decreases, the protocol chooses a better path for data transmission. In addition, the packet has a criterion extension zone to which the ETX information of the link between this node and any neighbor node is attached with a probability of $1/n$ (where n is the number of nodes in the route). This design allows the source and destination nodes to learn about other available links in the network. All nodes broadcast 300-byte ETX probe packets at random intervals (once every 10 s on average). The node calculates the packet loss rate of the communication link with each neighbor according to the reception in the first 3 min (approximately 18 probe packets). After SrcRR chooses a path, it must wait at least 5 s. Alternatively, when the new route is absolutely optimal, the protocol allows routing switches; as a result, the new route reduces the packet loss rate of the path quite well. The fundamental goal of this optimization process is to reduce the impact of routing conflicts.on ETX probe packet loss.

The MR-LQSR protocol is a multichannel WMN routing protocol developed by Microsoft. It is an improvement on the conventional DSR protocol. It not only must obtain the state information related to nodes on the path and their neighbor links, but also integrates the link-state information to evaluate the link quality and thus form its own routing criteria. By contrast, DSR treats nodes and links on the path as equals and simply sums the number of nodes as the criterion for the routing decision, thus realizing shortest-path routing. The MR-LQSR protocol assumes that all mesh routes in the WMN are static nodes and that each node has multiple non-interfering wireless transceivers. It employs a new routing performance criterion called the weighted cumulative transmission time (WCETT). WCETT is based on the expected transmission time (ETT), which is given by the following formula:

$$\mathrm{ETT} = \mathrm{ETX} \times \frac{\mathrm{S}}{\mathrm{B}} \tag{8.1}$$

where B is the link bandwidth, S is the size of the probe packet, and ETX is the expected number of transfers. Thus, the WCETT is defined as

$$\mathrm{WCETT} = (1-\beta) \times \sum_{i=1}^{n} ETT_i + \beta \times \max_{1 \le j \le k} x_j \tag{8.2}$$

$$x_j = \sum_{\text{The } i\text{th hop link operates on channel j}} ETT_i \tag{8.3}$$

The first term in (8.2) is the cumulative expected time of all links on a path, and the second is the sum of the ETTs of all links on a path over the channel that causes the maximum transmission delay (i.e., the bottleneck).

β is a variable factor and satisfies $0 \le \beta \le 1$; X_j is the number of times the jth channel is used on path p to capture interference in the data stream. In addition, $\max_{1 \le j \le k} x_j$ represents the maximum number of occurrences of the same channel on the path. Paths that are more diverse in the channel assignment of the link have lower in-stream interference, and their weights are smaller, reflecting the intensity of in-stream interference. The WCETT criterion considers link quality, channel variation, and minimum hops. Thus, MR-LQSR is a routing protocol that combines the cumulative transmission time with LQSR.

8.1.2.2 Flood Control Routing Protocols

Flood control routing protocols are used to reduce the control overhead. They include the optimized link-state routing (OLSR) protocol [11], localized on-demand link-state protocol [12], and mobile mesh routing protocol. Updating by flooding the entire network with routes may improve the scalability.

The OLSR protocol is a table-driven precedent routing protocol standardized by the MANET working group of the Internet Engineering Task Force. The protocol uses the classical shortest-path algorithm and calculates the routes in the network based on the hop criterion. However, the key concept of OLSR is to optimize the broadcast mechanism used to distribute the necessary link-state information across the network. Each node selects a multipoint relay (MPR) from among its neighbors. Although only the MPR broadcasts the message again, all two-hop neighbor nodes receive the broadcast message, and the node finds the MPR in this way. Forwarding broadcast messages through the MPR can significantly reduce only the number of broadcast messages. This optimal forwarding mechanism is used for all broadcasts in OLSR networks. OLSR uses two main control groups: the Hello group and topology control (TC) group. Hello groups are used to build a neighbor table for a node, which includes the address and delay or overhead of neighbor nodes. OLSR periodically broadcasts Hello messages to determine the state of neighbor nodes. Wireless links between nodes can be asymmetric links, symmetric links, or links to MPRs. Hello groups are also used to calculate the MPR of a node. They are broadcast only in the one-hop range and cannot be forwarded. By contrast, TC groups must be broadcast across the network. They contain instructions for selecting the node sending the TC group as the neighbor node of the MPR, and the node calculates the topology of the network according to the received TC group. In addition, link-state information is sent only to MPR selectors in OLSR networks because they are used to calculate the shortest path. This mechanism reduces the amount of link-state information that must be sent in the network. Each node periodically broadcasts Hello messages to detect local topologies. The Hello message (time to live: 1) containing the list of neighbors of the sending node cannot be forwarded. Each node in a WMN can learn its two-hop neighbors through the Hello mechanism and can also determine the bidirectionality of the links. OLSR labels each link with state information (symmetric or asymmetric). In addition, each node attempts to forward a packet in a Hello message. The information for Hello messages is stored in multiple repositories: the link, neighbor node, and two-hop neighbor node sets.

The localized on-demand link-state (LOLS) protocol is a routing protocol designed for scalable packet delivery in a static multihop mesh network. It assigns long-term and short-term overhead to a link; they represent the typical and current overhead, respectively. To reduce the control overhead, the short-term overhead is often used for sending to neighbor nodes, whereas the long-term overhead is used for sending to more distant nodes. LOLS uses the ETX or ETT to calculate routes. The protocol captures the relatively static topology of the WMN itself and the relative dynamic link state and obtains a long-term routing strategy based on the static topology. For the short-term link state, the resulting path is modified by local propagation. Local propagation using the short-term state requires only those links with reduced performance, whereas long-term state information is predetermined and becomes a fixed topology. The blacklist-aided forwarding (BAF) mechanism has been proposed in which each packet carries a blacklist and records the descending links. Each node has a blacklist cache, which is initially empty. When the performance of a link is degraded, a blacklist is added to the cache corresponding to the link. When the packet passes through this node, it also selects the next-hop node according to the target node. When a node reaches the blacklist, it removes the blacklist after the packet and continues forwarding according to the specified next hop. The BAF mechanism can be used for the local adjustment of the overall performance degradation due to poor short-term performance; it can also avoid loops.

The mobile mesh routing protocol (MMRP) was developed by the MITRE Corporation. The protocol is based on link-state information and uses a link-state algorithm.

The interface of a network node periodically broadcasts link-state information datagrams, and nodes that receive this information relay it within a specified time after the datagrams arrive. Thus, each node holds the topological information of the entire network. Using its topology database, each node can calculate the shortest propagation path to each node in the network. When sending data, the terminal finds the shortest propagation path to the destination node by querying the local routing table and sends it along this path to the destination node. Therefore, the MMRP is also routing-table-driven. The link-state information packet contains the interface information of the host, interface address of the linked neighbor node, and route weight. External routing information is also included in link-state information. This mechanism allows hosts to link to wired networks, enables terminals to connect to external networks, and improves the scalability. The routing protocol assigns a time to the routing message, like open shortest path first. When a node sends a routing message, it subtracts the message time from the estimated time it needs to relay. When the message time ends, the packet is discarded to prevent it from being relayed. The MMRP does not require a specified routing metric.

8.1.2.3 Communication-Aware or Tree-Based Routing Protocols

Communication-aware or tree-based routing protocols take into account the common communication matrix in WMNs. Because they assume that return access is a common application, they consider a tree network topology. These protocols include the ad hoc on-demand distance vector-spanning tree (AODV-ST) protocol [13] and the distributed protocol proposed by Raniwala and Chiueh [8].

In the AODV-ST protocol, which is based on the AODV protocol, a gateway periodically sends routes to each node in the network to update its routing table information. The gateway is the root of the tree. During the flooding process of the reverse link algorithm, each intermediate node establishes a reverse route to the source node. The set of grouped effective transmission paths constitutes a spanning tree from the source node to all other nodes in the (sub)network. The reverse of the spanning tree constitutes a directed acyclic graph from each node in the network to the source node. AODV-ST uses this algorithm to build a spanning tree from the gateway to all nodes in all subnets. The kernel is essentially unchanged and continues to follow the functionality that all threads have in the kernel. The protocol code compensates for the shortcomings of the basic AODV protocol such as the lack of high-throughput routing and an effective route maintenance mechanism and the large routing delay. It supports the ETX and ETT routing metrics, and the use of the spanning tree algorithm greatly reduces the routing delay and routing table maintenance overhead. Each gateway periodically broadcasts control messages and actively constructs a spanning tree. The spanning tree ensures that each relay node has the best path to each gateway. Because the path from the gateway to each relay node is actively maintained, the overall network route maintenance overhead is minimal. In addition, each node-to-gateway path in the spanning tree is updated periodically. Each node uses routing metrics to select the gateway with the best path as the default gateway. The path-finding mechanism of AODV-ST for communication between relay nodes is similar to that of basic AODV and uses a reactive path-finding mechanism. Therefore, AODV-ST is a hybrid routing protocol that actively updates the route between frequently used gateways and relay nodes and establishes the best route between relay nodes as needed.

8.1.2.4 Opportunistic Routing Protocols

There are no reports on opportunistic routing in China. Two opportunistic protocols are used internationally: extremely opportunistic routing (ExOR) [14] and resilient opportunistic mesh routing (ROMER) [15]. The ExOR protocol, which was developed by MIT, is a typical opportunistic routing protocol. First, when a source node transmits a packet, it contains a simple schedule that describes the priority of potential recipients who can forward the packet. The node calculates the timetable according to the transfer rate of the interactive node. The ExOR protocol then sends an acknowledgment packet to ensure that the recipient is in agreement regarding which node is the highest-priority recipient. ExOR includes a distributed media access control protocol that allows the receiver to simultaneously determine that only one receiver is forwarding the packet (thus avoiding conflicts) and an algorithm that predicts which packet will be most useful to forward ExOR reduces the total number of packets transmitted according to three factors: node density (the higher the better), path length (the longer the better), and the rate at which reception decreases with distance (the less steep the better). The ExOR protocol consists of three steps: (1) selecting forwarding candidates, (2) issuing confirmation packets, and (3) deciding whether to forward a received packet. Assume that each node in the network has a matrix that contains an approximation of the direct wireless transmission loss rate for each pair of nodes. This matrix can be established by link-state flooding, where nodes measure the packet loss rates and periodically update the flooding status. The first node of the ExOR forwarding sequence selects a subset of candidates from among all its neighbor nodes that can deliver the packet to a node closer to the destination node. The sender lists this set in the header and prioritizes it by distance. After transmission, each node that receives a packet looks for its address (that is, the recipient, if any) in the list of candidates for the packet header. Each recipient delays sending an acknowledgment packet for a period of time based on its location in the list. Each node looks at the set of acknowledgment packets it receives to determine whether to forward the packets. The forwarding node overrides the ExOR header with a new list of candidates and forwards the packet.

Many routing protocols have been proposed for WMNs. However, because of the unique transmission characteristics of UV light, conventional routing protocols for WMNs cannot be directly applied to UV WMNs. Therefore, it is necessary to develop a routing algorithm suitable for UV WMNs.

8.2 Wireless Mesh Network Based on UV Light

The conventional radio-frequency-based WMNs are facing great challenges resulting from spectrum scarcity and increasing data transmission. Infrared-based wireless networks have very stringent link requirements, and the light must essentially travel in a straight line with no obstacles. Most of these requirements are difficult to meet in practical applications. However, the rapid development of UV communication in recent years has made it possible to solve these problems.

8.2.1 Key Technologies for UV Wireless Mesh Networks

Research on UV WMNs is still in its infancy worldwide, and most research on UV communication and WMNs is still theoretical. Key technologies include the following.

(1) The main task of routing protocols is to select a reliable, fast, and inexpensive route between source and destination nodes. Like wireless multihop ad hoc networks, WMNs provide necessary routing capabilities through their routing protocols. However, because of differences in the sizes and uses of WMNs, there is no uniform standard for evaluating their routing algorithms, and the term "optimal routing algorithms" is seldom used. In addition, because of the unique transmission characteristics of UV light, routing protocols for conventional WMNs cannot be directly applied to UV-based WMNs.

(2) The purpose of quality of service (QoS) assurance is to satisfy all constraints, such as delay, bandwidth, and delay jitter, on the source, intermediate, and destination nodes when sending data in a distributed network and ultimately reach the minimum overhead or a specific level of service. Because WMNs are a type of wireless multihop network technology for civilian communication, as the demand for civilian communication clients for different communication services increases, it becomes important to ensure QoS in UV WMNs.

(3) Because UV communication is omnidirectional, nodes within the effective radius of a circle centered on the light source can receive information. Therefore, there is interference between any link in the network and the wireless links adjacent to its geographic location, resulting in unidirectional links and reducing network transmission capacity. Therefore, a reasonable topology design is needed to control the interference between wireless links.

(4) Network operation requires that a WMN is open when links are active, and thus nodes are vulnerable to attack. If vulnerabilities are not discovered in time during communication, the connectivity of network links and security of communication are often affected. Therefore, methods of preventing attacks by external malicious nodes and ensuring the security of UV WMNs is also highly important in research on WMNs.

8.2.2 Applications of UV Wireless Mesh Networks

Because they combine the advantages of UV communication and mesh networks, UV WMNs have broad application prospects for use in homes, businesses, and public places.

(1) Home wireless networks are an important application of mesh technology. They connect desktop computers, laptops and handheld computers, HDTVs, DVD players, game consoles, and other consumer electronic devices without the need for complex wiring and installation processes. Various home appliances both use the network and provide access services for other devices as part of the network infrastructure. This networking method can provide more capacity and coverage as communication capability is added to more home appliances. Another key benefit of mesh technology in the home is its ability to support bandwidth-intensive applications such as high-definition video.

(2) Most enterprise wireless communication systems use conventional cellular phone wireless links to provide point-to-point and point-to-multipoint transmission for users. WMNs allow network users to share bandwidth, eliminate the current bottleneck of single-hop networks, and achieve dynamic load balancing. Adding or adjusting APs) in a WMN is also easier and more flexible, and wireless APs are less expensive to install and use than wired APs. Especially for enterprises that need to move APs frequently, the multihop structure and flexible configuration of wireless mesh technology is very helpful for adjusting and upgrading network topology.

(3) Campus wireless networks are very similar to those of large enterprises, with some differences. First, campus WLANs are very large; they not only cover a large geographic area, but also have a large number of users and heavy traffic, because students use multimedia more frequently than typical enterprise users. Second, the requirements for network coverage are high. The network must be able to provide seamless roaming indoors, outdoors, and in auditoriums, dormitories, libraries, public spaces, etc. Third, load balancing is very important, because students often simultaneously use the network in a certain location, and network congestion may occur. Mesh networking not only enables easy upgrading and adjustment of the network structure, but also provides seamless roaming between outdoor and indoor locations.

(4) Hospitals are among the most difficult areas to install wireless networks because of the dense and complex structure of hospital buildings and the need to prevent electromagnetic radiation in some areas. Hospital networks have two main features. First, wiring is more difficult; wiring is typically routed through the walls of buildings, which clearly makes it more challenging to change the network topology. Second, the network must be highly robust; if important activities (such as surgery) occur in the hospital, any possible network failure will have catastrophic consequences. WMNs are an ideal solution to these problems. To adjust the wireless network topology, all that is necessary is to move existing mesh nodes or install new mesh nodes, which are very simple and

convenient to install. The robustness and high bandwidth of wireless mesh also make it suitable for deployment in hospitals.

(5) Mesh is ideal for remote locations where wiring is difficult or economically expensive and users need broadband wireless Internet access, such as tourist sites, resorts, and motels. Mesh can provide broadband services at these sites at the lowest cost.

(6) Mesh networks are undoubtedly the most economical and effective method of networking when rapid deployment or temporary installation is needed, for example, at exhibitions or fairs, or in disaster relief. For example, the use of mesh networking technology can minimize costs when a temporary network is needed for a few days for meetings or exhibitions.

8.2.3 Routing Protocols for UV Wireless Mesh Networks

Routing protocols enable a node to communicate with other nodes in a network through multiple hops. Most existing routing protocols for WMNs are based on radio communication, and many problems must be solved when they are applied to UV-based communication.

Because of the unique transmission characteristics of UV light in the atmosphere, routing protocols for conventional WMNs cannot be directly applied to UV-based WMNs. The ant colony algorithm is applied to find the best path in the network because of its positive feedback, distributed computing, and greedy heuristic search. QoS assurance in wireless networks has become more important with the continuous development of network technology and increasing user requirements. Reference [16] used two groups of ants to search for paths and finally chose the optimal path according to the QoS multicast constraints. Another study [17] considered the constraints on the delay, delay jitter, bandwidth, cost, and packet loss rate. The ant colony algorithm has been applied to QoS multicast routing, but it does not take into account loops between paths to each destination node. The pheromones found in these conventional algorithms are independent of each destination node, which slows convergence. In addition, in UV WMNs, the convergence rate is reduced by the use of NLOS communication. The differences in NLOS channel types, effects of atmospheric turbulence, and obstruction by hidden terminals or obstacles often result in the existence of many unidirectional optical links or asymmetric links. In this case, only unidirectional communication can occur between two nodes. Reference [18] used maintenance by a multihop reverse routing method to support unidirectional links, but forwarded packets must carry the entire reverse route, which increases the consumption of network resources. To study the unidirectional link problem in virtual backbone networks, a constant approximation algorithm has been proposed; it applies a strong connected dominating set to a strong connected dominating set and an absorptive set, but increases the cost [19]. Collaborative relay technology based on the classical AODV protocol has been used to expand coverage and solve the unidirectional link problem. However, this method can cause interference between communication links in UV WMNs because of the expanded coverage. The routing algorithm is studied in this chapter.

8.3 Ant Colony Algorithm

Humans are able to understand things by interacting with the natural world. Because the natural world exhibits many cases of adaptive optimization, human beings can perfectly solve seemingly complex optimization problems by applying the evolutionary mechanism of organisms and natural ecosystems. Since the mid-1950s, inspired by the mechanism of biological evolution, many bionic algorithms for solving complex optimization problems have been proposed; these algorithms simulate natural ecosystem mechanisms and include genetic, ant colony, particle swarm, simulated annealing, artificial fish swarm, and bee swarm algorithms. They provides feasible solutions to combinatorial optimization problems that are difficult to solve by conventional optimization algorithms. Some of these biomimetic optimization algorithms have shown great ability and potential for further development in solving nondeterministic polynomial (NP) problems and in practical applications [20].

The ant colony algorithm is a simulated evolutionary algorithm proposed by the Italian scholars Dorigo et al. [21] in the 1990s. Inspired by the bio-evolutionary mechanism, it simulates the behavior of ants in the natural world as they search for paths between food and the nest. It is a probability algorithm that finds the optimal path in a graph. It is also an important algorithm in group intelligence theory research. Preliminary studies show that it has many excellent properties: high intrinsic parallelism, strong robustness, an excellent distributed computer system, and easy integration with other methods.

8.3.1 Origin of Ant Colony Algorithm

Ants are an ancient social insect that first appeared 100 million years ago, at approximately the same time as dinosaurs. Their individual behavior is very simple. A single worker ant can make no more than 50 types of movements, most of which transmit information. However, the entire population of these simple individuals, the ant colony, shows a highly structured social organization and in many cases can perform complex tasks far beyond the ability of individual ants. Ants in an ant society play various roles. Ant colonies perform complex tasks by division of labor. As a type of social insect, ants can communicate with each other and transmit information in addition to implementing division of labor. They have unique information systems, including visual signals, sound communication, and distinctive pheromones, or exogenous hormones, which are volatile chemicals left along their routes [22].

In reality, each ant does not need to understand the entire world. Instead, it knows local information for a very small area and makes decisions on the basis of this information and a few simple rules. As a result, complex behaviors emerge in ant colonies. This description is the scientific interpretation of artificial life and complexity. The following is a detailed description of these simple rules in the simulation environment.

(1) Range: The range an ant can observe is a square world with a speed radius parameter (typically 3). The range an ant can observe is 3×3 square worlds, and it can move within this range.
(2) Environment: The environment in which ants live is a virtual world with obstacles, other ants, and pheromones. There are two types of pheromones: a food pheromone that is sprayed by ants who find food and a location pheromone that is sprayed by ants who find the nest. Each ant can perceive only the environment within its range. Furthermore, pheromones in the environment disappear at a certain rate.
(3) Foraging rules: Ants search for food within the range that they can perceive; if they find it, they go directly to it. Otherwise, they check for pheromones and identify the point in their range that has the most pheromones so that they can move in the direction of increasing pheromones. Each ant may make a mistake with low probability, rather than moving toward the point with the most pheromones. The ant looks for the nest in the same way, except that it reacts to the nest pheromone but not to the food pheromone.
(4) Moving rules: Each ant moves in the direction of the maximum amount of pheromones. When there are no pheromones around it, the ant moves in the direction of its original motion, and there is a small random disturbance in the direction of its movement. To prevent the ant from spinning in place, it remembers which points it has recently covered and will try to avoid them if it finds that it has recently passed the next point on its route.
(5) Obstacle avoidance rules: If an ant is obstructed in the direction in which it is moving, it behaves according to the rules of foraging if there are pheromones; otherwise, it randomly chooses one direction to move forward.
(6) Pheromone spreading rule: Each ant spreads the most pheromones when it first finds food or an ant nest, and as it travels farther away, it spreads fewer pheromones.

There is no direct link between individual ants, but according to these rules, each ant can interact with the environment, and each ant is linked to the others through pheromones. For example, when an ant finds food, it does not directly tell other ants where the food is. Instead, it spreads pheromones in the environment. When other ants enter the area, they sense the pheromones and find food according to their guidance.

When no ants find food, the environment contains no useful pheromones. Ants can find food relatively effectively owing to the rules by which they move. First, they try to maintain inertia so that they move forward as much as possible (starting with a randomly chosen direction forward), rather than spinning in place. Second, the motion of ants should be somewhat random. Although they have a fixed direction, they cannot move in a straight line like particles but exhibit random disturbances. Consequently, ants move with a certain purpose. Their heuristic is based on maintaining the original direction as far as possible. When they encounter obstacles, they change direction immediately. This behavior can be seen as a process of selection; that is, obstacles in the environment cause an ant to move in only one direction but not in others. Consequently, individual ants can find well-concealed food in complex environments such as mazes. Of course, when one ant finds food, the other ants can quickly find food by following the pheromones.

Because of the swarm biointelligence of ants, Dorigo and colleagues observed their feeding habits and found that ants always found the best way between their nests and food. It has been found that this group collaboration function of ants is mediated and coordinated by pheromones. This behavior is among the basic communication modes of ants and plays an important role in their life habits. A study of the feeding behavior of ants revealed that the entire ant colony cooperates using pheromones to form positive feedback so that ants on multiple paths can gradually converge onto the best one [23]. Thus, Dorigo et al. [24] first proposed the ant colony algorithm in 1991. It is a heuristic search algorithm based on population optimization, and its main feature is that it can find the optimal path using positive feedback and distributed collaboration. It takes full advantage of the collective optimization characteristic that biotic ant colonies use to search for the shortest path from an ant nest to food by simple information transmission between individuals. Because of the similarity between the optimization process and the solution of the traveler problem, an NP-complete optimal solution of the traveler problem has been obtained. In addition, the algorithm is also used to solve job-shop scheduling problems, quadratic assignment problems, and backpack problems, indicating its superior ability to solve combinatorial optimization problems.

In 1992, Dorigo further proposed Ant Colony System (ACS) [25] in his doctoral dissertation. He proposed three basic ant colony models based on different methods of calculating the pheromone increment: the ant-cycle, ant-quantity, and ant-density models. In addition, the effects of various parameters on the performance of the algorithm were discussed on the basis of many experiments, and the valid intervals of the main parameters of the algorithm were determined.

8.3.2 Research Progress on Ant Colony Algorithms Worldwide

With the rapid expansion of its applications, the ant colony algorithm has become an important subject in the field of intelligent computing. Although it is not as mature as genetic, simulated annealing, and other algorithms, this new optimization concept that simulates natural biological systems is expected to have a very bright future. It has attracted the attention of many researchers at home and abroad.

In the 1990s, Dorigo et al. proposed the first ant algorithm model, Ant System [21]. This algorithm was first used to solve the well-known traveling salesman problem (TSP) and yielded good results. Later, to overcome its drawbacks of slow convergence and a tendency to stop, Dorigo proposed a modified ant colony algorithm, Ant-Q [26]. This algorithm introduced an elite strategy, which combines a local pheromone update mechanism with a global pheromone update and uses a pseudo-random scale state transfer rule to replace the random scale selection rule in Ant System. Later, Dorigo proposed ACS, which is based on the Ant-Q algorithm; it not only has the same effectiveness in solving TSPs, but also is easier to implement. In 1997, Bullnheimer et al. proposed an order-based ant system (ASrank, a rank-based version of Ant System) [27]. After an iteration, the algorithm arranges the length of the ants' paths in order from short to long and assigns weights accordingly, where shorter paths have larger weights. Vittori et al. proposed an intelligent routing algorithm, Q-Agents [28]. This algorithm combines three learning strategies, Q-learning, double-reinforcement learning, and ant-based behavioral learning, and searches for the best route using a series of agents moving independently in the network. Bell et al. [29] used the heuristic method of the ant colony optimization (ACO) algorithm to solve vehicle routing problems (VRPs); this method simulates the decision-making process in ant colonies, similar to other adaptive learning and artificial intelligence technologies such as tabu search and simulated annealing and genetic algorithms. The ant colony algorithm used to solve conventional travel problems was modified to allow multipath search in VRPs [29]. Although the path found by the ant colony algorithm is always close to the shortest path, a large number of packets must be sent before the path is established, and the time delay is large. Therefore, in 2007 Kamali and Opatrny proposed a reactive routing algorithm, the position-based ant colony routing algorithm for MANETs. This algorithm combines the location information of the ant colony algorithm with the concept of nodes, and the time delay of shortest-path establishment is much smaller than that of the conventional ant colony algorithm [30]. In 2010, Kadono and Daisuke presented a new ACO routing algorithm based on path robustness, which can be applied to MANETs using the Global Positioning System (GPS). Each ant agent in the algorithm uses GPS information to evaluate the robustness of visited node paths and then determines the number of pheromones remaining on the path. The algorithm has a high data packet transfer rate [31] and low communication overhead. Research and improvement of ant colony algorithms in foreign countries is ongoing.

Compared with that in other countries, research on ant colony algorithms began relatively recently in China. Dr. Zhang Jihui and Xu Xin of Northeast University have performed the following research on ant colony algorithms. In 1999, they introduced a new method of solving the combinatorial optimization problem, the artificial ant colony algorithm [32]. This method simulates the process by which an ant colony searches for food and solves a more difficult combinatorial optimization problem. To address its long computation time, in the same year they also proposed a new ant colony algorithm with mutation characteristics [33]. This algorithm introduces a mutation mechanism into the basic ant colony algorithm and uses the 2-exchange method, which is simple and efficient; consequently, it not only has a faster convergence speed, but also conserves computing time. In 2000, in light of the many good qualities and some shortcomings of ant colony algorithms, they reviewed the current research and proposed an adaptive ant colony algorithm [34], which inspired related research. In 2001, Wang and Xie applied an ant colony algorithm combined with a heuristic algorithm to solve the multipoint routing problem and proposed a new multipoint routing algorithm based on adaptive ant colony system (AACS) [35]. Then, the ant colony algorithm was improved to overcome the drawbacks of slow convergence and easy trapping at local minima. In 2002, they improved ACO by retaining the optimal solution, adaptively changing the volatility coefficient, and introducing the crossover operator of the genetic algorithm; they also proposed a multicast routing algorithm model based on ACO with time-delay constraints [36]. In the same year, they proposed an adaptive ant colony algorithm [37] in which the volatility and other coefficients of the algorithm are changed adaptively. The algorithm can improve the global properties of the solution while ensuring rapid convergence. In 2005, Liu et al. applied the ant colony algorithm to route planning for unmanned aerial vehicles and presented an optimization algorithm for route planning [38]. In 2007, Liu et al. proposed an optimal job-scheduling algorithm based on the ant colony algorithm [39]. In 2009, Zhao et al. proposed an improved method based on various information requests, which enables the algorithm to intelligently select the routing direction [40] in a peer-to-peer network according to file type. In 2010, Zheng et al. proposed a path recovery algorithm for sensor networks based on ACO [41]. Most of these studies focused on improving and applying the theory underlying the algorithm.

8.3.3 Basic Principles and Models of Ant Colony Algorithm

8.3.3.1 Basic Principles of Ant Colony Algorithm

To study the feeding behavior of ants under artificial control, Deneubourg et al. designed the symmetric binary bridge experiment [23], in which there is a bridge between the nest and food. The bridge has two branches of the same length. In the ant colony experiment, Argentine ants were placed in the nest and could freely move back and forth between the nest and food. Each ant had to choose between two branches when moving from the nest to food or from food to the nest. Deneubourg et al. [23] recorded the temporal variation in the percentage of ants traveling through the upper and lower routes. The experiments showed that in a few minutes, almost all the ants converged onto a single path. After making a simple adjustment to this experiment, Goss et al. performed an asymmetric binary bridge experiment [42], where the two branches of the binary bridge have different lengths, and the other parameters are the same as in the symmetric experiment. In most experiments, more than 80% of the ants chose the shortest path from the nest to food. This phenomenon of ant colonies choosing the shortest feeding path from their nests to food can be explained by positive feedback (autocatalysis). Individual ants communicate indirectly through pheromones. When ants reach a bifurcation, they choose routes according to the concentration of pheromones they smell along the routes. They tend to choose routes with higher concentrations of pheromones. This route selection behavior has positive feedback because when an ant chooses a path, it leaves pheromones that increase the probability that other ants will choose the path in the future, and the pheromones left by other ants continue to increase the probability that the path will be selected again.

In the symmetric binary bridge experiment, there were initially no pheromones on either path; therefore, ants would randomly choose one of the two paths with the same probability when moving from the nest to food. The term "same probability" here refers to the statistical probability over a long period of time without the effect of pheromones. If at some point there are more ants on one path than on the other, the concentration of pheromone residues on that path will be higher than that on the other path, and the ants will tend to choose that path. Because of the positive feedback between ant routing and pheromone concentration, an increasing number of ants choose this path, and most ants eventually choose one of these paths. In the asymmetric binary bridge experiment, ants that choose shorter paths reached food more quickly because the paths have different lengths. Pheromones are distributed differently along the paths when the first ant arrives at the food and then returns to the nest and encounters a bifurcation again. Clearly, only the shorter paths have a higher concentration of pheromones; therefore, the ant tends to choose the shorter path. New pheromones are then added to the shorter path, making it more attractive to ants arriving later. As this process continues, pheromones are deposited significantly more quickly on the shorter path than on the longer path, and thus an increasing number of ants choose this path. Ultimately, most ants tend to choose the shorter route because of positive feedback.

As shown in Fig. 8.3, ants search for the shortest path. A represents the ant nest, F is the food source, and CD is an obstacle. Because of the presence of obstacles, F is reached from A through C or D, and A is reached from F. The distance between BC and CE is assumed to be 1, and that between BD and DE is 0.5. Each time unit is set to the time required for 30 ants to move from A to B and 30 ants to move from F to E; the amount of pheromone left behind by each ant is 1, and each ant moves at a speed of 1 per time unit. The residence time of the pheromone is set to 1, as shown in Fig. 8.3a.

At $t = 0$, because there are no pheromones on paths BC, CE, BD, and DE, ants on B and E select paths randomly. Statistically, they can be assumed to select BC, CE, BD, or DE with the same probability. Therefore, on average, 15 ants go to C and 15 go to D from B and E, respectively, as shown in Fig. 8.3b.

At time $t = 1$, 30 new ants arrive at B from A. They find a pheromone intensity of 15 on path BC, as pheromone was secreted by 15 ants that passed B going toward C, whereas that on path BD was 30. The pheromone intensity is the sum of the secretions of 15 ants from B to D and those of 15 ants that reached B by traveling from D to E, as shown in Fig. 8.3c.

Consequently, the probability that a path is chosen is skewed, and ants going toward D are twice as likely to go toward C, where 20 and 10 ants, respectively, went in each direction. Similarly, 30 ants moving from E to D would behave the same

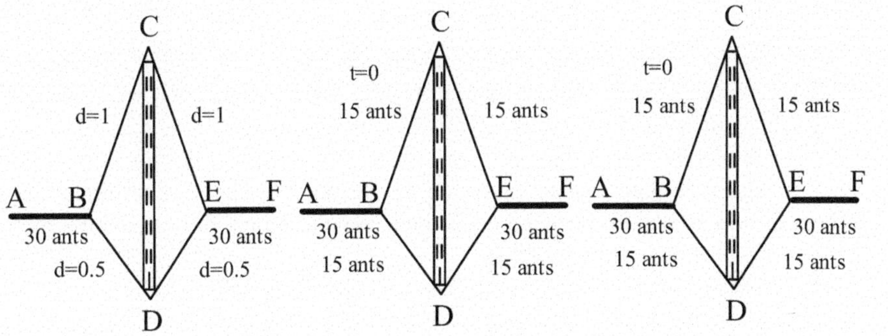

Fig. 8.3 Schematic diagram of ants finding the shortest path

way. Over time, ants choose path BDE with increasing probability, and eventually all ants choose the shortest path, ABDEF.

8.3.3.2 Differences Between Artificial and Real Ants

The ant colony algorithm introduces the concept of artificial ants to distinguish them from real ants in nature. The behavior of artificial ants is an abstraction of the behavioral characteristics of real ants, and artificial ants exhibit the most critical components of ant foraging behavior. However, to improve the effectiveness of ant colony algorithms, artificial ants have some characteristics that real ants do not have.

(1) Similarities between artificial and real ants:

   (a) Both artificial and real ants can cooperate to find a better solution to the problem. Each ant can construct a solution, but collaboration by the entire ant colony produces a better solution.
   (b) Both need to complete a common task, that is, to find the shortest path from the source node (ant nest) to the destination node (food source). They cannot jump; they can move only one step at a time to the adjacent state on the path until traversal is complete.
   (c) Both communicate indirectly using pheromones and use current information for path selection. That is, they can release pheromones on the path to change their environment. The strength of the pheromones weakens gradually with time. Through this volatilization mechanism, artificial and real ants gradually forget historical information; thus, they are not be limited to the experience of previous ants when choosing the path. Their routing strategy makes full use of only the current local information, but not future information. Therefore, the applied strategy is completely local in time and space.

(2) Differences between artificial and real ants:

(a) Real ants exist in three-dimensional space, and the solution of the problem generally lies in a plane; therefore, the three-dimensional space should be abstracted into a plane. However, the computer cannot directly and completely describe a continuous plane, and it must be discretized to form a discrete plane consisting of a group of points. Therefore, artificial ants live in a discrete world, and their movement is essentially a transition from one discrete state to another.

(b) Real ants have no memory of the path they have taken in the past. They choose the path only by evaluating the current environment. Artificial ants have an internal state that records their past behavior.

(c) Real ants release chemical pheromones on their paths, and artificial ants store digital state variables on their paths. The state variable uses an $n \times N$-dimensional pheromone matrix, where N represents the scale of the problem. It records the performance of the current and historical solutions of the artificial ant and can be read and written by other artificial ants who later follow the path.

(d) Real ants release pheromones while moving. Inspired by the characteristics of the problem space, artificial ants can update pheromones after establishing a feasible solution or every time they make a choice. In short, artificial ants do not update pheromones at arbitrary times and locations.

(e) To improve the performance of the algorithm, artificial ants have features that real ants do not, such as local optimization and the ability to predict the future or return to the original path.

8.3.3.3 Mathematical Model of Ant Colony Algorithm

The TSP is among the most well-known NP problems in combinatorial optimization, and it played an important role in the development of the ACO algorithm. The TSP was first selected by researchers as an ant colony algorithm to solve, and later researchers have tested many algorithms on the TSP. Therefore, the quality of the solution to the TSP has become a criterion for evaluating various ACO algorithms designed for static combinatorial optimization problems. The TSP is selected for three reasons. First, it is among the most frequently studied NP problems and is often used as a test object to compare the performance of algorithms. Second, the ant colony algorithm can be easily modified to solve the TSP. Third, the TSP has a clear solution process. The ant colony algorithm can be more intuitively described by solving the TSP, and thus the abstract algorithm description is clearer and easier to understand. This chapter illustrates the conventional ACS model by solving the TSP for n cities on a plane.

In this model, $G = \langle V,E \rangle$ is represented as a weighted graph, where $V = \{v_1, v_2, \ldots, v_n\}$ is the node set of the graph, $E = \{e_1, e_2, \ldots, e_n\}$ is the edge set of the graph, and D is the edge distance or cost. The salesman's residence and the city he is going to are denoted as nodes in V, the road between cities is regarded as edge E, and the weight on the edge is equal to the length D of the corresponding road. That is, for any city $v_{i,}v_j \in V$, the distance from v_i to v_j is denoted as $d_{ij} \in R_+$, where $d_{ij} = d_{ji}$ is assumed; that is, the symmetric TSP is considered. To simulate the behavior of real ants, the following symbols are first introduced:

$d_{ij}(i, j = 1, 2, \ldots, n)$ is the distance between cities i and j;

$b_i(t)$ is the number of ants $m = \sum_{i=1}^{n} b_i(t)$ in city i at time t;

$\tau_{ij}(t)$ is the intensity of the pheromone accumulated on path ij at time t;

$\eta_{ij}(t)$ is a heuristic function that can be interpreted as the visibility of path (i, j) or the heuristic information from city i to city j, which is given by the problem to be solved; $\eta_{ij}(t) = \frac{1}{d_{ij}}$ in the TSP.

Clearly, for ant k, with decreasing $\eta_{ij}(t)$, $p_{ij}^k(t)$ and $p_{ij}^k(t)$ increase. This heuristic represents the probability that ants move from node (city) i to node (city) j.

$p_{ij}^k(t)$ represents the probability that ant k move from node i to node j at time t, $j \in allowed_k$. Each ant should behave according to the following rules.

(1) The next path is selected with the appropriate probability according to the pheromone intensity on the path.
(2) The path traveled in this cycle is not chosen as the next path. A tabu table is used to control this behavior.
(3) When a cycle is completed, the intensity of the pheromone is released according to the length of the entire path, and the pheromone intensity on the traveled path is updated.

Initially, the pheromone intensity is the same along each path; that is, $\tau_{ij}(0) = \mathrm{C}$ (where C is a constant). Ant $k(k = 1, 2, \ldots, m)$ determines the direction of transfer according to the pheromone function of each path during its movement. Ant k at time t at a node can select only one of the target nodes to move toward at that time. Ant k at time t at node i chooses node j as the target node with the probability

$$p_{ij}^k(t) = \begin{cases} \frac{[\tau_{ij}(t)^\alpha \times \eta_{ij}(t)^\beta]}{\sum_{u \in allowed_k} [\tau_{iu}(t)^\alpha \times \eta_{iu}(t)^\beta]} & if \quad \mathrm{j} \in \mathrm{allowed_k} \\ 0 & else \end{cases}. \tag{8.4}$$

In this formula, $allowed_k = \{V - tabu_k\}$ denotes the nodes that ant k may select next. V is the set of neighbor nodes of the node, and $tabu_k$ is the tabu table, which contains the set of nodes that ant k will travel through. In addition, α is an information heuristic factor that indicates the relative importance of tracks; it reflects the relative importance of pheromones accumulated by ants as they perform ant colony search. As α increases, the ant becomes more likely to choose the same path as other ants have chosen; that is, the collaboration between ants becomes stronger. β is the expected heuristic factor, which indicates the relative importance of visibility; it reflects the relative importance of heuristic information in guiding ant colony search. As β increases, the state transition probability approaches the result of the greedy rule. With time, previously deposited pheromones gradually disappear. After n moments, the ant completes a cycle, and the pheromones on each path are adjusted according to the following formula:

$$\tau_{\mathrm{ij}}(t+n) = \rho \times \tau_{\mathrm{ij}}(t) + \Delta\tau_{\mathrm{ij}} \tag{8.5}$$

$$\Delta\tau_{\mathrm{ij}} = \sum_{k=1}^{m} \tau_{ij}^{k} \tag{8.6}$$

where ρ represents the pheromone volatilization coefficient, and $1 - \rho$ represents the pheromone residue factor; $\Delta\tau_{ij}$ represents the pheromone increment on path ij in this cycle. The initial time $\Delta\tau_{ij} = 0$; τ_{ij}^{k} represents the pheromone intensity released by the kth ant on path ij in this cycle.

According to the pheromone updating strategy and the value of $\Delta\tau_{ij}$, Dorigo proposed three basic ant colony algorithm models: the ant-cycle, ant-quantity, and ant-density models [25].

In the ant-cycle model,

$$\Delta\tau_{ij}^{k} = \begin{cases} \frac{Q}{L_k} \ \textit{If the kth ant passes through this cycle}(i,j) \\ 0 \ \ \textit{else} \end{cases} \tag{8.7}$$

where Q is a constant; d_{ij} represents the distance between nodes i and j. In the ant-density model,

$$\Delta\tau_{ij}^{k} = \begin{cases} \mathrm{Q} \ \textit{If the kth ant passes through this cycle}(i,j) \\ 0 \ \ \mathrm{else} \end{cases} \tag{8.8}$$

The difference between the three models is that the ant-quantity and ant-density models use local information; that is, the ant updates the pheromone on the path after completing one step. In the ant-quantity model, $\Delta\tau_{ij}^{k}$ is related to the distance between *cities* i and j; that is, the residual pheromone strength increases with decreasing intercity distance. In the ant-density model, $\Delta\tau_{ij}^{k}$ is a constant, Q, independent of the path; The ant-cycle model uses the global pheromone information; that is, the ant updates the pheromone on all paths after completing a cycle. $\Delta\tau_{ij}^{k}$ is related to the total path length. Because the ant-cycle model exhibits good performance in solving the TSP, it is typically used as the basic model in the ant colony algorithm.

8.3.4 Description of Ant Colony Algorithm

Taking the TSP as an example, suppose that there are m ants in n cities, and $NC_{\max}$ is defined as the maximum number of cycles. The ant-cycle model is used to update the pheromone increment on the path, and we stipulate that each ant chooses not to revisit cities it has passed through before completing a cycle. With time, the previously deposited pheromones gradually disappear. After *n* times, the ants complete a cycle. In this way, each ant closes a loop after *n* migrations. Figure 8.4 shows a flow chart of the ant colony algorithm.

The main steps of the ant algorithm are as follows.

(1) At initialization, time *t* and the number of cycles NC are set to 0; $\tau_{ij}(0)$ on each path is a fixed constant, $\Delta\tau_{ij} = 0$. The position of each ant is set to the initial position.

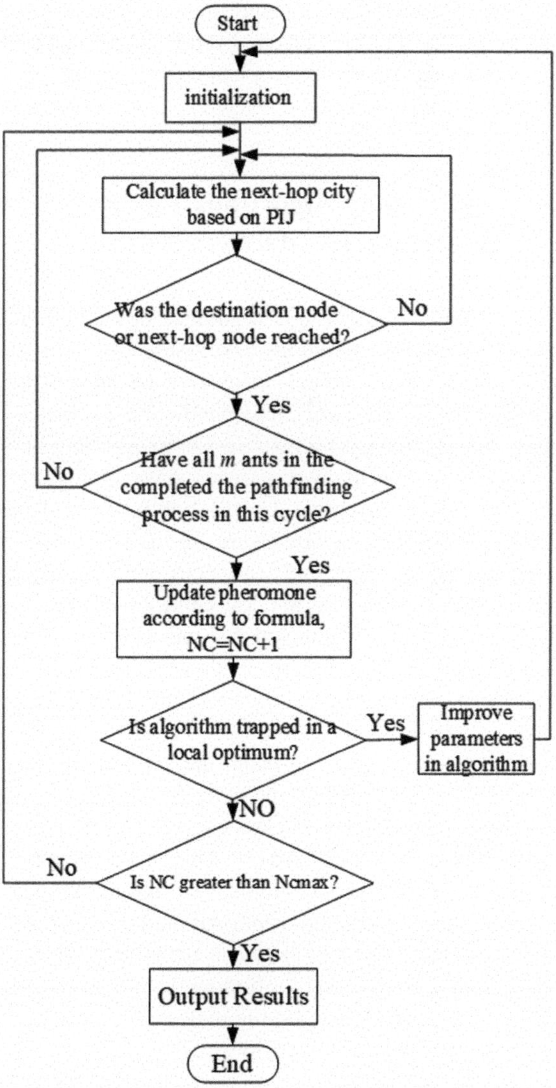

Fig. 8.4 Flowchart of ant colony algorithm

(2) Equation (8.4) is used to calculate the next-hop node for each ant; this step is repeated until the ant reaches its destination node or there is no next-hop node to choose.
(3) If each ant has completed step (2) in this cycle, step (4) is performed; otherwise step (2) is repeated.
(4) If each ant has traveled a complete path, (8.5), (8.6), and (8.7) are used to perform global updates of the pheromones on the traversed path, and NC is set to NC + 1.
(5) The algorithm determines whether it is trapped in a local optimum. If so, step (6) is executed; otherwise, the algorithm proceeds.
(6) The parameters of the algorithm are changed.
(7) Steps (2)–(5) are repeated until NC exceeds NC_{max} or the solution has not improved significantly in recent generations.
(8) The results are output.

8.4 Ant Colony Algorithm–Based QoS Multicast Routing Algorithm for Wireless Networks

The implementation of QoS assurance in wireless networks is increasingly important owing to the continuous development of network technology and increasingly stringent user requirements. However, because of high node mobility and limited resources such as energy and bandwidth [43], it is more difficult to implement multi-constrained QoS on wireless networks than on wired networks.

In recent years, many researchers have applied the ant colony algorithm to wireless network QoS routing [43–46]. For example, the average latency and maximum depth of the propagation tree were used as a cost measure, which was minimized after merging [43]. Another study [44] considered the dynamic characteristics of wireless self-organizing networks and low control packet cost to distribute network resources fairly. Reference [45] showed that more business communication requirements can be satisfied and the lifetime of the entire network can be extended by controlling the energy of nodes and using ant agents to discover and maintain paths. Reference [46] combined the characteristics of a wireless self-organizing network and an on-demand routing algorithm using the ant colony algorithm. However, these studies generally did not consider the node power. Because the nodes in wireless networks have limited power, it is necessary to consider the rational use of node power to extend the lifetime of the entire network. This chapter presents an algorithm for a QoS multicast routing based on ant colony algorithm for wireless networks (QMR-ACA-WN), which is based on the conventional ant colony algorithm. The algorithm has three main features: (1) loops between paths to each destination node are deleted; (2) pheromones are reserved for finding the previous destination node to identify subsequent destination nodes; (3) the node power is considered. The improved algorithm is applied to a wireless network.

8.4.1 QoS Multicast Routing

QoS multicast routing is used to find the optimal path in a distributed network, which requires that all destination nodes be traversed from the source node and all constraints must be satisfied to reach the minimum overhead or a specific service level [17]. To facilitate the analysis of routing problems, wireless networks are considered as undirected and weighted connectivity graphs.

The wireless network is denoted as $G\langle V, E\rangle$, where V is the node set, E is the bidirectional link set; $s \in V$ for multicast source nodes, and $M \subseteq \{V - \{s\}\}$ is the multicast destination node set. In addition, R_+ is a positive real number set, and R^+ is a non-negative real number set.

For any link $e \in E$, four measures are defined:

(1) Delay function, $delay(e)$: $E \to R_+$: The delay is the time it takes for a group to reach the destination node from the source node through the network.
(2) Delay jitter function, $delay(e)$: $E \to R^+$: The delay jitter is the variation in packet transmission time. The delay and delay jitter affect the quality of network transmission.
(3) Bandwidth function, $bandwidth(e)$: $E \to R_+$: Bandwidth is the highest data rate from one point in the network to another in a unit time. It is crucial to reducing the end-to-end latency.
(4) Cost function, $\cos t(e)$: $E \to R_+$: The link cost can be represented by the link distance, remaining link bandwidth, etc.

For any network node, $n \in V$, five measures are defined.

(1) Delay function: $delay(n)$: $V \to R_+$
(2) Delay jitter function: $delay\_jitter(n)$; $V \to R^+$
(3) Packet loss rate function: $packet\_loss(n)$: $V \to R^+$. The packet loss rate is the percentage of packets lost to data loss and errors during data transmission.
(4) Cost function: $\cos t(n)$: $V \to R_+$. The node cost can be represented by the node queue length, node processing speed, etc.
(5) For a given source node $s \in V$, the destination node set M and multicast tree $T(s, M)$ composed of s and M have the following relationships assuming the power function $power(n)$: $V \to R_+$.

$$delay(P_T(s,d)) = \sum_{e \in P_T(s,d)} delay(e) + \sum_{n \in P_T(s,d)} delay(n)$$

$$bandwidth(P_T(s,d)) = \min\{bandwidth(e), e \in P_T(s,d)\}$$

$$\cos t(P_T(s,d)) = \sum_{e \in P_T(s,d)} \cos t(e) + \sum_{n \in P_T(s,d)} \cos t(n)$$

$$delay\_jitter(P_T(s,d)) = \sum_{e \in P_T(s,d)} delay\_jitter(e) + \sum_{n \in P_T(s,d)} delay\_jitter(n)$$

$$packet\_loss(P_T(s,d)) = 1 - \prod_{n \in P_T(s,d)} (1 - packet\_loss(n))$$

$$power(P_T(s,d)) = \sum_{n \in P_T(s,d)} power(n)$$

where $P_T(s,d)$ is the path from source node $T(s,M)$ to destination node S on multicast tree *D*, where the destination node $d \in M$.

The goal of QoS multicast routing is to find a multicast tree that satisfies the constraints in the QoS multicast routing problem defined below. The constraints of the QMR-ACA-WN algorithm are

(1) Bandwidth constraint: $bandwidth(P_T(s,d)) \geq B$
(2) Delay constraint: $delay(P_T(s,d)) \leq D$
(3) Delay jitter constraint: $delay\_jitter(P_T(s,d)) \leq DJ$
(4) Packet loss rate constraint: $packet\_loss(P_T(s,d)) \leq PL$
(5) Power constraint: The next-hop node with higher power in $power(n)$ is selected from among the available next-hop nodes.
(6) Cost constraint: The smallest multicast trees $\cos t(T(s,M))$ that meet the above criteria are selected.

Here *B*, *D*, *DJ*, and *PL* are the bandwidth, latency, delay jitter, and packet loss rate constraints, respectively, of path $P_T(s,d)$. This chapter considers a network structure with eight nodes. As shown in Fig. 8.5, each edge in a node is described by the quaternion (d, dj, b, c), and a node is described by the five-tuple (d, dj, pl, c, p), where *b*, *d*, *dj*, *pl*, *c* and *p* b, d, dj, pl, c, p represent the bandwidth, latency, delay jitter, packet loss rate, overhead, and power, respectively.

8.4.2 QoS Multicast Routing Based on Ant Colony Algorithm

In the QMR-ACA-WN algorithm, each node maintains a pheromone table, which acts as a conventional routing table to record the transition probability of its neighbor nodes. When an ant reaches a node, it chooses the next-hop node according to the transition probability of neighbor nodes in the pheromone table. After several iterations, the path with the most accumulated pheromones is the optimal path, that is, the required path.

8.4.2.1 Transition Probability

The probability that an ant will transfer from one node to another at any given time is given by

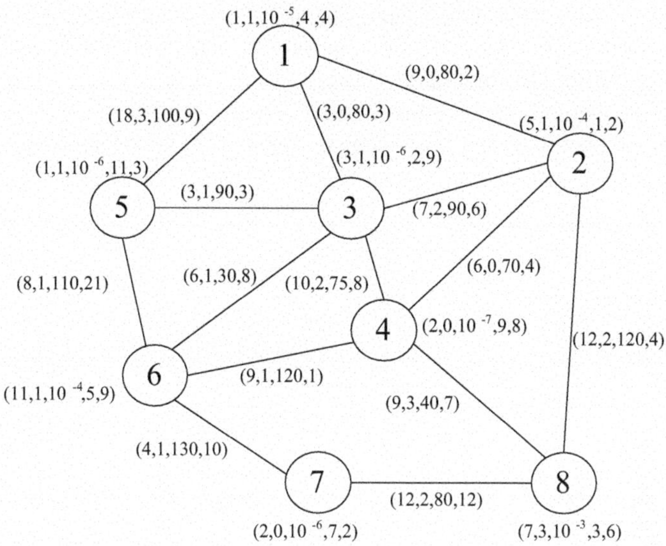

Fig. 8.5 Model of eight-node network structure

$$P_{ij}^{k}(t) = \begin{cases} \frac{\tau_{ij}^{\alpha}(t)\cdot\eta_{ij}^{\beta}(t)}{\sum_{s\in allowed_k}\tau_{is}^{\alpha}(t)\cdot\eta_{is}^{\beta}(t)}, & j \in allowed_k \\ 0, & \text{else} \end{cases} \tag{8.9}$$

The learning described by (8.9) represents the intensity of pheromones accumulated on the path $\tau_{ij}(t)$ at time t, and $\eta_{ij}(t)$ represents the probability that ants move from node i to node j. Here, $\eta_{ij}(t) = \frac{1}{d+dj}$, and $allowed_k = \{V - tabu_k\}$ is the set of nodes from which ant k may select next. V is the neighbor node set of the node, and $tabu_k$ is the tabu table, which contains the node set that ant k travels through. In addition, α is an information heuristic factor; it indicates the relative importance of the track and reflects the relative importance of the pheromones accumulated by moving ants in guiding the ant colony search. β is an expected heuristic factor; it indicates the relative importance of visibility and reflects the relative importance of heuristic information in guiding the ant colony search.

8.4.2.2 Node Power

Nodes in a wireless network are typically powered by batteries. They have limited power and no backup power supply. If node power is not considered, it is very likely that nodes in the network will be exhausted and stop operating because of overuse at low power, which reduces the lifetime of the entire network. We consider the node power when updating the pheromone on the path for ants and select a node with high power as the next-hop node according to the transition probability formula when later ants traverse the path. This choice avoids the use of nodes with low power, prolongs the lifetime of the entire network, and improves the suitability of the conventional ant colony algorithm for wireless networks.

The specific methods are as follows. After Δt moments, m ants have completed a cycle, and the pheromone (i, j) on the path $\tau_{\mathrm{ij}}(t + \Delta t)$ that the ants travel is

$$\tau_{\mathrm{ij}}(t + \Delta t) = (1 - \rho) \times \tau_{\mathrm{ij}}(t) + \Delta\tau_{\mathrm{ij}} \tag{8.10}$$

$$\Delta\tau_{\mathrm{ij}} = \sum_{k=1}^{m} \tau_{ij}^{k} \tag{8.11}$$

$$\tau_{\mathrm{ij}}^{\mathrm{k}} = \frac{Q \times P(j)}{L_k} \tag{8.12}$$

where ρ represents the pheromone volatilization coefficient, and $1 - \rho$ represents the pheromone residual factor; $\Delta\tau_{ij}$ ij represents the pheromone increment on path ij in this cycle. The initial time $\Delta\tau_{ij} = 0$, and τ_{ij}^{k} represents the pheromone left by the kth ant on path ij. Q is constant, and $P(j)$ represents the energy of the next-hop node j. Larger values of P indicate more pheromones on path ij and a larger value of $P_{ij}^{k}(t)$, which is the sum of the d and dj values of all nodes and edges experienced by the kth ant. Updating the pheromones on the path using these formulas enables the ants to change the transition probability and thus select routes with high power, low delay, and low delay jitter.

8.4.2.3 Deletion of Loops Between Paths to Each Destination Node

In the QoS multicast routing algorithm based on the conventional ant colony algorithm, loops are most likely to occur when ants are finding paths to each destination node. For example, when the path to destination node 6 is 1-5-6 and the path to destination node 7 is 1-3-6-7, loops are generated when these two paths form a multicast tree, as shown in Fig. 8.6.

In the QMR-ACA-WN algorithm, loops are deleted by selecting the path of subsequent destination nodes. When the selected next-hop node is on the path to a previously visited destination node, the size of the delay on two paths to the same node is compared with that on the path to a previously visited destination node. The path with a large delay is deleted, and the path with a small delay is selected as a useful path for both routes to avoid the generation of loops. The steps in the algorithm for deleting loops are as follows.

(1) When the destination node sought is not the first destination node in the destination node set, the algorithm determines whether the currently sought node is on the path of a destination node that was previously visited.

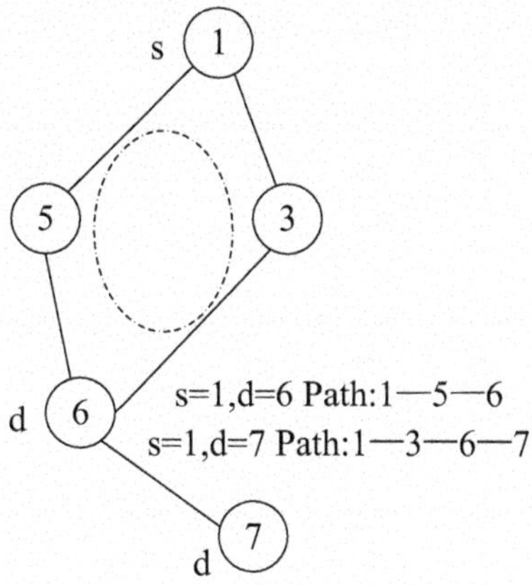

Fig. 8.6 Loops in QoS multicast routing algorithm based on conventional ant colony algorithm

(2) If the node is on the path to a previously visited destination node, the delays on the two paths from the source node to the node are compared, and the path with the smaller delay is selected as the best path for the two routes.
(3) Otherwise, the algorithm returns to step (1) and continues to find the next node until all the paths to the destination nodes are found.

8.4.2.4 Implementation Steps of QMR-ACA-WN Algorithm

In the conventional ant colony algorithm based the QoS multicast routing algorithm, loops may be generated when the paths to destination nodes are sought, and the processes of finding the pheromones to reach each destination node are independent of each other. Here, the pheromones used to find the previous destination node are reserved for finding later destination nodes, and the loops generated in identifying the path to each destination node are deleted, taking into account the power consumption of the nodes. Finally, the improved algorithm is applied to a wireless network. A flowchart of the QMR-ACA-WN algorithm is shown in Fig. 8.7, and the program is presented in Appendix F.1.

The algorithm is implemented as follows.

(1) The network is initialized; each node maintains a pheromone table to record the transition probability for its neighbors.
(2) Edges that do not satisfy constraint (1) in Sect. 8.4.1 are deleted according to the bandwidth of each edge (d, dj, b, c). Nodes that do not satisfy constraint (4) and the edges connected to them are also deleted according to the packet loss rate on each node (d, dj, pl, c, p).
(3) The source node is added to the $tabu_k$ table, where $m = 1, 2, \ldots$, is the number of ants in the network.
(4) The next-hop node is selected according to the transition probability. If the destination node that is sought is not the first destination node in the destination node set, the loop deletion algorithm described above is used to determine

Fig. 8.7 QMR-ACA-WN algorithm flowchart

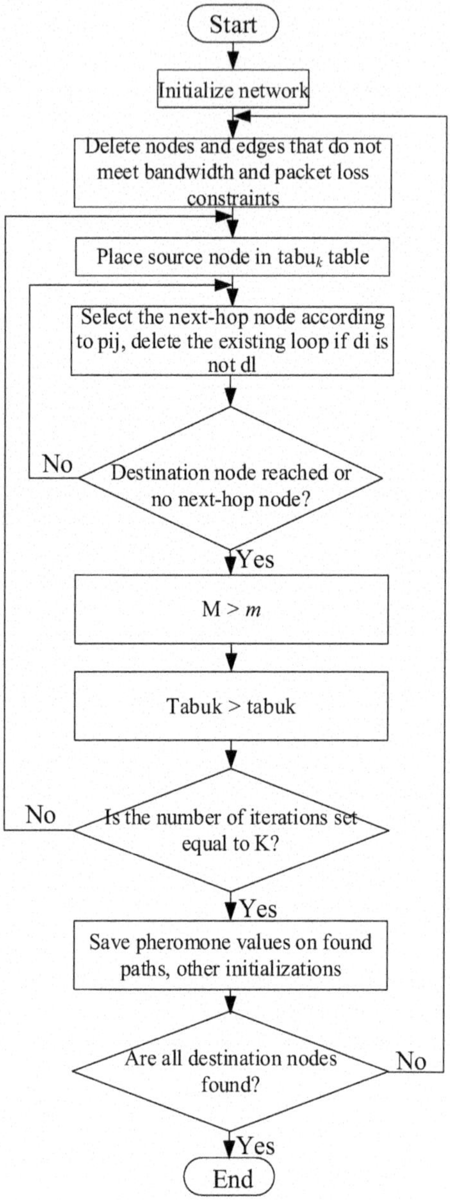

where there is a loop between the arrival destination nodes. If a loop is found, it is deleted; the next-hop node with a smaller delay on the two paths is selected, and the node is added to the $tabu_k$ table. This step is repeated until m ants have found the destination node or there is no next-hop node.

(5) After Δt moments, m ants have completed a cycle, the pheromone is changed to $\tau_{ij}(t+\Delta t)$ on each path (i, j) an ant passes through according to (8.10), (8.11), and (8.12).
(6) The $tabu_k$ table is emptied, and the current value of the node energy is extracted. The transition probability is updated according to (8.9), (8.10), (8.11), and (8.12). The algorithm returns to step (3), and this step is repeated until k iterations have been completed.
(7) When an ant finds a path, the pheromone values on the path are retained, and the pheromones on the other paths are initialized. The algorithm returns to step (2) until a path has been found to each destination node.

8.4.3 Analysis of Simulation Results

We simulated the algorithm proposed in [17] and the QMR-ACA-WN algorithm. Here the cost represents the expected delay, and the goal of this algorithm is to find the multicast tree with the minimum expected delay.

8.4.3.1 Considering Node Power

The initialization diagram of the simulation is shown in Fig. 8.5. The simulation parameters are as follows: $B = 70$ kbps, $D = 50$ ms, $DJ = 12$ ms, $PL = 0.001$, $\alpha = 1$, $\beta = 1$, $\rho = 0.5$, and $Q = 100$. The number of iterations k is 20, and the number of ants per iteration m is 10. The source node $s = 1$ and the destination node set $M = \{5, 6\}$ were simulated 30 times in succession using the algorithm proposed in the literature [17] and the QMR-ACA-WN algorithm. The results are shown in Table 8.2, where the expected delay, delay, and delay jitter are the values on the path to the first destination node, 5, and the expected delay of the multicast tree is the total expected delay to the destination node set M. The main procedure for considering the node power is shown in Appendix F.1.

Because the network has only eight nodes, the path to the destination node set can be found quickly. For $k = 20$ iterations, $k < 5$ is chosen to compare the convergence speed of the two algorithms. With increasing number of convergences, the convergence speed of the algorithm increases. As the mean number of successful convergences decreases, the algorithm converges closer to previous iterations, and the convergence speed increases. For $k < 5$, the numbers of successful convergences of the previous algorithm [17] and the QMR-ACA-WN algorithm are 28 and 30, respectively. The number of convergences of the expected delay in the previous algorithm [17] is 9, whereas that of the QMR-ACA-WN algorithm is 24, and the average values are almost equal; the QMR-ACA-WN algorithm converges almost 67% more quickly, and the average number of successful convergences is one less. For the delay jitter, the QMR-ACA-WN algorithm converges 29 times, and the previous algorithm [17] converges 21 times. By contrast, the convergence speed is also improved, and

Table 8.2 Comparison of results with and without considering node power

Parameter	Algorithm	QoS multicast routing algorithm based on ant colony algorithm [17]	QMR-ACA-WN algorithm
Expected delay	Number of successful convergences	28	30
	Convergences for $k < 5$	9	24
	Average number of successful convergences	6.67	3.2
	Average value (ms)	25.1765	25.1675
Delay	Number of successful convergences	28	30
	Convergences for $k < 5$	9	24
	Mean number of successful convergences	6.67	3.2
	Average value (ms)	14.6875	14.5375
Delay jitter	Number of successful convergences	28	30
	Convergences for $k < 5$	21	29
	Mean number of successful convergences	4.6	2.1
	Average value (ms)	5.05	5.065
Expected delay of multicast tree	Number of successful convergences	28	30
	Convergences for $k < 5$	17	25
	Mean number of successful convergences	5	3.17
	Average value (ms)	60.07	60.27
Average power of multicast tree (mW·h)		25	29.8

the mean number of successful convergences is reduced by one-half. The average expected delays of the entire multicast tree are very similar, whereas the QMR-ACA-WN algorithm has 8 more successful convergences, and the average number of successful convergences decreases by 1.83. The average power of the multicast

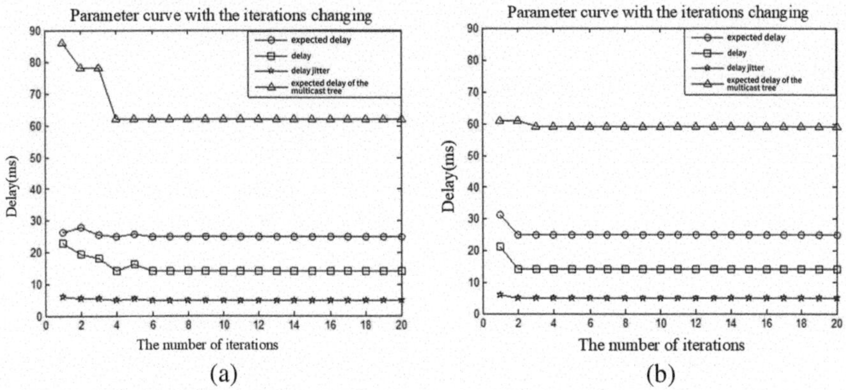

Fig. 8.8 Parameters versus number of iterations for **a** QoS multicast routing algorithm based on ant colony algorithm [17] and **b** QMR-ACA-WN algorithm (which considers node power)

tree in the previous algorithm [17] is lower than that in the QMR-ACA-WN algorithm, indicating that the latter chooses the next-hop path with higher-power nodes, thus prolonging the lifetime of the entire network. Therefore, it can be concluded that the expected delay, delay, and delay jitter of the QMR-ACA-WN algorithm and the convergence speed of the expected delay of the multicast tree are significantly improved. The simulation results are shown in Fig. 8.8.

8.4.3.2 Deletion of Loops Between Paths to Each Destination Node

In the topology in Fig. 8.5, assume the source node is node 1 and the destination node set is {6, 7}. For the loop illustrated in Fig. 8.6 for the conventional algorithm, the loop algorithm between the paths to each destination node is deleted, and the path with the smaller delay is selected as the optimal path for both paths according to the sum of the delay of nodes and paths on the two paths from node 1 to node 6. The delay on path 1-3-6 is less than that on path 1-5-6. Therefore, the path to node 6 is replaced with 1-3-6, and the path to node 7 is 1-3-6-7, as shown in Fig. 8.9. Thus, the loop is deleted, and the total delay of the path is reduced. See Appendix F.1 for the main procedure for deleting loops between paths to each destination node.

8.4.3.3 Retaining Pheromones for Subsequent Destination Nodes

The initialization diagram of the simulation is shown in Fig. 8.5. The simulation parameters are as follows: $B = 70$ kbps, $D = 50$ ms, $DJ = 12$ ms, $\mathrm{PL} = 0.001$,$\alpha = 1$,$\beta = 1$, $\rho = 0.5$, and $Q = 100$. The number of iterations k is 20, and the number of ants per iteration m is 10; the source node s is 1, and the destination node set M is {5, 6}. The algorithm in the literature [17] and the QMR-ACA-WN algorithm were

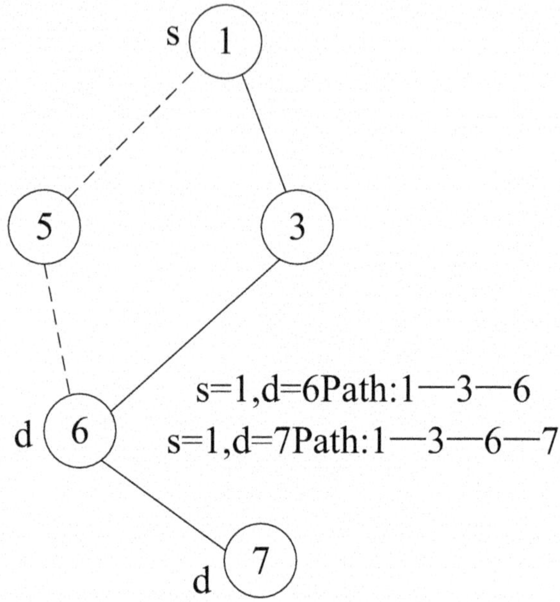

Fig. 8.9 Diagram of loop deletion procedure in QMR-ACA-WN algorithm

simulated 30 times. The results are shown in Table 8.3, where the expected delay, delay, and delay jitter are the values on the path to the first destination node, 6, and the expected delay of the multicast tree is the total expected delay to the destination node set M. The main procedures for reserving pheromones for finding subsequent destination nodes are given in Appendix F.1.

When $k < 5$, the numbers of successful convergences of the previously reported algorithm [17] and QMR-ACA-WN algorithm are 29 and 30, respectively. The average values of the expected delay, delay, and delay jitter for the two algorithms are almost the same; the number of convergences of the expected delay, delay, and delay jitter are 15 and 29, respectively. The convergence of the QMR-ACA-WN algorithm is significantly accelerated, and the average number of successful convergences is significantly reduced. The previously reported algorithm [17] has six fewer convergences of the expected delay of the multicast tree than the QMR-ACA-WN algorithm, and the average number of successful convergences also decreased, by 1.7, whereas the average expected delays of the multicast tree are very similar. Therefore, it can be concluded that the expected delay, delay, delay jitter and expected delay convergence speed of the multicast tree in the QMR-ACA-WN algorithm are significantly improved. The simulation results are shown in Fig. 8.10 [47].

Table 8.3 Comparison of results with and without considering pheromone retention

	Algorithm		
Parameter		QoS multicast routing algorithm based on ant colony algorithm [17]	QMR-ACA-WN algorithm
Expected delay	Number of successful convergences	29	30
	Convergences for $k <$ 5	15	29
	Average number of successful convergences	5.2	1.37
	Average value (ms)	61.01	61.56
Delay	Number of successful convergences	29	30
	Convergences for $k <$ 5	15	29
	Average number of successful convergences	5.2	1.37
	Average value (ms)	34.5075	34.025
Delay jitter	Number of successful convergences	29	30
	Convergences for $k <$ 5	15	29
	Average number of successful convergences	5.2	1.37
	Average value (ms)	8.3225	8.095
Expected delay of multicast tree	Number of successful convergences	29	30
	Convergences for $k <$ 5	19	25
	Average number of successful convergences	4.4	2.7
	Average value (ms)	60.665	61.2

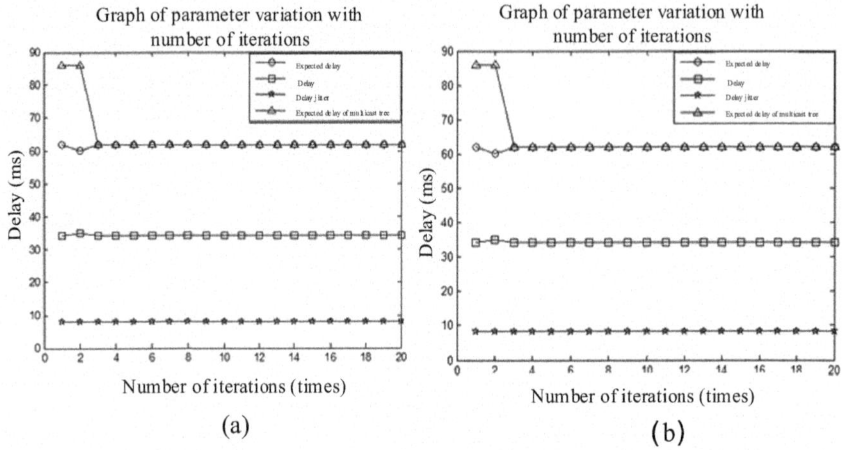

Fig. 8.10 Parameters versus number of iterations for **a** QoS multicast routing algorithm based on ant colony algorithm [17] and **b** QMR-ACA-WN algorithm (which considers pheromone retention)

8.5 Unidirectional Link Routing Algorithm Based on Ant Colony Algorithm for UV Wireless Mesh Networks

We propose a unidirectional link-routing algorithm based on the conventional ant colony algorithm for UV WMNs. The algorithm regards ants as mobile agents, that is, control packets, which are divided into three categories: forward ants (Fants), backward ants (Bants), and broadcast backward ants (BBants). The route is determined by the interaction between ant agents. The processes are described in Appendix F.2.

8.5.1 Topology with Unidirectional Links

Nodes in a UV communication network can operate in both line-of-sight (LOS) and NLOS modes. To ensure good real-time communication, a long communication distance, and a high communication rate, the LOS communication mode should be used as much as possible during communication. When LOS communication is impossible, optimal NLOS communication can be ensured by adjusting the transmission and reception elevation angles in the system. Here, the NLOS(C) communication mode is used. In the actual network topology, NLOS scattering communication can be flexibly deployed by changing geometric parameters of the system such as the elevation angle, light source divergence angle, and field of view of the transmitting light source and receiving device to meet various needs and maintain efficient communication. We arrange the network topology according to the following rules.

(1) The communication nodes in the topology adopt multi-interface nodes.

(2) In a UV WMN, multiple links are typically in operation simultaneously. Mutual interference between adjacent links can be eliminated by setting different interfaces between nodes and their neighbor nodes and varying the transmitting and receiving directions and angles of nodes.
(3) To avoid the exposed terminal problem, the algorithm determines whether neighbor nodes are within the transmission coverage according to the divergence angle and field of view to obtain a reasonable arrangement of neighbor nodes. When the communication coverage area of multiple communication nodes is appropriately adjusted, parallel transmission in the same frequency band can be performed without collision, and the spatial multiplexing of the network can be improved.
(4) In LOS communication mode, the emission angle of the light source is as small as possible to avoid interference with other communication links.

The UV WMN can be represented as a directed graph composed of several nodes and several directional links, where V represents the node set, and E represents the directional link set between nodes in the network. For any node $G\langle V, E\rangle$, V indicates a bidirectional link between nodes i and j; that is, the nodes are within communication range of each other. In addition, $(i, j) \in E$ indicates that node j is within the communication range of node i; that is, node i can communicate with node j, but the node j cannot communicate with the node i. Similarly, $\langle i, j\rangle \in E$ indicates that node i is within the communication range of node $\langle j, i\rangle \in E$, and thus $(i, j) = \langle i, j\rangle \cup \langle j, i\rangle$.

According to these principles for the UV WMN, we designed a network topology with four nodes, as shown in Fig. 8.11, where nodes (x, y, z) are represented by the three-dimensional coordinates of objects in actual communication. Node 1 is the source node S, node 4 is the destination node D, and there is a unidirectional link between nodes 1 and 2. There is NLOS communication between these two points. During NLOS communication, the divergence angle $\theta_1 = 35°$, and the transmitter elevation angle $\phi_1 = 3°$. To enable information from node 1 to bypass obstacles and reach node 2, $\phi_2 = 60°$, where ϕ_2 is the angle between the line connecting the transmitting node and the highest node near the obstacle and the horizontal plane, and the receiver elevation angle θ_2 is 60°.

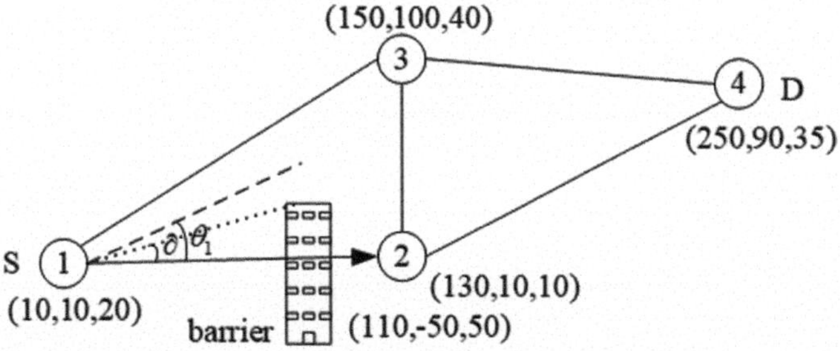

Fig. 8.11 Diagram of four-node network topology

8.5.2 Route Establishment in ULR-ACA-UVWMN Algorithm

In the route establishment process, the optimal path from the source node to the destination node is identified. The algorithm for unidirectional link routing based on the ant colony algorithm for UV WMNs (ULR-ACA-UVWMN algorithm) sends Fants from the source node to find the path to the destination node. After reaching the destination node, it generates Bants to update the pheromone on the path that Fants traveled. If there is a unidirectional link in the path, it generates BBants to find the next-hop node path that cannot be reached. A flow chart of the ULR-ACA-UVWMN algorithm is shown in Fig. 8.12. The steps are as follows.

(1) The network is initialized, and each node maintains a pheromone table to record the transition probability for neighbor nodes.

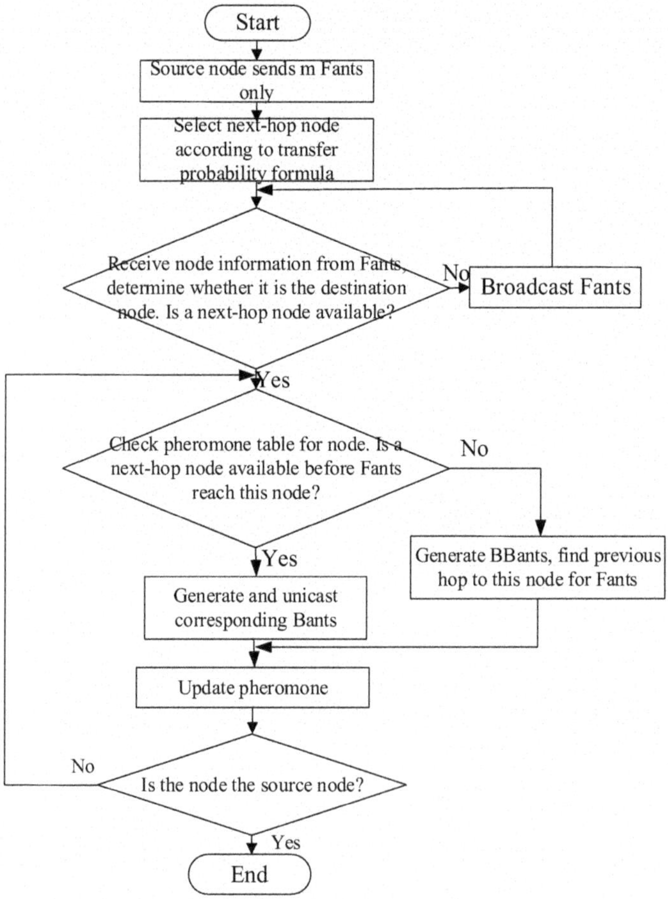

Fig. 8.12 Flowchart of ULR-ACA-UVWMN algorithm

(2) Only Fants are generated at the source node; the source node is listed in the table, where $k = 1, 2\ldots, m$, m is the number of Fants in the network.
(3) The next-hop node is selected according to the transition probability; it is added to the table, and this step is repeated until Fants find only the destination node.or no next-hop node is available.
(4) After Δt time, only Fants complete a cycle to determine whether each Fant has found the destination node. If they have, the corresponding Bant is generated.
(5) If there is a path in the pheromone table with the previous hop for Fants to reach the node, Bants are unicast along the path, and the transition probability is updated according to the transition probability and pheromone increment formulas. Otherwise, BBants are produced to find the path with the previous hop before the Fants arrive at the node. After they arrive at the previous hop, the pheromone is updated according to the information of the corresponding Fants. The intermediate node repeats this step until the source node is reached.
(6) The $tabu_k$ table is cleared. The algorithm returns to step (2) and repeats this step until the next iteration is completed.

8.5.3 Analysis of Simulation Results

We simulated and analyzed the ant colony algorithm with only bidirectional links and the ULR-ACA-UVWMN algorithm. The main program of the ULR-ACA-UVWMN algorithm is presented in Appendix F.1 [48].

8.5.3.1 When Only Unidirectional Links Can Reach the Destination Node

The initial conditions of the simulation are shown in Fig. 8.13. The simulation parameters are as follows: $\alpha = 1$, $\beta = 1$, $\rho = 0.5$, and $Q = 100$. The number of iterations of Fants, k, is 20, and the number of Fants per iteration, source node, and destination node are simulated differently for the ant colony algorithm with only bidirectional links and the ULR-ACA-UVWMN algorithm. The simulation results show that the path to the destination node, 3, cannot be found for the former algorithm because there is a unidirectional link can reach the destination node. As a result, node 3 loses contact with the entire network. When the topology graph is simulated using the ULR-ACA-UVWMN algorithm, the path to node 3 is found to be 1-3, and when node 3 needs to communicate with node 1, its path is 3-2-1, which ensures normal communication between nodes 1 and 3.

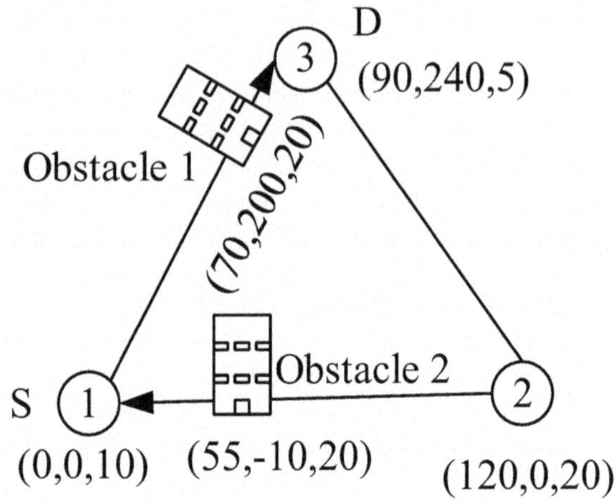

Fig. 8.13 Topology of destination node with only unidirectional links

8.5.3.2 When Both Unidirectional and Bidirectional Links Can Reach the Destination Node

A network topology with four nodes and only one unidirectional link was simulated. The initial conditions are shown in Fig. 8.11. The simulation parameters are the same as above, the source node $S = 1$and the destination node $D = 4$. The ant colony and ULR-ACA-UVWMN algorithms supporting only bidirectional link were simulated. The simulation results are shown in Fig. 8.14.

A network topology with eight nodes and two unidirectional links was simulated. The initial conditions are shown in Fig. 8.15. The simulation parameters are the same as above, the source node $S = 1$ and the destination node $D = 8$. The ant colony and ULR-ACA-UVWMN algorithms supporting only bidirectional links were simulated. The simulation results are shown in Fig. 8.16.

Figures 8.14 and 8.16 show that the ant colony algorithm supporting only bidirectional links to the optimal path has a significantly lower convergence rate than the ULR-ACA-UVWMN algorithm. When the performance parameters of unidirectional links in the network topology are better than those of bidirectional links, the time delay of unidirectional links is less than that of bidirectional links. Because the ULR-ACA-UVWMN algorithm combines unidirectional and bidirectional links, the total delay on the optimal path is much smaller than that of the conventional ant colony algorithm that supports only bidirectional links. Therefore, the ULR-ACA-UVWMN algorithm always chooses paths with good performance, for example, those with small delay, small delay jitter, and high bandwidth, and the convergence speed of the optimal path is significantly faster [49].

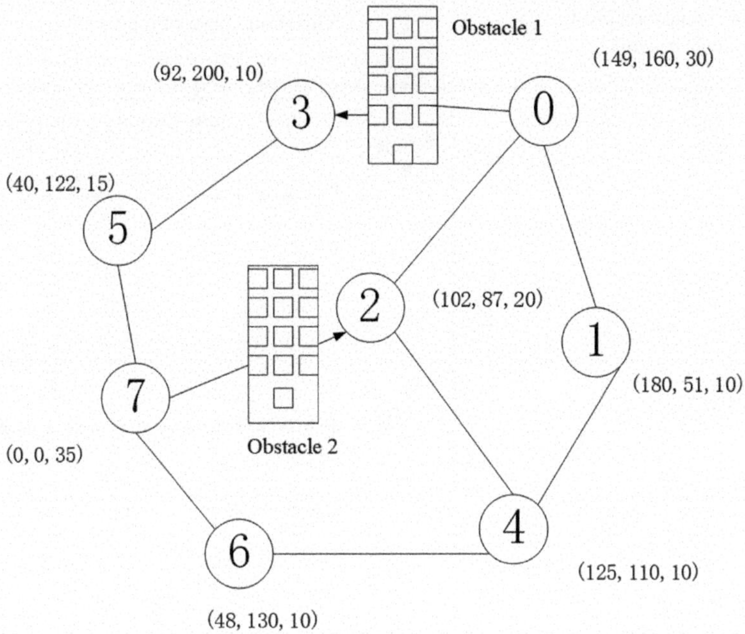

Fig. 8.14 Simulation results for four-node network

Fig. 8.15 Diagram of eight-node network topology

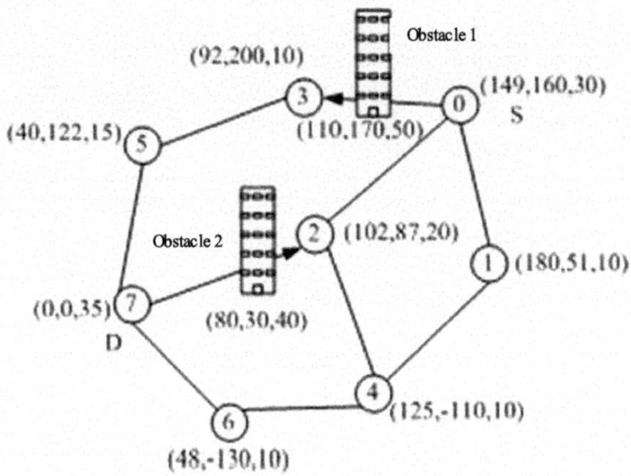

Fig. 8.16 Simulation results for eight-node network

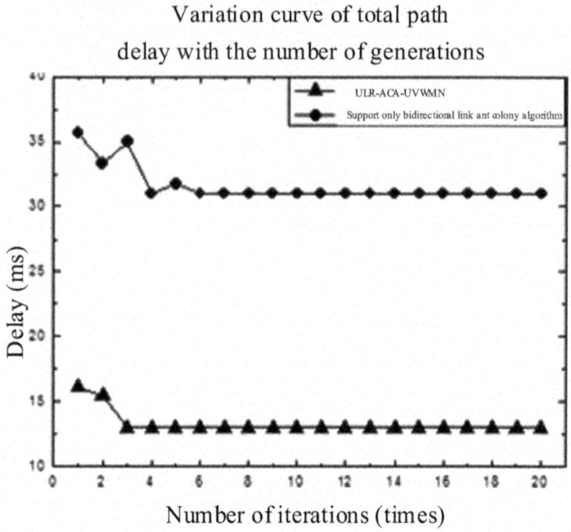

References

1. Akyildiz IF, Wang XD (2005) A Survey on wireless mesh networks. IEEE Commun Mag 43(9):S23–S30
2. Rayner K (2003) Mesh wireless networking. IEE Commun Eng 1(5):44–47
3. Chen CC, Chekuri C, Klabjan D (2009) Topology formation for wireless mesh network planning. In: Proceedings-IEEE INFOCOM, Rio de Janeiro, pp 2671–2675
4. Bayer N, Roos A, Karrer RP et al (2006) Towards carrier grade wireless mesh networks for broadband access. In: 2006 1st workshop on operator-assisted (wireless mesh) community networks, Berlin, pp 1–10
5. Fowler T (2001) Mesh networks for broadband access. IEEE Rev 47(1):17–22
6. Nortel Network (2005) Wireless mesh network—the new wide-area broadband wireless access solution. In: 2005 China wireless technology conference, Beijing, pp 1–3
7. Campista MEM, Esposito PM, Moraes IM et al (2008) Routing metrics and protocols for wireless mesh networks. IEEE Netw 22(1):6–12
8. Raniwala A, Chiueh TC (2005) Architecture and algorithms for an IEEE 802.11-based multi-channel wireless mesh network. In: Proceedings—IEEE INFOCOM, vol 3, Miami, FL, pp 2223–2234
9. Aguayo D, Bicket J, Morris R (2005) SrcRR: a high throughout routing protocol for 802.11 mesh networks (DRAFT). Massachusetts Institute of Technology Technology Report, Boston, pp 1–8
10. Draves R, Padhye J, Zill B (2004) Routing in multi-radio, multi-hop wireless mesh networks. In: Proceedings of the annual international conference on mobile computing and networking, MOBICOM, pp 114–128
11. Clausen T, Jacquet P, Laouiti A et al (2001) Optimized link state routing protocol for ad hoc networks. In: IEEE international multi topic conference 2001 IEEE INMIC 2001, Lahore, pp 62–68
12. Nelakuditi S, Lee S, Yu YZ et al (2005) Blacklist-aided forwarding in static multihop wireless networks. In: 2005 Second annual IEEE communications society conference on sensor and ad hoc communications and networks, vol 25, pp 252–262

13. Ramachandran KN, Buddhikot MM, Chandranmenon G et al (2005) On the design and implementation of infrastructure mesh networks. In: Proceedings of the IEEE workshop on wireless mesh networks, Santa Clara
14. Biswas S, Morris R (2005) ExOR: opportunistic multi-hop routing for wireless networks. Comput Commun Rev New York 35(4):133–44
15. Yuan Y, Yang H, Wong SHY et al (2005) ROMER: resilient opportunistic mesh routing for wireless mesh networks. IEEE Work Wirel Mesh Netw (WiMesh) 9:26
16. Li KW, Tian J (2009) Application of improved ant colony algorithm to the QoS multicast routing. In: 200 International workshop on education technology and training and international workshop on geoscience and remote sensing, ETT and GRS, vol 2, Shanghai, pp 780–784
17. Gu J, Hou X, Song J, Li L (2002) The QOS broadcasting problem based on the ant algorithm is solved. J Hebei Univ Technol 31(4)19–24
18. Ramasubramanian V, Mosse D (2008) BRA: a bidirectional routing abstraction for asymmetric mobile ad hocnetworks. IEEE/ACM Trans Netw 16(1):116–129
19. Yamada K, Umebayashi K, Kamiya Y, et al (2010) A study on routing protocol suitable for directional Links. In: 2010 IEEE radio and wireless symposium, New Orleans, LA, pp 328–331
20. Haibin D (2005) Aquin algorithm and its application. Science Press, Beijing
21. Dorigo M, Maniezzo V, Colorni A (1996) Ant system: optimization by a colony of cooperating agents. IEEE Trans Syst Man Cybern Part B Cybern 26(1):29–41
22. Li S, Chen Y, Li Y (2004) Aquin algorithm and its application. Harbin University of Technology Press, Harbin
23. Wu Q, Wang L (2004) Smart ant colony algorithm and application. Shanghai Science and Technology Education Press, Shanghai
24. Colomi A, Dorigo M, Maniezzo V (1992) Distributed optimization by ant colonies. In: Proceedings of the first European conference of artificial life, Paris, pp 134–142
25. Dorigo M (1992) Optimization, learning and natural algorithm. Ph.D., thesis, DEI, Politecnico di Milano, Italy
26. Gambardella LM, Dorigo M (1995) Ant-Q: a reinforcement learning approach to the traveling salesman problem. In: Proceedings of the 12th international conference on machine learning, Palo Alto, pp 252–260
27. Bullnheimer B, Hartl RF, Strauss C (1999) A new rank-based version of the ant system: a computational study. Cent Eur J Oper Res Econ 7(1):25–38
28. Vittori K, Araujo AFR (2001) Agent-oriented routing in telecommunications networks. IEICE Trans Commun E84-B(11):3006–3013
29. Bell JE, McMullen PR (2004) Ant colony optimization techniques for the vehicle routing problem. Adv Eng Inform 18(1):41–48
30. Kamali S, Opatrny J (2007) POSANT: a position based ant colony routing algorithm for mobile ad-hoc networks wireless and mobile communications. In: Third International conference on wireless and mobile communications, 2007. ICWMC'07, Guadeloupe, pp 21–21
31. Kadono D, Izumi T, Ooshita F et al (2010) An ant colony optimization routing based on robustness for ad hocnetworks with GPSs. Ad Hoc Netw 8(1):63–76
32. Zhang J, Xu X (1999) A new evolution algorithm-ant colony algorithm. Syst Eng Theory Pract 19(3):84–87
33. Wu Q, Zhang J, Xu X (1999) Anton algorithm with mutation characteristics. Comput Res Dev 36(10)240–245
34. Zhang J, Gao Q, Xu X (2000) Adaptive ant colony algorithm. Control Theory Appl 17(1)1–3
35. Ying W, Jianying X (2001) A multi-point new algorithm based on the ant colony system. Comput Eng 27(1):55–56
36. Ying W, Jianying X (2002) A multimedia multimedia multi-broadcast route algorithm based on ant colony algorithms. J Shanghai Jiaotong Univ (Chin Ed) 36(4):526–528
37. Ying W, Jianying X (2002) A self-adaptive ant colony calculation and its simulation research. Syst Simul J 14(1):31–33
38. Changan L, Guangping L, Heping W (2005) The application of the antiquaria algorithm in the planning of drones. Firepower Command Control 30(6):22–24

39. Liu Z, Li Y, Li S (2007) Job-shop multi-resource constrained job shop scheduling based on ant colony algorithm. J Syst Simul 19(1)216–220
40. Hong Z, Weizhi X, Chenxi Z (2009) Research on unstructured P2P search based on ant colony algorithm. Comput Technol Dev 19(2):31–34
41. Zheng W, Liu S, Kou X (2010) An ant colony optimization path recovery algorithm for sensor networks. J Xi'an Jiaotong Univ 44(1):83–86
42. Goss S, Aron S, Deneubourg JL et al (1989) Self-organized shortcuts in the argentine ant. Naturwissenschaften 76(12):579–581
43. Sabari A, Duraiswamy K (2008) Ant based multicast routing algorithm with multiple constraints for mobile ad hocnetworks. In: International conference on security technology, Hainan Island, pp 35–40
44. Hussein OH, Saadawi TN, Lee MJ (2005) Probability routing algorithm for mobile ad hoc networks' resources management. IEEE J Sel Areas Commun 23(12):2248–2259
45. Shokrani H, Jabbehdari S (2009) A novel ant-based QoS routing for mobile ad hoc networks. In: First international conference on ubiquitous and future networks, Hong Kong, pp 79–82
46. Gunes M, Sorges U, Bouazizi I (2002) ARA–the ant-colony based routing algorithm for MANETs. In: International conference on parallel processing workshops, Vancouver, pp 79–85
47. Zhao T, Ke X, Wu C et al (2009) QoS multicast routing algorithm based on ant colony algorithm in wireless networks. J Xi'an Univ Technol 25(4):404–409
48. Marina MK, Das SR (2002) Routing performance in the presence of unidirectional links in multihop wireless networks. In: Proceedings of the international symposium on mobile ad hoc networking and computing (MobiHoc), Lausanne, pp 12–23
49. Wu C (2011) Research on routing protocol in ultraviolet optical communication network based on ant colony algorithm. Master's Thesis of Xi'an University of Technology, Xi'an

Chapter 9
Ultraviolet Wireless Sensor Networks

Wireless sensor networks (WSNs) are a special type of ad hoc network that can be used in areas where wiring and power supply are difficult, in areas that are inaccessible (for example, because of environmental pollution or war), and in temporary situations (such as when natural disasters occur or fixed communication networks are blocked). They do not require fixed network support and can be rapidly deployed; they have strong invulnerability and can be widely used in military, industrial, transportation, environmental protection, and other fields. Compared with conventional wireless networks, WSNs have severe node energy constraints, and energy exhaustion is the main determinant of WSN performance and the network operation cycle. Many factors affect the energy consumption of nodes, the most important of which is the routing algorithm. However, sensor networks have unique characteristics, and the routing algorithms used in earlier ad hoc networks cannot be applied in them. Therefore, it is necessary to design energy-saving routing algorithms for WSNs.

This chapter compares and analyzes the design concepts and solutions of existing routing protocols for WSNs. On this basis, a directed flood routing protocol and directed diffusion (DD) routing protocol suitable for ultraviolet (UV) WSNs are proposed. Finally, the energy consumption of UV multihop communication is modeled, and the mathematical expression for calculating the optimal number of hops is derived.

9.1 Wireless Sensor Networks

Like mobile ad hoc networks (MANETs), WSNs are multihop wireless transmission networks with movable nodes and limited resources; however, there are also significant differences. MANETs are generally composed of dozens to hundreds of nodes. The nodes have a continuous energy supply or a backup energy source. The network is oriented toward data transmission. The design goal is to improve the

X. Ke, *UV Light Self Organizing Network Theory*, Optical Wireless Communication Theory and Technology, https://doi.org/10.1007/978-981-97-7854-6_9

quality of service (QoS) as much as possible to ensure end-to-end data transmission. WSNs integrates four technologies: sensors, embedded technology, networking, and wireless communication. They are new wireless self-organizing networks without infrastructure.

9.1.1 Overview of Wireless Sensor Networks

A WSN is composed of multiple sensor nodes with functions such as wireless communication, data processing, and cooperation to realize distributed wireless data collection [1]. The node that broadcasts the query is called the sink node; it is not only the injection node of the task, but also the return node of the obtained data. The nodes that perform the response task, that is, collect data according to the query and return the result, are called source nodes. The communication architecture of WSNs is shown in Fig. 9.1.

Because many sensor nodes are typically used, they can only be deployed randomly, so their locations cannot be predetermined [2]. To transmit data to sink nodes, the nodes must be connected through wireless channels in a self-organizing network topology that supports coordination between sensor nodes, local data collection, and data interaction between nodes through multihop communication to complete global tasks. The nodes transmit data generated by the sensors to task management nodes through the Internet or a satellite network [3].

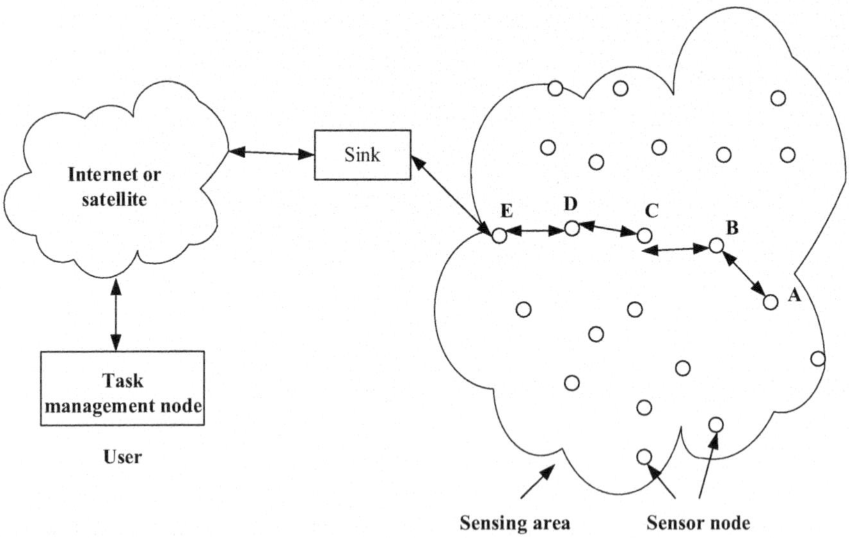

Fig. 9.1 Communication architecture of WSN

9.1.2 Key Technologies for Wireless Sensor Networks

The key technical issues of WSNs are as follows.

(1) Routing problem

The WSN layer establishes an energy-efficient route to reliably relay data from sensor nodes to sink nodes and maximize the network lifetime. The routing problem in WSNs is extremely challenging, mainly because WSNs have several unique characteristics compared to contemporary communication networks and wireless ad hoc networks. First, it is impossible to establish network-wide addressing for WSNs with a large number of sensor nodes. Therefore, classical IP-based protocols cannot be applied to sensor networks. In addition, relative to typical communication networks, almost all applications of sensor networks require sensing data streams to be delivered from multiple areas (data sources) to a central node. Multiple sensor nodes may be located near the observed phenomenon and thus may generate the same data. Consequently, the data redundancy in the network is very high, and the routing protocol must reduce the number of packets sent over the wireless medium using data fusion and improve the energy and bandwidth utilization. Finally, the transmit power, energy supply, processing capacity, and storage capacity of sensor nodes are very limited; thus, network resources should be carefully managed. Owing to these unique requirements and constraints, new problems have arisen in research on WSN routing. We should design a routing protocol that is reliable, considers energy efficiency, and can ensure balanced network energy consumption.

(2) Energy problems

Nodes in a sensor network are typically powered by batteries, and the battery capacity is very limited. In WSNs with thousands of nodes, it is very difficult or even impossible to replace the batteries. If the nodes in the network cannot work because of energy exhaustion, problems occur such as changes in the network topology and the re-establishment of routes; the network may even become disconnected, interrupting communication. Therefore, conserving battery energy in WSNs without affecting the functionality has become the core problem in the design of hardware and software for WSNs. First, because most WSNs are designed for a particular purpose, removing unnecessary functions can save energy for the chosen functions. Second, special protocols that improve the energy efficiency of sensor networks can be designed, or special technologies can be used. These protocols and technologies involve all levels of the network [4]. For example, the physical layer can use ultra-wideband wireless communication technology, and the media access control (MAC) layer can use suitable nodes; the access protocol for switching between dormant and working states in the network layer can use energy as a routing metric. In addition, a cross-layer design can be used to improve the energy efficiency of the network.

(3) Security concerns

Two main security methods are used: secure multicast and key management. Secure multicast may be used for WSNs in a hostile environment. To prevent information providers from injecting false information into the network, it is necessary to implement secure multicast based on source authentication in the WSN. Key management is used in WSNs to minimize communication, because the energy required for communication can exceed that required for computing. Sensor network design should also consider data fusion technology and other issues to reduce data redundancy. The preconfigured key scheme usually calculates the session key using prestored secret information. Because of the limited node storage and energy, the preconfigured key management scheme must conserve storage space and reduce the communication overhead.

(4) Clock synchronization

Communication protocols and applications in sensor networks, such as MAC protocols based on time-division multiple access (TDMA) and time-sensitive monitoring tasks, require clock synchronization between nodes. Elson and Estrin [5] presented a simple and practical clock synchronization method. Briefly, the node records the time using its own clock and then corrects it using a reference time broadcast by a third party; the accuracy depends on the measurement of the interval between the two times. This synchronization mechanism is applied with sufficient precision to monitor a sequence of events from different nodes. Most research on time synchronization has focused on two methods: minimizing the dependence of the synchronization algorithm on the time server and channel quality, and shortening the critical path along which synchronization errors may occur. The synchronization algorithm must be efficient.

(5) Service quality

When a source sends packets to a destination, the network guarantees to meet service performance constraints such as end-to-end delay, bandwidth, and packet loss rate according to predefined QoS requirements. To guarantee QoS, the first task is to find the necessary resources between the source and destination nodes to meet the QoS requirements of the route. Second, once the route for a specific packet is selected, the necessary resources must be reserved for the packet (such as bandwidth and cache space in the router). QoS routing can combine these tasks; thus, QoS guarantees become QoS routing problems.

9.1.3 Key Technologies for UV Wireless Sensor Networks

Research on UV WSNs worldwide is still focused on basic theoretical studies. However, UV WSNs have many unique characteristics that differ from those of other networks and require in-depth research. Research hotspots in UV WSNs focus mainly on the following topics.

(1) Energy-saving communication protocol

The low emission power of UV light enables sensor nodes deployed in the physical environment to maintain a sufficiently long lifetime, which greatly extends the network lifetime. During the lifetime of the network, some sensor nodes may become useless, or new nodes may enter the network. Therefore, it is necessary to arrange the network activity time of sensor nodes so that redundant nodes can be in sleep mode for as long as possible to conserve energy. Another way to reduce energy consumption is to use different types of UV communication to adjust the communication distance of sensor nodes, which can not only minimize the sensing distance, but also cover the target node and conserve energy while ensuring network connectivity.

(2) Congestion control

Wired networks typically use end-to-end and network layer mechanisms for congestion control. However, because simultaneous transmission on different wireless links results in interference, the congestion control methods used in wired networks cannot solve the congestion problem in wireless networks. WSNs communicate mainly by broadcasting, which easily causes congestion. When the packet arrival rate exceeds the packet service rate, congestion easily occurs on sensor nodes close to the central node. The directionality of UV communication can be used to cause packets with different rates to choose different paths to reach the central node. The traffic carried by sensor nodes closer to the central node is then shared by other nodes, and congestion does not easily occur. Congestion on wireless links caused by competition and interference can be alleviated by using the UV non-line-of-sight (NLOS) communication mode NLOS(C) to reduce the number of hops from the source node to the destination node. When fewer nodes participate in data forwarding, competition for the channel decreases, and congestion is less likely. In addition to controlling congestion, an appropriately designed transport protocol can also guarantee QoS in terms of energy efficiency, throughput, packet loss rate, and end-to-end delay.

(3) Routing protocol

Conventional routing protocols rarely consider the energy consumption of nodes when selecting routes. However, nodes in WSNs have limited energy, and prolonging the lifetime of the entire network has become an important goal in the design of routing protocols for sensor networks. Therefore, it is necessary to consider the energy consumption of nodes and the balanced use of network energy. To conserve energy, UV WSNs usually adopt multihop communication. In addition, nodes cannot store a large amount of routing information or perform highly complex routing calculations because of their limited storage and computing resources. To realize efficient and energy-saving routing mechanisms, omnidirectional UV communication can be used to obtain network topology information, whereas directional UV communication can be used for data transmission so that nodes on the transmission path are working, and other nodes are in a sleep state. Alternatively, a mixed routing mechanism of single-hop and multihop communication can be used to ensure energy saving and also reduce the packet loss rate of the network.

(4) Data aggregation

To reduce the number of packets sent on a wireless link, it is necessary to accumulate the highly redundant data generated by the sensor nodes. Data aggregation is an energy-efficient way of gathering important data from sensors with minimal latency. Data latency is very important for many applications such as environmental monitoring, and the freshness of data is an important factor for latency-sensitive applications. Therefore, the development of energy-efficient data aggregation algorithms is critical to improving network lifetime.

(5) Channel access and MAC layer protocol

The data link layer is responsible for access control and establishing reliable communication links between nodes. It is difficult to adapt the conventional MAC protocol based on a competition mechanism to the needs of WSNs. Because this protocol requires multiple handshakes, the probability of data conflict, which results in energy waste, is high. Therefore, the main goal of MAC protocols for UV WSNs should be energy conservation. In addition, they should have good scalability and collision avoidance ability and should use pre-planning mechanisms, such as TDMA, to conserve node energy.

(6) Security mechanism

Most routing protocols do not consider security requirements and thus are vulnerable to attack, which can cause an entire WSN to collapse. Many secure routing protocols have been proposed to ensure the confidentiality and reliability of sensing information [6–11]. Most of them use mechanisms such as link layer encryption and authentication, multipath routing, identity authentication, and authentication broadcast to reject forged routing information from external sources. However, the system overhead of these security mechanisms is relatively high, which is not conducive to the energy saving of sensor nodes. UV communication has better regional confidentiality, making it difficult for communication to be intercepted and improving the confidentiality of information transmission. In addition, the directionality of UV wireless communication enables selective data forwarding by the routing process in WSN networks. It also reduces unnecessary energy consumption by network nodes and prolongs the lifetime of the entire network.

9.2 Wireless Sensor Network Routing Protocols

Because WSNs differ from other types of wireless networks, especially MANETs, researchers worldwide have proposed many routing algorithms for WSNs. In addition, in view of the inherent characteristics of WSNs, researchers have also proposed many new technologies, such as data fusion technology (data aggregation), intranet processing technology (in-network processing), and so on. The proposed WSN routing protocols can be roughly divided into three categories: flat, hierarchical, and location-based routing protocols.

9.2.1 Classification of Routing Protocols for Wireless Sensor Networks

9.2.1.1 Flat Routing Protocols

(1) Flooding protocols: These are conventional broadcast routing protocols. They do not need to maintain the topology of the network, nor do they require routing calculations. A node that receives a message forwards it to all neighbor nodes by broadcasting. Although the implementation is simple, it is prone to problems with implosion and overlap.

(2) Sensor protocol of information via negotiation (SPIN) [12]: This protocol is a routing algorithm that can avoid implosion and overlap through a negotiation mechanism and resource-adaptive control. The negotiation is conducted using two types of messages, announcing data (ADV) and agreeing to receive (REQ); a third type, the data message (DATA), is used to send the data required for negotiation to the designated node.

The protocol effectively overcomes the shortcomings of flood routing protocols and improves resource utilization. Topology changes do not affect neighbor nodes beyond two hops. However, it does not consider energy saving and data transmission with multiple channels. Therefore, improved routing protocols based on SPIN have been developed, such as point-to-point SPIN, energy-control SPIN, route-loss SPIN, and broadcast channel SPIN.

(3) DD [13]: This protocol is data-centric, and its routing mechanism was designed for sensor networks by Estrin et al. [13]. Other data-centric routing protocols have been developed by improving the DD protocol or extending its basic idea. The main feature of this protocol is that the concept of the gradient is introduced. After the gradient is established through a network control message, the optimal path is selected to transmit the message.

The DD routing protocol is an on-demand routing protocol; thus, it can reduce the number of control messages sent and reduce energy consumption somewhat. However, sink nodes in DD must send many interest messages to request data, increasing the control overhead.

(4) Rumor routing (RR) [14]: RR is an improvement on the flooding routing protocol. To save energy, RR uses the principle of randomness. When a node detects valid data, it adds the event to the local list and then forwards it to a randomly selected neighbor node until the message becomes invalid. When the sink node needs data, it sends a broadcast request. After the node finds data that meet the request conditions from the local list, it returns the data by the reverse route of the request.

RR can reduce the number of times the sink node requests forwarding. In the case of emergencies and many requests, the RR protocol uses more than 10% less energy than the flood routing protocol. RR can solve the implosion problem, but not the overlap problem.

(5) Minimum-cost forwarding algorithm (MCFA) [15]: This algorithm is a direction-aware algorithm. The node does not need to maintain a routing table and instead maintains only a minimum-cost table to reach the sink node. The table has only two columns: the minimum cost estimate and transmission path. Every time an intermediate node receives a packet, it checks whether it is on the path with the least cost. If it is, it transmits the packet and broadcasts it to its neighbors; otherwise, it does nothing.

The minimum cost table is established starting from the sink node, which broadcasts the message with zero cost; the new node sets the minimum cost to infinity. After receiving the overhead packet sent by the sink node, the node calculates the minimum overhead of the link, adds this value to the value in the packet, updates its own overhead table, and forwards the packet with the updated overhead field. In this way, all nodes in the entire network can find the path with the least cost.

The MCFA obtains the minimum-cost path in a simple way, which can conserve network resources. However, the nodes on the least-cost path quickly consume energy and stop operating because of the heavy load, and the network topology changes rapidly.

(6) Energy-aware routing (EAR) [16]: This energy-saving routing protocol uses a reactive routing mechanism. It is also based on the DD mechanism; it maintains a set of routes and determines the packet forwarding path by calculating a probability. This calculation requires the node to know the energy status of all nodes in the network. Nodes calculate the energy consumption between source and destination pairs and forward the packet to the next-hop node with the lowest energy consumption according to the energy consumption criterion.

The EAR routing protocol reduces energy consumption by approximately 21.5% and extends the network lifetime by 44% compared to the DD routing protocol. However, nodes using the EAR routing protocol must collect local information, and routes must be established nearly twice as often as for the DD mechanism.

(7) Random walk routing protocol (RWRP) [17]: This algorithm was the first to implement both multipath routing and statistical load balancing. The Bellman–Ford algorithm is used to estimate the distance between nodes and obtain the location information of the network. After the location information is known, the neighbor node closest to the destination is selected as the next-hop node on the packet forwarding path by a probability calculation. If the probability calculation is good enough, the load balancing of network traffic can be guaranteed.

Because the RWRP adopts multipath routing, it consumes a large amount of computing and storage resources. It is more suitable for large static WSNs. However, it is based on two assumptions; One is that nodes can automatically go to sleep as needed. The other is that the network has a grid-like node distribution, which is commonly referred to as a grid network. It is difficult to satisfy the latter condition in practical applications; therefore, the practical applicability of this protocol is not high.

9.2.1.2 Hierarchical Routing Protocols

(1) Low-energy adaptive clustering hierarchy (LEACH) [18]: This protocol is a low-power adaptive hierarchical routing algorithm. A node is randomly selected as the cluster head and starts to send broadcast requests. Common child nodes select a cluster to join according to the signal strength. The cluster assigns a time slot to each common child node using TDMA and broadcasts the request. Common child nodes send data to the cluster head in the specified time slot.

LEACH adopts data compression and data fusion technology. The cluster head combines multiple messages by compressing data and uses data fusion to reduce the number of messages and energy required to send messages. This protocol can prolong the network lifetime and has good scalability. However, although it includes the concept of periodic reselection of clusters, the entire network is easily paralyzed because all the clocks in the network cannot be synchronized.

(2) Power-efficient gathering in sensor information system (PEGASIS) [19]: This routing algorithm is an improved version of the LEACH protocol. The PEGASIS routing protocol selects a node as the starting node in the network to establish an optimal loop chain. The starting node sends the fused data information to the sink node. Because of the heavy load on the starting node, all network nodes take turns serving as the starting node of the loop chain for balance.

In each round, there is only one starting node and on sink node for data transmission, and the use of data fusion reduces energy consumption. Network simulation results show that the PEGASIS protocol outperforms the LEACH protocol in prolonging the network lifetime. However, when the optimal return link is established in each round, each node must have the location information of other nodes in the network, which consumes too many resources. In addition, the monitoring data of all nodes in each round can be sent to the sink node in only one message, and PEGASIS is suitable only for application scenarios with highly correlated monitoring data.

(3) Threshold-sensitive energy efficient network (TEEN) protocol [20]: This protocol adopts a mechanism for delayed data transmission by nodes, and multiple data are packaged and aggregated for transmission after the identification time, which reduces the number of packets sent and energy consumption. The threshold is used to control the frequency of aggregate sending, which is broadcast to other nodes by the cluster head node. The network can flexibly adjust the threshold to meet the needs of practical applications.

However, if a common node does not receive the threshold information broadcasted by the cluster head, the node will not send any data, which may easily cause network paralysis because of unidirectional links and poor network quality. In addition, the TEEN protocol is based on a time criterion, and time synchronization and poor real-time performance are inherent problems. The synchronization problem can be solved by using the TDMA method, but there is no effective countermeasure for applications with high real-time requirements.

(4) Self-organizing protocol (SOP) [21]: In this protocol, fixed nodes act as routers to form a backbone network. A tree-like route is generated by a local Markov cycle algorithm in a process called self-organization. There can be only one hop between the detection and routing nodes. After the detection node sends the detected data to the router, the router forwards the data packet to the sink node through the spanning tree.

    Some nodes in the SOP differ from probe nodes and act as pure routers. These routers are static in the network and constitute a complete network. Each sensor node chooses a router as a subnet. The router communicates with the sink node after data fusion. The SOP requires less energy per broadcast than SPIN. However, because SOP is not an on-demand routing protocol in the self-organization phase, it introduces additional energy and control overhead.

(5) Virtual grid architecture routing [22]: This protocol is an energy-saving routing protocol that is used mainly for static or low-mobility network nodes. It requires that the distribution of cluster head nodes in the network has an overlapping symmetrical grid shape. The grid structure allows the cluster head to learn a fixed linear virtual topology. The main function of the cluster head is data fusion, and the function of the nodes in a cluster is to observe the same area.

    Two algorithms are typically used to select cluster heads. One is the integer linear program algorithm based on the greedy algorithm; the other is the clustering-based aggregation heuristic routing algorithm based on clustering to extend the network lifetime by conserving energy.

    Data fusion is performed at two levels, local and global. All cluster head nodes must perform local data fusion; selected cluster heads perform global data fusion and send the globally fused data to sink nodes.

(6) Hierarchical power-aware routing [23]: This protocol needs to know the energy status of nodes in the entire network; it selects the path with the most energy remaining and relatively few hops as the packet transmission path. This path is called the max–min path. The max–min path algorithm is used to approximate the energy consumed by the path, and the path with maximum energy remaining in the nodes is selected. Then the Dijkstra algorithm is used; it is a typical single-source shortest-path algorithm that finds the shortest path from one node to all other nodes. The optimal path is calculated, and then the paths obtained by these two algorithms are optimized and synthesized to obtain the maximum and minimum paths. Nodes in the cluster can arbitrarily select other nodes as relay nodes, whereas relay nodes outside the cluster can be selected only from the backbone network formed by the cluster heads.

    This algorithm is an effective combination of conventional distance vector routing and energy-saving routing; it can reduce the energy consumption of nodes. However, the computational load of the nodes increases, and each node must store the remaining energy information of all other nodes in the entire network.

(7) Two-tier data dissemination (TTDD) [24]: This algorithm assumes that nodes are static and the location information of each node is known. A network can have multiple sink nodes, and sink nodes can move arbitrarily in the network.

The nodes in the network are divided into several areas in the form of a virtual grid. When an event occurs in the monitoring area, multiple nearby nodes select a node to trigger a data report message. A cluster head node sends the data report message to adjacent nodes in the other four grids outside the grid, and the adjacent node forwards it to the other three adjacent nodes of the grid; finally, the reported data message is sent to all nodes in all grids. Therefore, regardless of where the sink node is in the network, it can receive the reported data message from the nearest node.

However, the TTDD protocol requires nodes in the network to have a certain storage capacity to ensure that the reported data can be cached when multiple events are triggered simultaneously. It is suitable for applications with little data collection and few sudden events.

9.2.1.3 Location-Based Routing Protocols

(1) Geographic adaptive fidelity (GAF) [25]: This protocol is a location-based energy-aware routing algorithm; it is used mainly in ad hoc networks but is also applicable in WSNs. The network is divided into fixed grid regions. Depending on the time, nodes are selected in turn to be in the working state while other nodes in the grid enter the sleep state. Working nodes detect events and report data, whereas the other nodes can minimize energy consumption until they are awakened to the working state. If the grid division is effective, the GAF algorithm will not affect the reliability of routing. However, nodes must be equipped with a GPS unit, which is convenient for effectively dividing the grid.

(2) Geographic and energy-aware routing (GEAR) [26]: This protocol is an energy-aware location-based routing protocol. Its core function is to limit the scope of DD interest message transmission and to propagate interest only to a certain area rather than the entire network. Therefore, GEAR can conserve more energy than DD. GEAR must maintain a table that stores an estimate of the energy cost from a neighbor to the destination node. When a data message is sent, the local energy cost estimate value is added to the message so that the node receiving the message can revise the energy cost table.

(3) Most forward within radius (MFR), distance routing (DIR), and geographic distance routing (GEDIR) [27]: These routing protocols are all based on distance and direction. The key idea is that message transmission is divided into upstream and downstream traffic. According to the known location information, a packet is delivered to a neighbor according to a certain criterion. The MFR algorithm, like the greedy algorithm, selects the path with the fewest hops. DIR selects the forwarding path to the destination node with the smallest direction angle as the next-hop forwarding path. GEDIR is a two-hop range-based greedy algorithm.

    Because they apply the concept of the greedy algorithm, MFR and GEDIR do not generate routing loops. By contrast, DIR may generate routing loops unless a timestamp is added to the message and a list of received messages is maintained in the node.

(4) Greedy adaptive routing [greedy other adaptive face routing (GOAFR)] [28]: The GOAFR algorithm is a combination of positioning routing and the greedy algorithm; it selects the nearest neighbor to the destination node as the forwarding node. Because it is easy to generate nodes that have no neighbors but are closest to the destination node, sometimes forwarded data cannot be delivered to the sink node.

(5) Power-saving technique for multihop ad hoc wireless networks (SPAN) [29]: SPAN is a wireless ad hoc network algorithm for energy-saving coordination and topology maintenance. Network nodes attempt to be in the sleep state and obtain only the local information of neighbors and their neighbors through link maintenance. If a node finds that one of its two neighbors cannot communicate, it voluntarily becomes a coordinator and switches from the sleep state to the working state. To avoid multiple nodes becoming the coordinator simultaneously, a random back-off method is used to issue announcements.

SPAN is, strictly speaking, an energy-saving routing protocol for self-organizing networks. It can be used not only in MANETs but also in WSNs. Simulation results showed that a network that uses SPAN requires 60% less energy than a WSN without a sleep mechanism.

9.2.2 Design Requirements for Routing Protocols for Wireless Sensor Networks

In contrast to conventional networks, WSNs face many new challenges in information processing, such as data-centricity, managing the computing power and storage capacity of a single node, frequent changes in network topology, and robustness to algorithm. Adaptability make high demands of the routing algorithm. The most important point is that the energy of sensor network nodes is extremely limited, and all routing algorithms must reduce node energy consumption as much as possible to prolong the lifetime of the network and the entire system. Nodes are generally powered by batteries, and once nodes are deployed and begin operation, it is impossible or relatively costly to replace their batteries because of the environment in which nodes work.

Energy consumption cannot be accurately estimated. It is necessary to ensure the longest network lifetime while also ensuring that nodes fail because of energy exhaustion at approximately the same time to reduce the frequency of network topology changes.

Time synchronization cannot be precise. Data collection and reporting in WSNs are strictly differentiated according to time. However, delays occur during the forwarding of message and data packets. Link and network delays are difficult to estimate accurately, and thus communicating nodes cannot achieve precise time synchronization.

9.2.3 Transmission in Wireless Sensor Network Data Service

A major difference between WSNs and MANETs is the data transmission method. A sensor network is data-centric, and MANETs are used primarily to transmit information.

In contrast to MANETs, where all nodes have equal status, nodes in WSNs can have different statuses. Network nodes are divided into sink nodes that receive and store data and source nodes that observe the environment and collect data. Nodes use different data transmission methods, and the programming implementation also affect the algorithms of the routing protocol. The algorithm for implementing the routing protocol in the data-centric mode needs to be determined according to the data transmission methods. The data transmission methods used in WSNs can be divided into the following categories.

(1) No-question, no-answer method: The network has only one task, and the required data are obtained by a fixed method with a certain time parameter. When the source node transmits regularly to the sink node, the sink node does not need to send an $req(n)$. The source node sends data $[rep(n+f(n))]$ to the sink node according to a time rule. Alternatively, the sensor node sends data $[rep(n, time)]$ only when the detected data change abruptly.
(2) One-question, one-answer method: The sink node sends an $req(n)$, and each source node responds with the corresponding $rep(n)$.
(3) One-question, multiple-answer method: The sink node sends an $req(n)$, and each source node responds periodically to the sink node with $rep(n + mK)$, where K is a periodic response parameter, and m is an integer (0, 1, 2 ...). until the sink node sends a $stop(n)$ to the source. The node then stops sending data to the sink node.
(4) Multiple-questions, one-answer method: When the real-time requirements are not high, the sink node requests data frequently using $req(n)$, $req(n + 1)$, ..., $req(n+i)$, and the source nodes can cache all the queried data and then cache it to a certain amount of data. The data are concentrated in a response, $rep(n, n+1, \ldots, n+i)$, which is returned to the sink node during one response.
(5) Multiple-questions, no-answer method: The collected data have no valid changes in multiple queries $req(n)$, $req(n + 1)$, ..., $req(n + i)$, and the node does not respond to $rep(n)$, $rep(n+1)$, .. , $rep(n+i-1)$. Valid data are not sent until the collected data change and a query $req(n + i)$ arrives.

These five methods are based on the query-response mechanism, but when they are applied in sensor networks in different environments, factors such as data transmission delay, data efficiency, and transmission energy efficiency are important. The algorithms corresponding to these five methods are shown in Fig. 9.2.

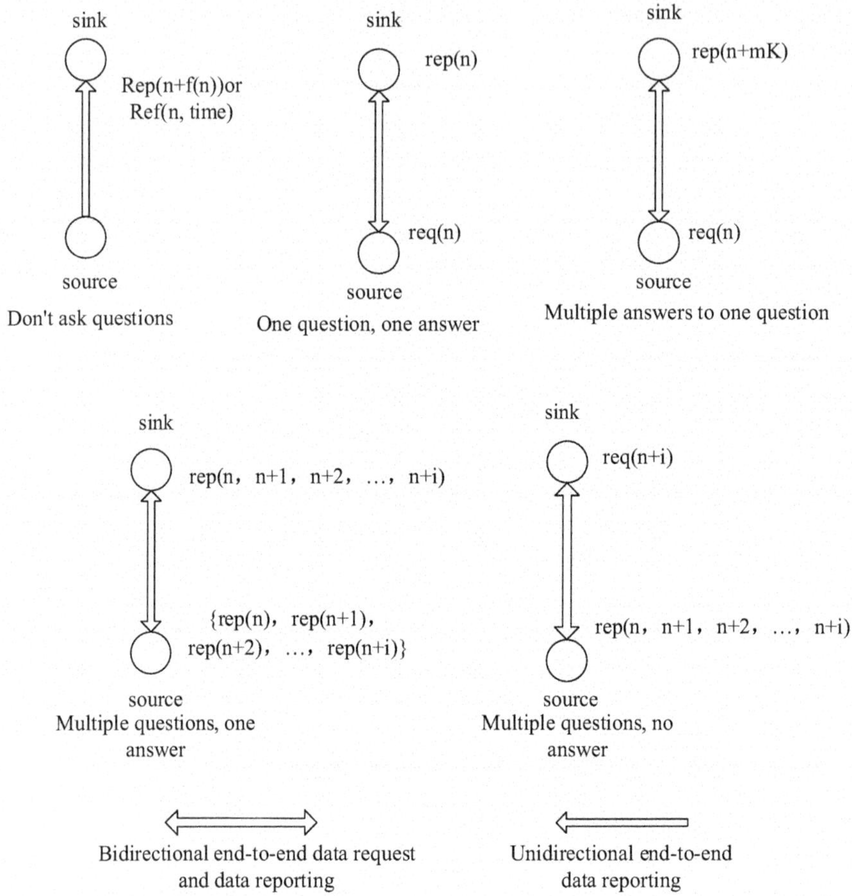

Fig. 9.2 Data service delivery mode based on data-centric network [30]

9.3 Routing Protocol for UV Wireless Sensor Networks

Nodes in WSNs are generally powered by batteries, and it is difficult to replace the batteries after they are deployed and begin operation. Therefore, the routing protocols of WSNs must reduce the energy consumption of nodes as much as possible to prolong the lifetime of the entire network. Because UV communication has the advantages of low emission power, a wide frequency band, a high transmission rate, small size, and NLOS communication [31], UV communication can be combined with a WSN using the UV communication method proposed in the literature [32]. Three communication methods are used for networking; they are highly suitable for low-security sensing systems with short distances and limited energy.

9.3.1 Flood Routing Protocol for UV Wireless Sensor Networks

9.3.1.1 Conventional Flood Routing Algorithm

The flood routing algorithm is a conventional broadcast routing protocol. It does not need to maintain the network topology or perform routing calculations. The process is as follows. When a source node needs to send a data packet to a destination node, the source node first transmits a copy of the data packet to each of its neighbor nodes through the network. Its neighbor nodes broadcast a copy of the packet to all of their neighbors except for the node that just sent them a copy of the packet, and so on until one of these situations occurs: (1) the packet is delivered to the destination node; (2) the packet is sent to the destination node, and the set life cycle becomes zero; (3) all nodes have a copy of the packet.

Figure 9.3 illustrates the implementation of the conventional flooding routing algorithm [33], where S represents the source node, D represents the destination node, and A, B, C, E, F, G, and H represent the other nodes in the network. Now suppose that the source node S wants to send a data packet to the destination node D; it needs to find a transmission path from itself to D. First, S sends a routing request to all its neighbor nodes: A, B, C, and E. When nodes C and E receive the routing request, they forward the data packets to all their neighbor nodes, namely, nodes F, G, and H. Therefore, G, which is a neighbor node of both C and E, receives the data packets from both of them. In this model, G forwards the first arriving data packet to all its neighbor nodes and discards the same data packet when it arrives again later. Node D receives the data packet from node G, which completes data packet transmission between the source and destination nodes.

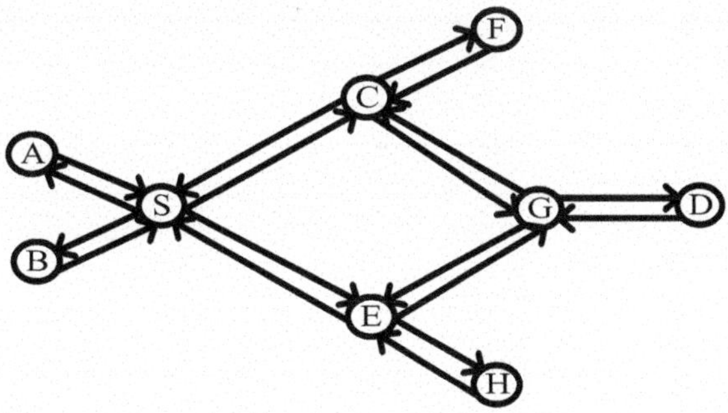

Fig. 9.3 Implementation of conventional flood routing algorithm [33]

9.3.1.2 Directed Flooding Routing Algorithm

The conventional flooding model wastes energy, because every node in the network has to forward packets regardless of whether it is on the final forwarding path. Consequently, the network is flooded with many useless packets, consuming considerable node energy and affecting the lifetime of the entire network. To address the shortcomings of this model, we improved it and obtained a directional flooding routing algorithm based on node location information. In this algorithm, flooding routing appears as the selective forwarding of data packets, which is performed within the coverage of the sending node, whereas nodes outside the coverage do not participate. That is, each node in the network does not send the data packet to all its neighbor nodes, but forwards it only to nodes within the coverage of the node. The location information changes its coverage area, and thus routing proceeds toward the destination node.

The NLOS(B) communication mode is used between nodes in a solar-blind UV WSN [32]. After each node discovers its neighbors, regular information exchange between nodes can reveal to each node the network status and topology. Each node has a set of transceiver equipment, and there are six solar-blind light-emitting diodes (LEDs) [34, 35] on the transmitting equipment. The area covered by each LED is a sector with an opening angle of 60°. During the process, all the LEDs operate at the same time, and only the LEDs covering the routing area operate during data transmission. The implementation of the directional flooding routing algorithm is illustrated in Fig. 9.4. The farthest coverage distance of a node is R. Suppose the source node S wants to route messages to the destination node D. Nodes A, B, and C establish routes as follows. First, the six LEDs of each node operate simultaneously for neighbor node discovery. They divide the coverage area into six blocks. The farthest coverage distance of each block is R, and the coverage angle of the fan-shaped area is 60°. By calculating the angles between the source node S and nodes A, B, and C and the horizontal direction, the source node S knows the coverage area in which its neighbor nodes are located. If the angle is within $\pm\pi/6$, the node is in coverage area No. 1 of S. As shown in Fig. 9.4, nodes C, A, and B are in coverage areas No. 1, 3, and 6, respectively, of S. After all the neighbor nodes of each node and the coverage area of each neighbor node are determined, the nodes exchange information. The source node S knows that the destination node D is a neighbor node of node C; thus, S covers the area in which node C is located, No. 1, by operating the corresponding LED. In the same way, node C operates the No. 2 LED covering the destination node D so that the data packet can be transmitted from the source node to the destination node.

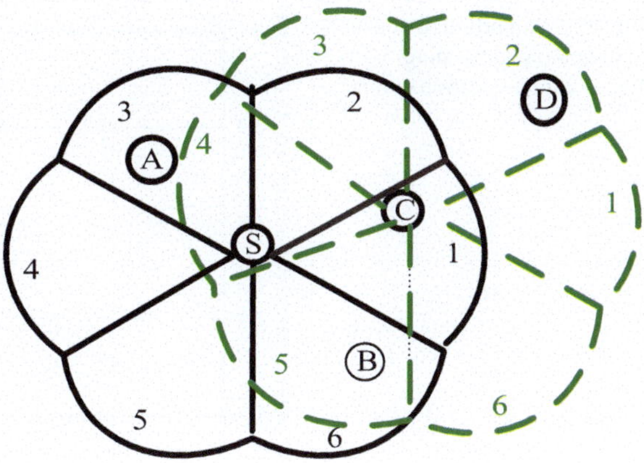

Fig. 9.4 Implementation of directional flood routing algorithm

9.3.1.3 Simulation and Analysis

(1) Performance comparison between conventional and directed flooding

The simulation environment is shown in Fig. 9.5, and the simulation script is presented in Appendix G.2. Ten nodes are configured in an area of 80×80 m^2, and the initial energy of each node is 10 J. NLOS(B) communication is used between nodes; the transmitter elevation angle is $\pi/3$, the divergence angle is $\pi 59/200$, the receiver elevation angle is $\pi/2$, and the receiver field of view is $\pi/3$. Each node's transmitter is equipped with six solar-blind LEDs, each of which covers a sector with an opening angle of 60°. We conducted computer simulations of the two algorithms to compare and analyze the proportion of nodes participating in flooding [the node ratio (NR)] and the average residual energy of nodes. Figure 9.6 shows the NR versus the communication radius.

The NR is 1 for the conventional flooding routing algorithm, because this algorithm broadcasts the message to all its own neighbor nodes, and then the neighbor nodes broadcast it to all their neighbor nodes. Therefore, all nodes in the network receive the message during transmission. The NR of the improved flood routing algorithm is significantly lower than that of the conventional flood routing algorithm, because message routing occurs only in a specific routing area that is directed toward

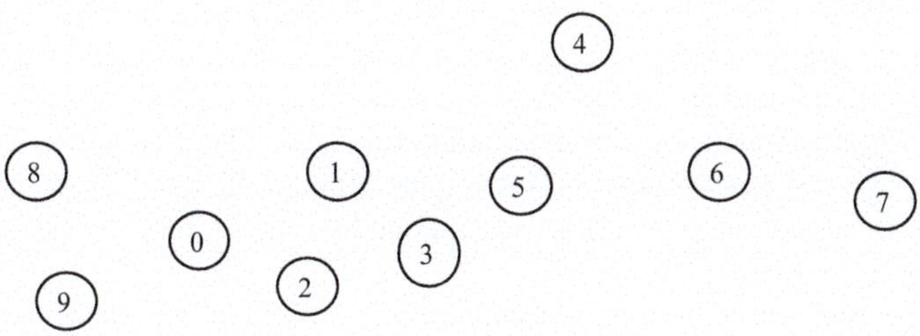

Fig. 9.5 Diagram of 10-node distribution used in simulation

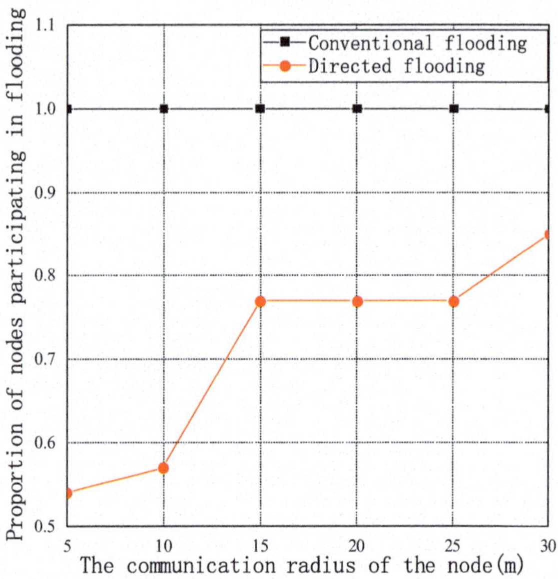

Fig. 9.6 Proportion of nodes participating in flooding versus communication radius

the destination node, and nodes outside the routing area do not receive messages. Therefore, the number of nodes on the transmission path is reduced, and the ratio is less than 1. Figure 9.6 also shows that with increasing communication radius, the NR of the improved model increases. The reason is that with increasing communication radius, the coverage of the nodes increases, and the node density of the sensor network becomes very high; nodes that were not initially covered may be covered, and the number of nodes participating in forwarding increases. When fewer nodes participate in flooding, the average residual energy of the node clearly increases. Figure 9.7 shows the average residual energy of the node over time. The procedure for calculating the average residual energy is given in Appendix G.6. The directional flooding routing algorithm is more energy efficient than the conventional flooding algorithm.

(2) Performance comparison of NLOS(A), NLOS(B), and NLOS(C) communication modes with directed flooding

The NLOS(B) mode can be converted into the NLOS(A) and NLOS(C) modes by adjusting the sum. To understand the effect of directional flooding on the performance of these three communication modes, the end-to-end delay, throughput, delay jitter, and energy consumption are simulated and compared. The simulation scenario is shown in Fig. 9.8. See Appendix G.2 for the simulation script. Eight nodes are configured in an area of $80 \times 80\ \text{m}^2$; each node has one set of transceiver devices, and the initial energy is 10 J.

As shown in Fig. 9.9, the end-to-end delay of the NLOS(C) mode is the smallest, followed by those of the NLOS(B) and NLOS(A) modes. Because the NLOS(A) mode covers the smallest distance, the number of hops from the same source node to

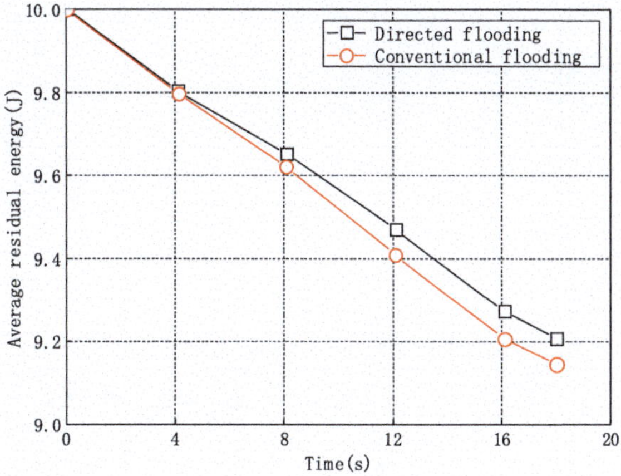

Fig. 9.7 Average residual energy of nodes as a function of time

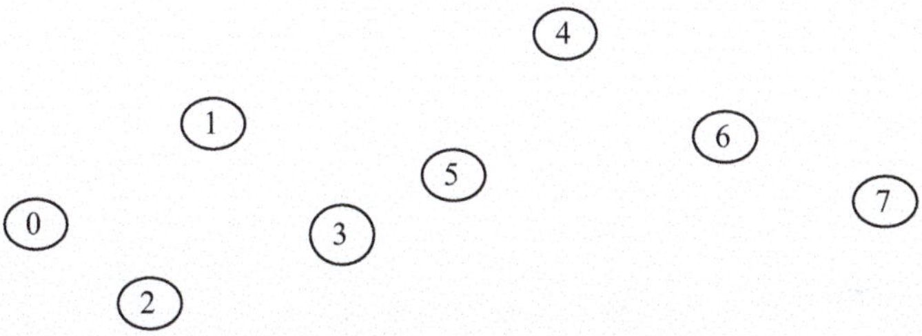

Fig. 9.8 Diagram of eight-node distribution used in simulation

the same destination node is larger than those of the NLOS(B) and NLOS(C) modes. To avoid broadcast collisions, each hop is assigned a random delay at the MAC layer; therefore, with increasing number of hops, the total random delay and end-to-end delay increase. Figure 9.10 illustrates the effect of multihop transmission on the delay jitter. With increasing number of hops, the probability of collision between packets increases, as well as the delay jitter. That is, the NLOS(A) mode has the largest jitter, followed by the NLOS(B) and NLOS(C) modes. The simulation scripts are presented in Appendices G.3 and G.4.

Figure 9.11 shows the throughput versus time for the three modes; the simulation script is presented in Appendix G.6. Overall, the throughput reaches a maximum at the beginning of the simulation and then tends to be stable at the end. NLOS(C) has the highest throughput, followed by NLOS(B) and NLOS(A). The main reason is that the NLOS(C) mode covers a long distance and reaches the destination node through the minimum number of hops, and fewer nodes participate in packet forwarding. Therefore, fewer nodes compete for the channel, and the throughput is higher than those of NLOS(B) and NLOS(A). Figure 9.12 compares the average residual energy for the three communication methods. The procedure for calculating the average

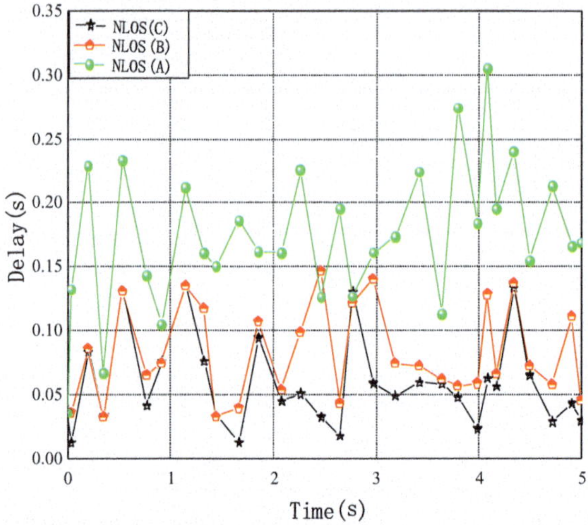

Fig. 9.9 End-to-end delay of three communication modes

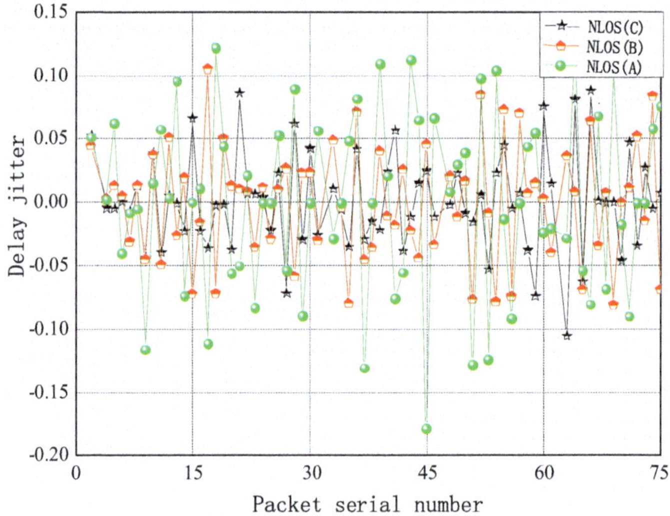

Fig. 9.10 Delay jitter in three communication modes

residual energy is given in Appendix G.6. The NLOS(A) mode has the lowest energy consumption, because data packets are sent from the same source node to the same destination node. The destination node uses the NLOS(A) mode for forwarding more often than the NLOS(B) and NLOS(C) modes. The energy required to send data is proportional to the square of the distance. For smaller distances, less energy is consumed; therefore, the long-distance transmission in the NLOS(C) mode is

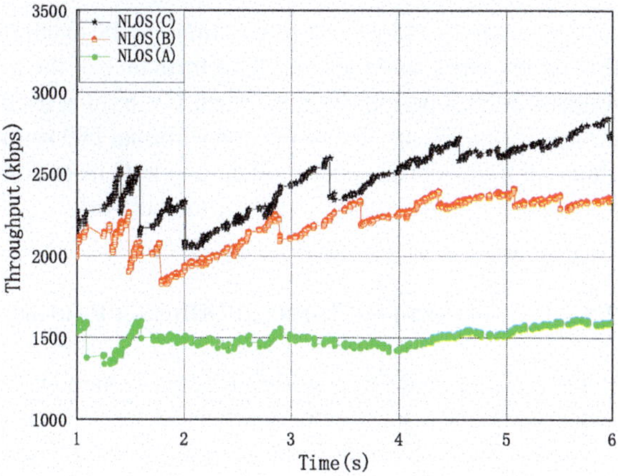

Fig. 9.11 Throughput of three communication modes

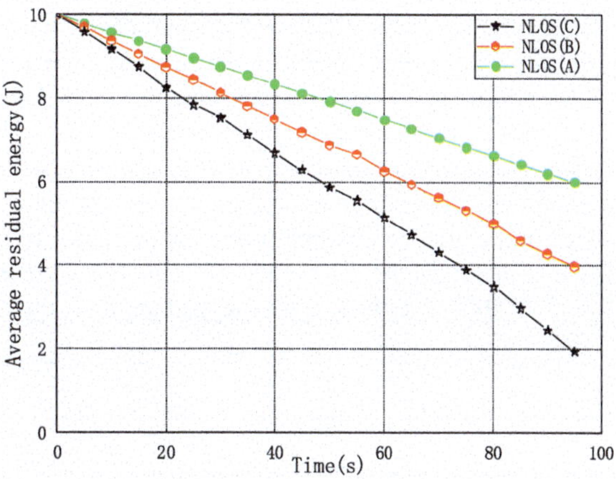

Fig. 9.12 Average residual energy of three communication modes

the most energy-intensive. Although multihop forwarding can conserve a certain amount of energy, if the number of hops is too large, the routing and MAC loads of intermediate nodes increase, offsetting the energy gain obtained by multihop forwarding. Therefore, there is an optimal number of hops that yields the best network performance. The simulation scripts are given in Appendices G.5 and G.6.

9.3.2 Directed Diffusion Routing Protocol for UV Wireless Sensor Networks

The conventional DD routing protocol relies mainly on two flooding processes to complete the entire data transmission process. Although its robustness is good, when nodes send more data, the interest diffusion and data backhaul of the broadcast method result in a large end-to-end time. In the event of delays, the energy consumption is very large, and the overall performance of the network decreases sharply after it operates for a period of time. When UV light is used as the information carrier, its directivity can be applied to the conventional DD routing protocol, and the node can choose the appropriate communication mode from among the three modes according to the performance requirements of the network.

9.3.2.1 Conventional Directed Diffusion Routing Algorithm

In the conventional DD algorithm, data are named using attribute pairs. Perception tasks are injected into the network as an interest in some data and diffuse through the network. As the interest propagates, several gradients are established to extract specific events about which users care, that is, actual data that matches their interests. The obtained data are returned to the sink node along different paths according to the existing gradient. The DD algorithm then selects one or several paths with the lowest cost from several return paths for data transmission according to certain criteria [36].

The following is a detailed introduction to the DD algorithm using an example. For example, to monitor the distributions of rare wild animals, a large number of sensor nodes are placed in a forest. Then a sink node is selected, from which the whereabouts of the animals in the area are sent out as an interest. When the sensor nodes in the area receive the task, they periodically return the observation results. The command can be: Please report to me every 20 ms all nodes in the area [− 50, 100, 200, 500] that observe four-legged animals; this command is valid for 10 s.

(1) Data naming

The observation task in the above example is described as follows:

type = four-legged animal	(type of animal)
interval = 10 ms	(frequency of reporting)
duration = 10 s	(validity period of the order)
rect = [−50, 100, 200, 500]	(observed area)

The returned matching data are also named according to the same rules. For example, if a sensor node finds an elephant, it sends response data in the following format:

type = four-legged animal	(type of animal)
instance = elephant	(name)
location = [125, 220]	(node location)
intensity = 0.6	(signal strength)
confidence = 0.85	(credibility)
timestamp = 01:20:40	(timestamp)

(2) Propagation of interest

The sink node initializes an interest with a specific type of information and periodically broadcasts the interest to its neighbors to accommodate the unreliability of the WSN and prevent the loss of interest. When the sink node periodically sends the interest of a task to the neighbor node, the timestamp attribute must be changed. The value of the period depends on the trade-off between network reliability and network load for the specific application.

Each sensor node has an interest cache, and each interest value corresponds to an entry in the cache. Each entry consists of several fields, which essentially correspond to the fields defined by the interest. Among them, the value of the timestamp field indicates the time when the node most recently received the interest. The interest entry contains a list of gradients, each of which records the flags of the neighbors that sent the interest to the node and the value of the data rate corresponding to the interest. Obviously, the data rate is the inverse of the interval. Each gradient also contains a "duration" field that indicates the validity period of the interest. The duration is generally greater than the network delay; otherwise, the interest is discarded before it is received by the responding node. Figure 9.13 shows a flowchart of interest propagation.

(3) Dissemination of data

When a sensor node in a certain area receives an interest it executes the corresponding function and collects environmental data. If no interest is received, it can be in a semi-sleep state to save energy. When a sensor node finds a target, it searches its interest cache to find matching interest entries (here matching means that the node position is within the rect range of the interest, and the target observed by the node belongs to the type of interest). When a node finds a matching interest entry, it computes the highest data rate among all its output gradients. Then, the observation node selects the neighbor node with the highest data rate in the gradient as the destination node and starts to transmit data.

By analyzing its own data cache, a node can calculate the data rate of the data messages it receives. Before forwarding a data message, the node checks the data rate corresponding to each gradient in the corresponding interest entry. For a gradient whose required data rate is equal to or higher than the data rate received by the node, the node forwards all the received data messages to the target neighbor node in the gradient record. If the required data rate of the gradient is lower than the data rate received by the node, the node can forward only a portion of the received data message. For example, if the date rate of a data message received by a node is 80 events/s, and the data rate required for a gradient to a neighbor node H is 40 events/s, the current node can forward 50% of the data message to node H. Figure 9.14 shows a flowchart of the data transfer process.

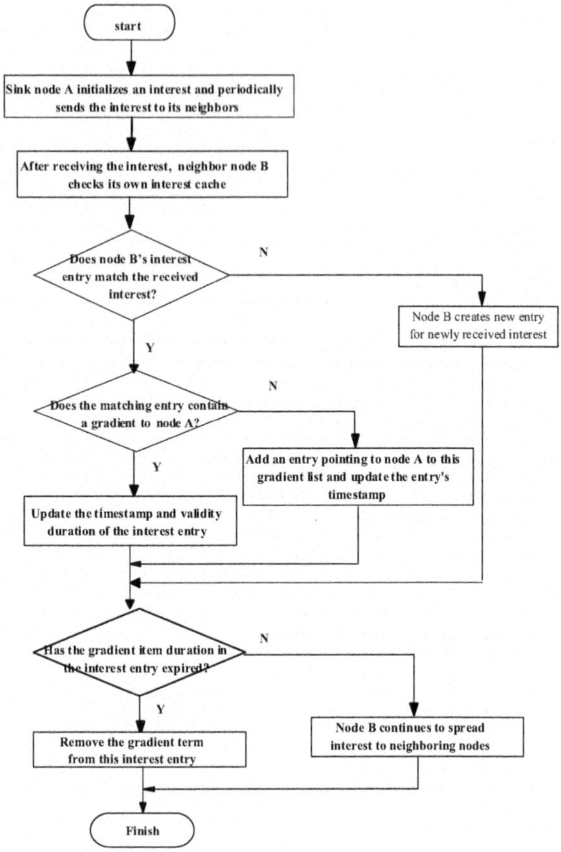

Fig. 9.13 Flow chart of interest propagation

9.3.2.2 UV Directed Diffusion Routing Algorithm

As shown in the previous section, the DD algorithm uses two floods, interest diffusion and exploration data diffusion, to establish a route, and flooding generates many redundant messages and consumes a large amount of network resources. To address this problem, a UV DD (UVDD) algorithm based on the location information of each node in the network is proposed. In the algorithm, interest propagation and data dissemination are performed by the selective forwarding of data packets. Packets are forwarded within the coverage of the sending node, and nodes outside the coverage do not participate in forwarding. In particular, when low-rate exploration data reach the sink node, the sink node strengthens the node sending the data. At this time, only the UV LEDs covering the nodes to be strengthened are operated, and the other LEDs are in the sleep state. The same method is used for high-rate data backhaul from the source node to the sink node. This process avoids the directionality and blindness of interest propagation and data dissemination and reduces unnecessary energy consumption by nodes in the network.

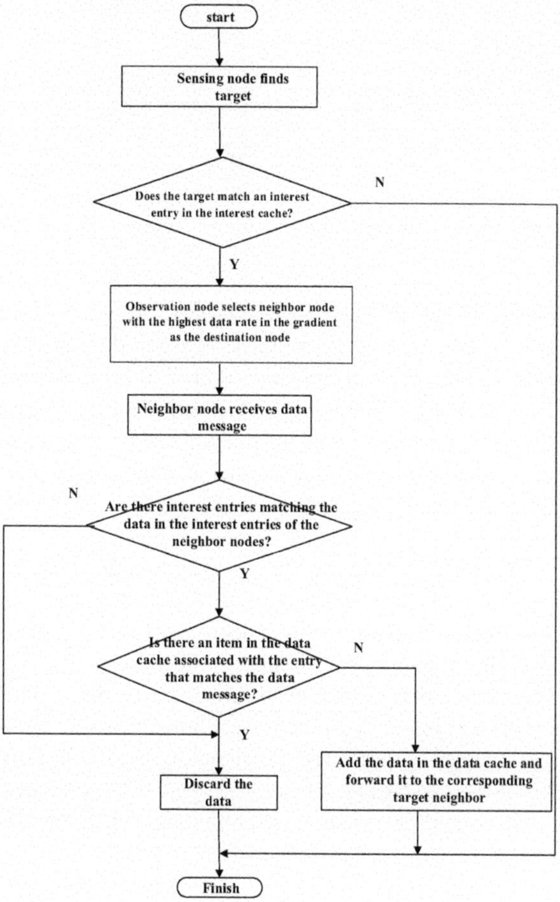

Fig. 9.14 Flowchart of data transmission

9.3.2.3 Simulation and Analysis

(1) Performance comparison of three UV communication modes using UVDD algorithm

Because the NLOS(B) mode can be converted into the NLOS(A) and NLOS(C) modes by adjusting θ_1 and θ_2, the performance of these three modes with directional flooding was evaluated by comparing the end-to-end delay, throughput, delay jitter, and energy consumption. The simulation scenario is shown in Fig. 9.15. See Appendix G.1 for the underlying transmission model. Eight nodes are configured in an area of $80 \times 80\ \text{m}^2$. Each node has one set of transceiver devices, and the initial energy is 10 J.

Figure 9.16 compares the end-to-end delay of the three communication modes when the UVDD routing algorithm is used. The NLOS(C) mode has the lowest latency, followed by the NLOS(B) mode. The time delay of NLOS(A) is the largest.

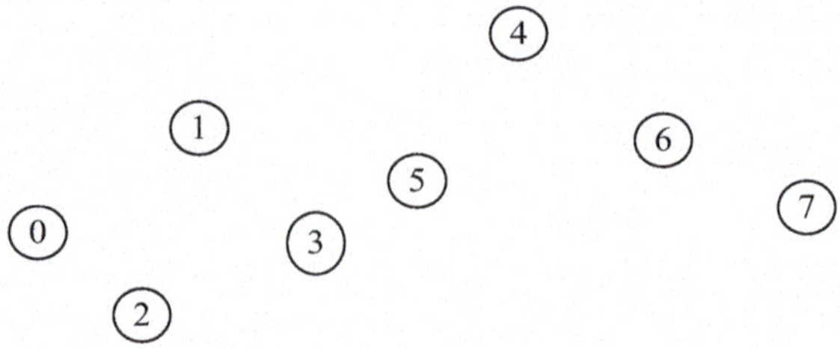

Fig. 9.15 Diagram of eight-node distribution used in simulation

The main reason is that the NLOS(C) mode has the smallest coverage distance, and the fewest hops are required to reach the same destination node from the same source node. With decreasing number of hops, the total random delay of the MAC layer decreases. The delay at the end is small. Figure 9.17 illustrates the effect of multihop transmission on the delay jitter. For fewer hops, the probability of collision between data packets and the delay jitter are smaller. Thus, the delay jitter of the NLOS(C) mode is the smallest, followed by those of the NLOS(B) and NLOS(A) modes.

Figure 9.18 compares the throughput of the three communication modes when the UVDD routing algorithm is used. The throughput reaches a maximum at the beginning of the simulation; it then decreases sharply and finally stabilizes. The NLOS(C) mode has the highest throughput, followed by the NLOS(B) and NLOS(A) modes. The main reason is that the NLOS(C) mode covers a long distance, and packets reach the destination node in the fewest hops. Because there are fewer forwarding nodes, fewer nodes compete for the channel, and the throughput is larger than that of the NLOS(B) and NLOS(A) modes. Figure 9.19 compares the average energy consumption for the three communication modes when the UVDD algorithm is used, where

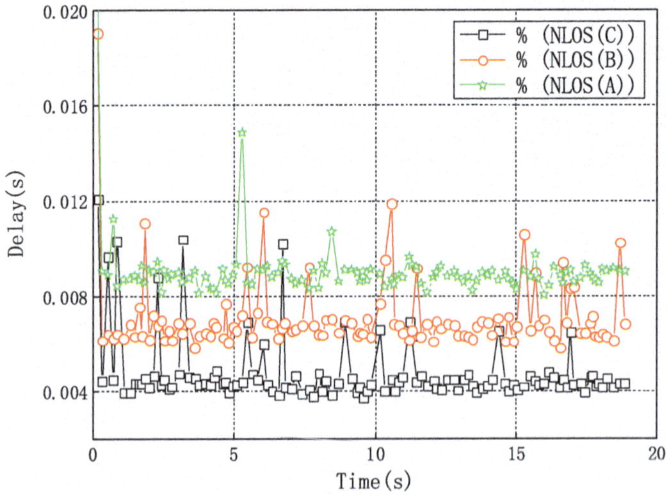

Fig. 9.16 End-to-end delay of three communication modes

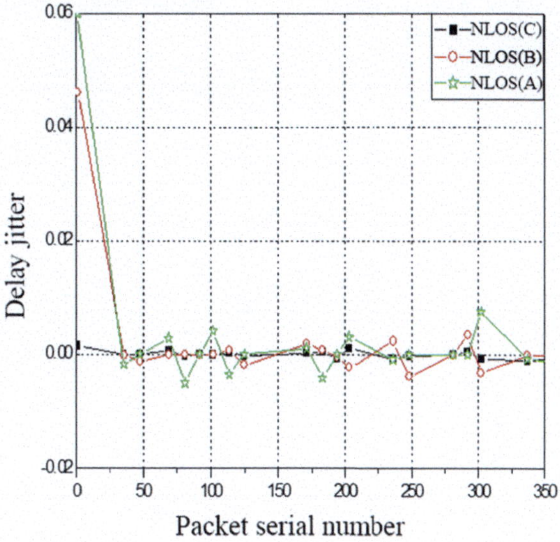

Fig. 9.17 Delay jitter of three communication modes

the energy required to send data in UV communication is proportional to the square of the distance. Therefore, the NLOS(A) mode consumes the least energy, followed by the NLOS(B) mode, and the NLOS(C) mode is the most energy-intensive for long-distance transmission.

(2) Performance comparison of three communication modes after path enhancement

In the DD schemes discussed above, the central node repeatedly diffuses low-rate events, which facilitate path establishment and maintenance; they are called exploratory events. When the source node detects a matching target, it immediately sends probe events along multiple paths to the central node. When a low rate is used

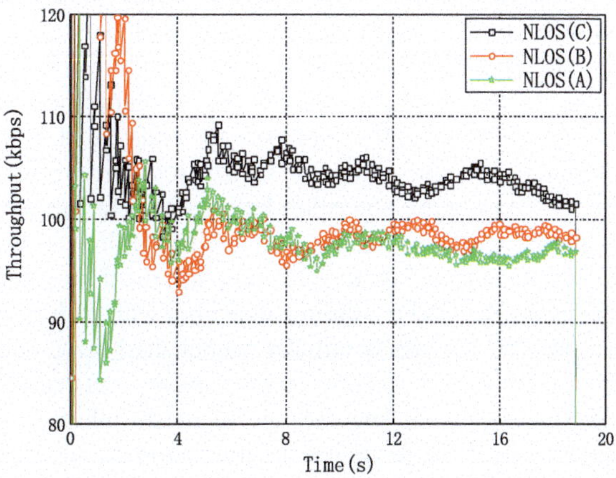

Fig. 9.18 Throughput of three communication modes

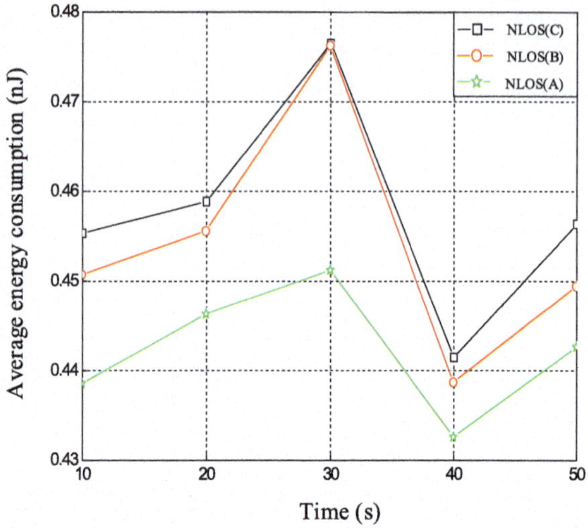

Fig. 9.19 Average energy consumption of three communication modes

to transmit data after the path is established, the collision probability of multihop transmission increases, resulting in a large data delay. Therefore, it is necessary to increase the data transmission rate, strengthen the path, and realize end-to-end low-latency data transmission. The method of path enhancement is as follows. When data from multiple paths are transmitted to the source node, each intermediate forwarding node establishes a gradient for each data packet and records the arrival time of each packet. The source node selects the node corresponding to the earliest data packet to perform path enhancement and sends the path enhancement information. The intermediate node then selects the next node to be strengthened according to the same rules. When the path enhancement information is propagated throughout the network, a high-speed data transmission channel is established between the source and sink nodes.

To evaluate the performance of the UVDD routing algorithm for the three communication modes, we simulated and analyzed the end-to-end delay, delay jitter, and throughput. The simulation scenario in Fig. 9.15 was used. See Appendix G.1 for the underlying transmission model. Path establishment and data return use omnidirectional and directional UV communication, respectively.

Figures 9.20 and 9.21 compare the end-to-end delay and delay jitter of the three communication modes for the UVDD routing algorithm with path enhancement. As for the DD routing algorithm before path enhancement, the delay and delay jitter of the NLOS(C) mode are the smallest, followed by those of the NLOS(B) and NLOS(A) modes. Figure 9.22 compares the throughput of the three communication modes for the UVDD routing algorithm with path enhancement. The throughput of the NLOS(C) mode is initially higher than those of the NLOS(B) and NLOS(A) modes. However, when the network throughput decreases sharply, the throughput of the NLOS(C) mode becomes the smallest, followed by that of the NLOS(B) mode, and the NLOS(A) mode has the maximum throughput.

Fig. 9.20 End-to-end delay after path enhancement for three communication modes

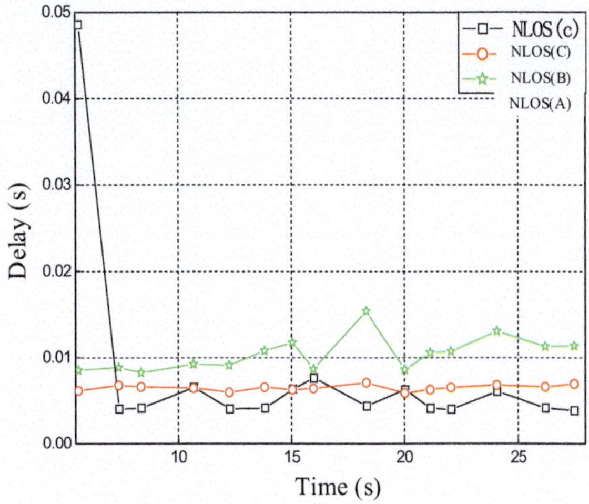

Fig. 9.21 Delay jitter after path enhancement for three communication modes

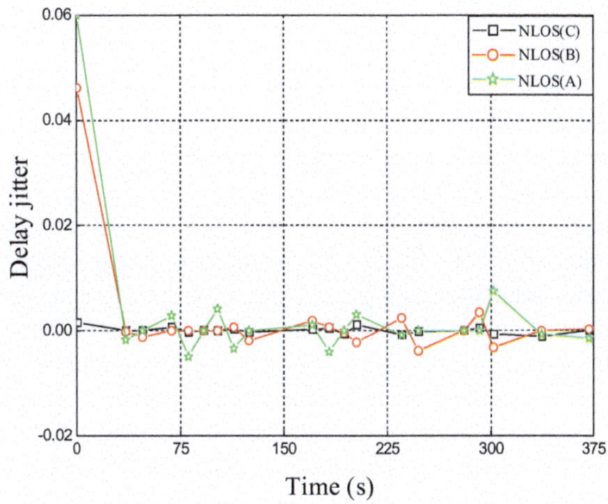

Fig. 9.22 Throughput after path enhancement for three communication modes

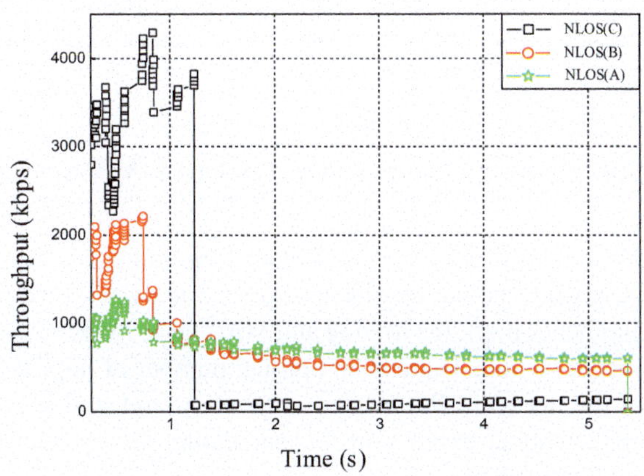

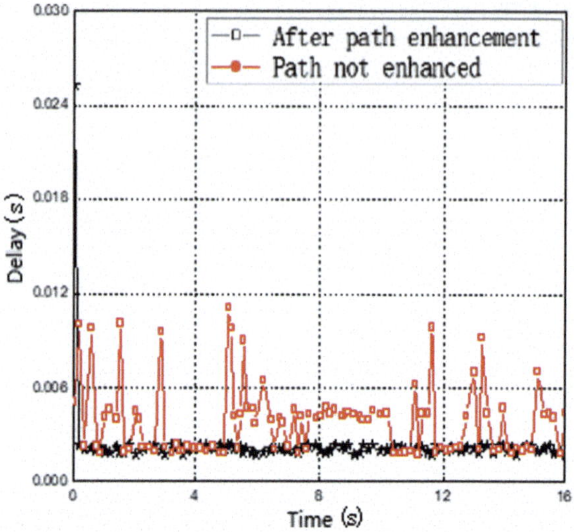

Fig. 9.23 End-to-end delay before and after path enhancement

Figure 9.23 compares the end-to-end delay before and after path enhancement. The end-to-end delay is smaller after path enhancement than before path enhancement, because the data transmission rate after path enhancement is higher, and the data return path after path enhancement is a single enhanced path. The probability of data packet collision is smaller than that in multipath transmission before path enhancement; therefore, the end-to-end delay is small. Figure 9.24 compares the delay jitter before and after path enhancement. With increasing data transmission rate, the delay jitter of data packets transmitted from the same source node to the same destination node decreases. Figure 9.25 shows the throughput before and after path enhancement. The throughput reaches a maximum at the beginning of the simulation and tends to be stable at the end. The throughput is higher after path enhancement than before path enhancement. The reason is that after path enhancement, when data are returned from the sink node to the source node, the data rate is higher, and data are returned by single-path transmission, where collisions are less likely than in the multipath transmission used before path enhancement. Therefore, the throughput is high.

9.4 Energy Saving Analysis of UV Wireless Sensor Network

Although the use of UV multihop communication can reduce energy consumption, increasing the number of hops does not always save more energy. With increasing number of hops, the routing and MAC layer loads increase, which is not conducive to energy saving. There is an optimal number of hops that optimizes the performance of the network. We discuss single-hop and multihop energy saving and obtain a mathematical expression for calculating the optimal number of hops in multihop communication.

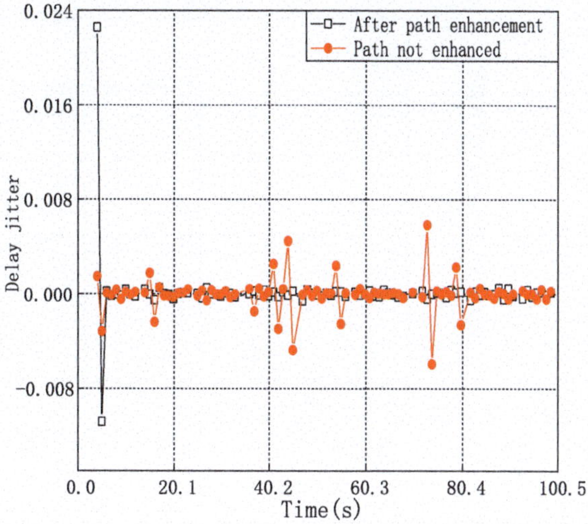

Fig. 9.24 Delay jitter before and after path enhancement

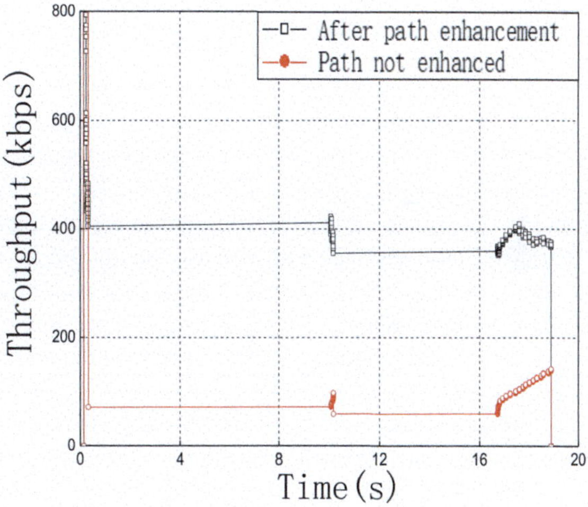

Fig. 9.25 Throughput before and after path enhancement

9.4.1 Energy Consumption Analysis of UV Wireless Sensor Network

9.4.1.1 Energy Consumption Distribution in Wireless Sensor Networks

In WSNs, the modules that consume energy in nodes include sensor, processor, and wireless communication modules [37]. A typical energy consumption distribution for a sensor node is shown in Fig. 9.26. To prolong the lifetime of WSNs, the energy usage rate of the communication module should be improved.

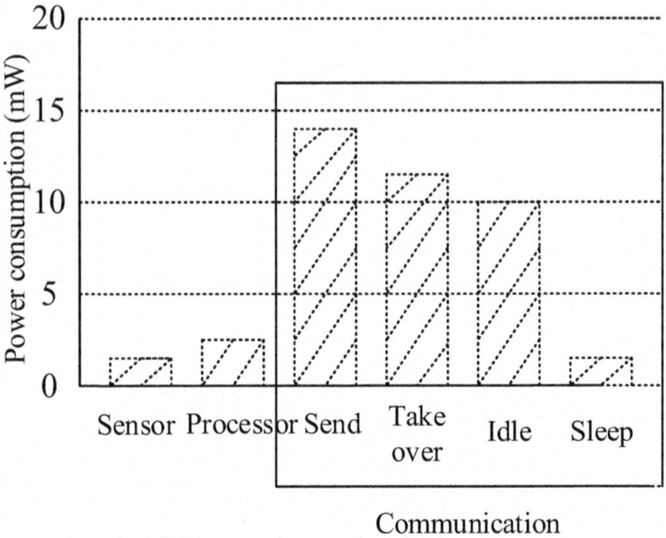

Fig. 9.26 Energy consumption distribution of sensor nodes [38]

The communication module in a sensor node consumes energy mainly for sending, receiving, idling, and sleeping, where sending consumes the most energy. The energy consumed by the transmission of UV light in line-of-sight communication is proportional to the square of the transmission distance; thus, it is necessary to minimize the transmission distance and increase the number of transmissions to save energy. However, as the number of hops increases, the routing and MAC layer loads of intermediate nodes increase, and they consume more energy. Therefore, it is necessary to find the optimal number of hops to minimize the energy consumption of each node in the network. Another problem in multihop communication is the unequal energy consumption of network nodes, which results in rapid energy consumption by nodes close to sink nodes. When nodes fail because of energy exhaustion, the connectivity and coverage of the WSN cannot be guaranteed. This chapter assumes that all sensor nodes consume the same amount of energy when idle and sleeping and reports an energy-saving analysis of sending and receiving.

9.4.1.2 Energy Consumption Model

The energy consumption of data transmission between nodes i and j is defined as follows:

$$E_t(i,j) = \alpha \cdot f_{ij}$$
$$E_r(j,i) = \beta \cdot f_{ij}$$

where $E_t(i, j)$ represents the energy consumed by node i to transmit data to node j at the rate f_{ij} per unit time, $E_r(j, i)$ represents the energy consumed by node j to receive data at the rate f_{ij} per unit time, α represents the total energy consumed by node i to send 1 bit, and β Indicates the energy consumed by node j to receive 1 bit.

$$\alpha \text{ is defined as } \quad \alpha = \begin{cases} a + b \cdot \mathrm{d}_{\mathrm{i,j}}^2 & d_{\min} \le d_{i,j} \le d_{\max} \\ a + b & 0 \le d_{i,j} \le d_{\min} \end{cases}$$

where d_{ij} is the distance between nodes i and j, $d_{\max}$ is the maximum distance at which node i can transmit, and $d_{\min}$ is the minimum distance at which the node can transmit. Here we assume that the energy node i consumes to send 1 bit is equal to that used by node j to receive 1 bit; that is, $\alpha = \beta$, and $b \cdot d_{ij}^2$ represents the energy consumed to transmit 1 bit on the path. The unit of α and β is nJ/bit, and the unit of b is $(pJ \cdot bit^{-1})/m^2$.

9.4.1.3 Energy Consumption Analysis

Assuming that the distance between the source and sink nodes is fixed at $d_{\max}$ and the distance per hop between the source and sink nodes is $d_{\min}$, the number of hops required between the source and sink nodes to minimize the energy consumption of the entire network should be calculated. Figure 9.27 shows a one-dimensional linear network model. Assuming $d_{\max} = k\ d_{\min}$, the optimal number of hops is discussed in the following two cases.

(1) Source node 1 transmits data directly to sink node $k + 1$

In single-hop transmission, the energy consumed for transmitting 1 bit is

$$E_{s\_hop} = E_t(1, \mathrm{k} + 1) + E_r(\mathrm{k} + 1, 1) = 2\mathrm{a} + b \cdot (kd_{\min})^2 \tag{9.1}$$

In multihop transmission, the energy consumed for transmitting 1 bit is.

$$E_{m\_hop} = \sum_{i=1}^{k} E_t(i, i + 1) + \sum_{j=2}^{k+1} E_r(j, j - 1) = 2ka + bkd_{\min}^2 \tag{9.2}$$

When $E_{s\_hop} \le E_{m\_hop}$, a single hop uses the least energy, and $k \le 2a/(bd_{\min}^2)$.

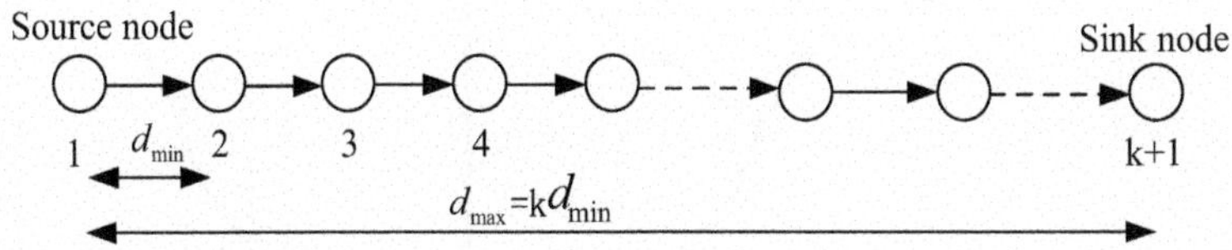

Fig. 9.27 One-dimensional linear network model

(2) Source node 1 transmits data to sink node $k + 1$ over a k-hop distance $d_{\min}$

When $k \geq 2a/(bd_{\min}^2)$, there is an optimal number of hops N that minimizes E_{m_hop}. We denote the distance from the source node to the sink node as R; then $d_{i,i+1} = R/N$. This result is substituted into (9.2) to obtain $E_{m\_hop} = 2aN + bR^2/N$, and the optimal number of hops N_{opt} that minimizes E_{m_hop} is $R \cdot \sqrt{\frac{b}{2a}}$. It can be seen from $N_{opt} = R \cdot \sqrt{\frac{b}{2a}}$ that the optimal distance d_{opt} between two nodes is $\sqrt{\frac{2a}{b}}$, and the optimal distance depends only on the parameters a and b. For a known network, these two constants are known.

9.4.2 Simulation Analysis of Average Energy Consumption with Different Hops

The UV sensor node model is shown in Fig. 9.28. Each node has a set of transceiver equipment, and there are six solar-blind LEDs on the transmitting equipment. The area covered by each LED is a sector with an opening angle of 60°. Omnidirectional UV communication is used in topology discovery. Six LEDs operate simultaneously, and B and C, two neighbor nodes of S, can be covered. Directional UV communication is used to transmit data between nodes S and D. Node S allows only the No. 1 LED covering node C to operate. Similarly, node C allows only the No. 1 LED covering node D to operate, which reduces its unnecessary energy consumption.

The simulation environment is shown in Fig. 9.29. See Appendix G.1 for the underlying transmission model. Eleven nodes are configured in an area of 600 × 600 m^2, and the initial energy of each node is 10 J. Node 1 is the source node, and node 11 is the sink node; $d_{\max} = 550$ m, and $d_{\min} = 55$ m. That is, data can be relayed from the source node to the sink node through a distance $d_{\min}$ with up to 10 hops. For $a = 50$ nJ/bit and $b = 8.2$ (pJ bit^{-1})/m^2, the optimal number of hops according

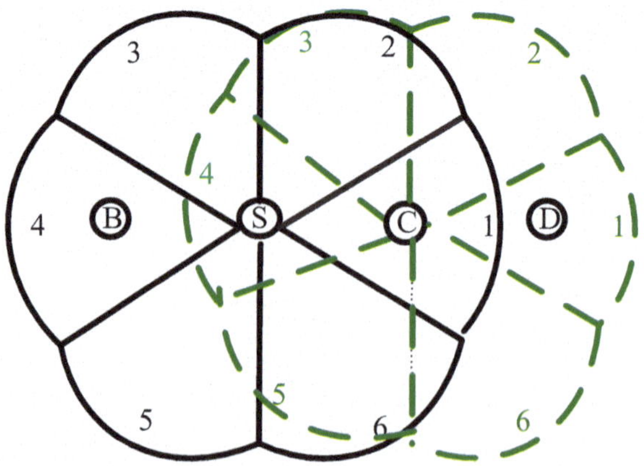

Fig. 9.28 UV sensor node model

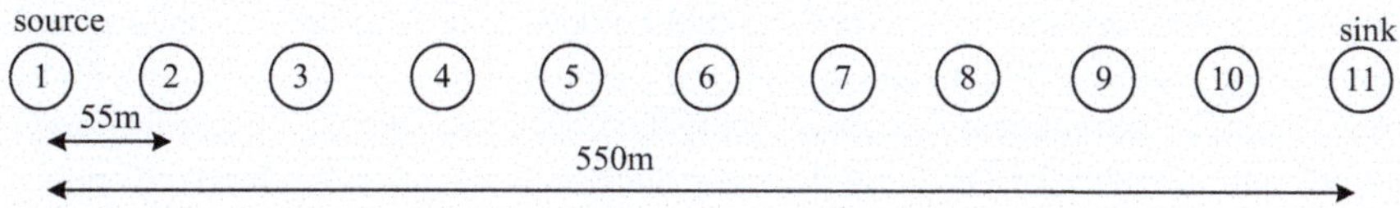

Fig. 9.29 Diagram of linear topology distribution with 11 nodes

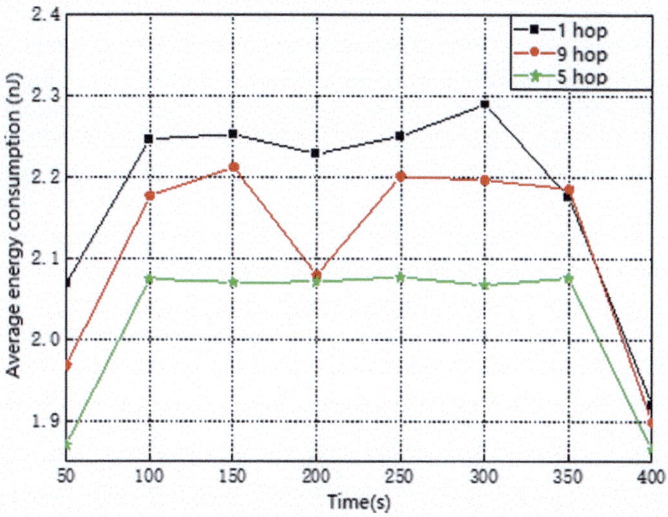

Fig. 9.30 Average energy consumption of one-hop, nine-hop, and five-hop communication

to the above analysis is $N_{opt} = R \cdot \sqrt{\frac{b}{2a}} = 5$. We simulated single-hop, multihop, and optimal-hop communication, and compare and analyze the energy consumption of the three communication methods.

As shown in Fig. 9.30, five-hop communication uses less energy than one-hop and nine-hop communication, which is consistent with the analysis in Sect. 9.4.1.3.

9.4.3 Average Energy Consumption, Packet Loss Rate, and Delay at Different Sink Node Speeds

To understand the impact of node speed on network performance, we simulated and compared the average energy consumption, packet loss rate, and end-to-end delay of sink nodes moving at different speeds. The model of the UV sensor node and simulation scenario are the same as those in 9.2, and a combination of omnidirectional and directional UV communication is used in the simulation. As interest propagates, the six LEDs at each node operate simultaneously, and omnidirectional UV communication is adopted, which can ensure that when the network topology changes, each node in the network knows its own neighbor nodes and establishes a data return transmission gradient. When data are relayed from the source node to the sink node, directional UV communication is used. Only the LEDs covering the forwarding node operate, and the other LEDs are in the sleep state, which can also reduce the energy consumption of the node [39].

At 2 s, the sink node starts to move at speeds of 0, 10, 20, 30, 40, 50, 60, and 70 m/s as the hop count increases from one to the optimal hop count. As shown in Fig. 9.31, as the node speed increases, Energy consumption is not gradually decreasing, and the average energy consumption of the entire network is minimum at a node speed of 50 m/s. The reason is that the single-hop distance is 55 m. When the sink node moves with the optimal number of hops at different speeds, it also broadcasts interest packets to its neighbor nodes. When it moves at 50 m/s, the average distance from the sink node to all neighbor nodes is smaller than those to other nodes. Because the energy required to send data is proportional to the square of the distance, less energy is consumed at smaller distances; consequently, the average energy consumption of the entire network is lowest when the sink node moves at 50 m/s.

Figure 9.32 compares the packet loss rate of sink nodes moving at different speeds. The packet loss rate increases gradually with increasing speed between 3 and 6 s. The ratio of the packet loss rate when the network topology is stable with that at a sink node speed of 50 m/s is lowest; the ratio is larger for other sink node speeds. The reason is that packet loss occurs mainly during data transmission from the source node to the sink node. The data transmission path is based on the gradient established when interest propagates, but the movement of the sink node may create a path in the gradient. Some nodes become unreachable, and the node loses packets until the sink node broadcasts interest again. Network nodes update their neighbor nodes

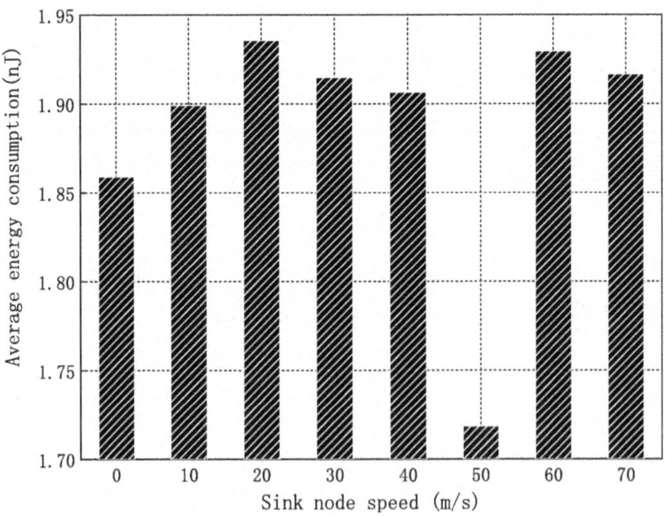

Fig. 9.31 Average energy consumption of entire network when sink node moves at different speeds

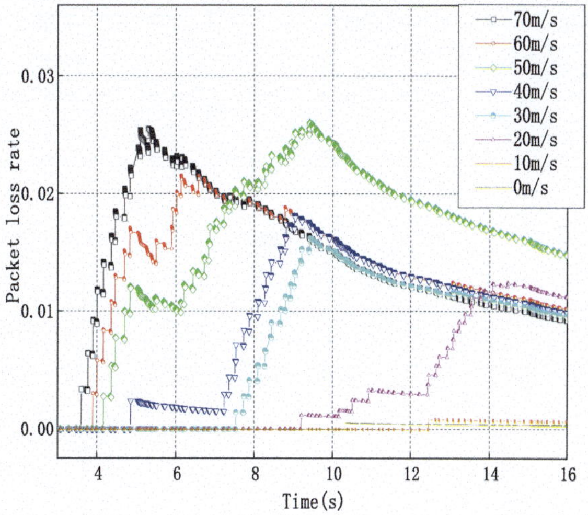

Fig. 9.32 Packet loss rates at different sink node speeds

again, and the packet loss rate decreases after a new gradient is established [40]. The establishment time of the new gradient varies with sink node speed, which is clearly shown in Fig. 9.32

Figures 9.33, 9.34, 9.35, 9.36, 9.37 and 9.38 compare the end-to-end delay when the sink node speed is 50 m/s with those at sink node speeds of 0, 10, 20, 30, 40, and 60 m/s, respectively As shown in Fig. 9.33, when the sink node moves, the end-to-end delay at a sink node speed of 50 m/s is smaller than that when the sink node does not move. The reason is that the movement of the sink node reduces the number of hops from the source node to the sink node, and to avoid broadcast collisions, each hop has a random delay at the MAC layer. At smaller values, the end-to-end delay is smaller. When the network topology changes, the difference in end-to-end delay is largest between sink node speeds of 50 and 0 m/s and smallest between sink node speeds of 50 and 60 m/s. When the topology is stable, the end-to-end delay is roughly the same. The reason is that when the sink node moves at different speeds, the distance between the sink and source nodes decreases with increasing speed; the number of hops, and thus the end-to-end delay, decrease. When the topology is stable, the number of hops from the source node to the sink node is fixed, and the end-to-end delay does not change greatly.

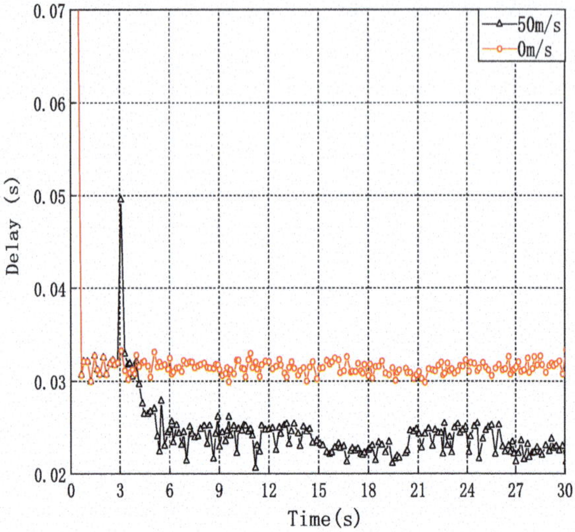

Fig. 9.33 End-to-end delay at sink node speeds of 0 and 50 m/s

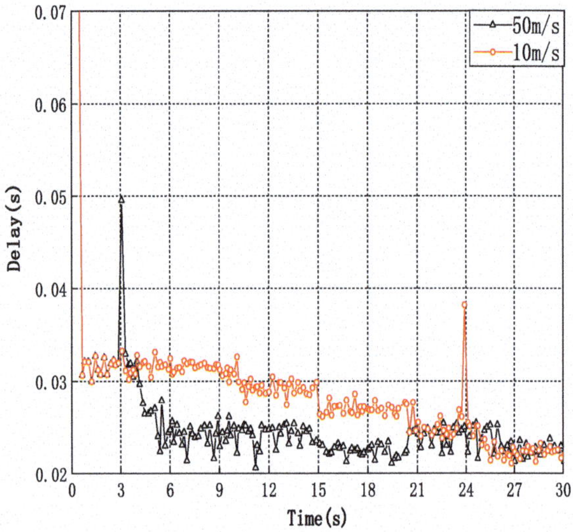

Fig. 9.34 End-to-end delay at sink node speeds of 10 and 50 m/s

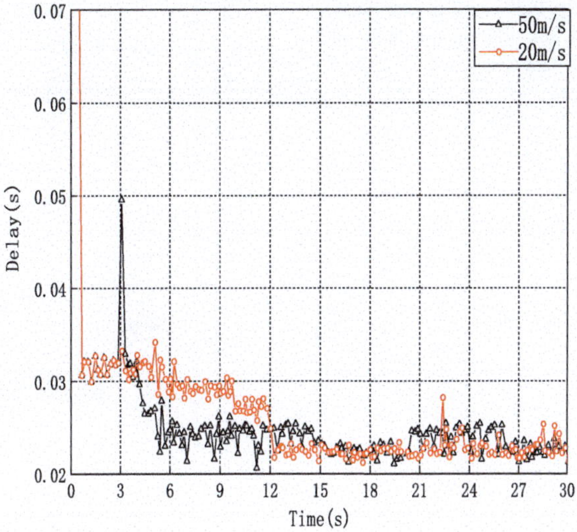

Fig. 9.35 End-to-end delay at sink node speeds of 20 and 50 m/s

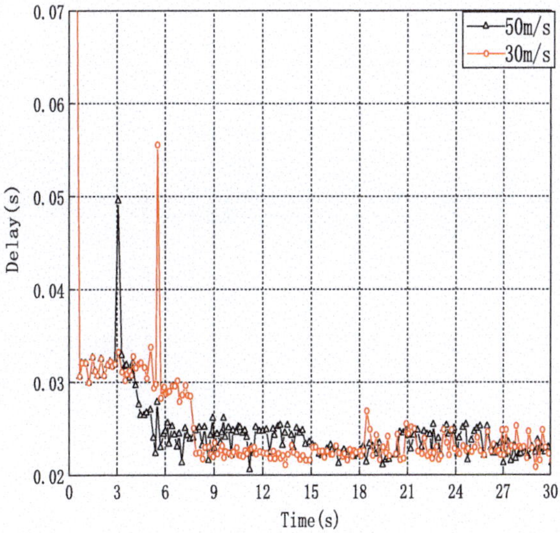

Fig. 9.36 End-to-end delay at sink node speeds of 30 and 50 m/s

Fig. 9.37 End-to-end delay at sink node speeds of 40 and 50 m/s

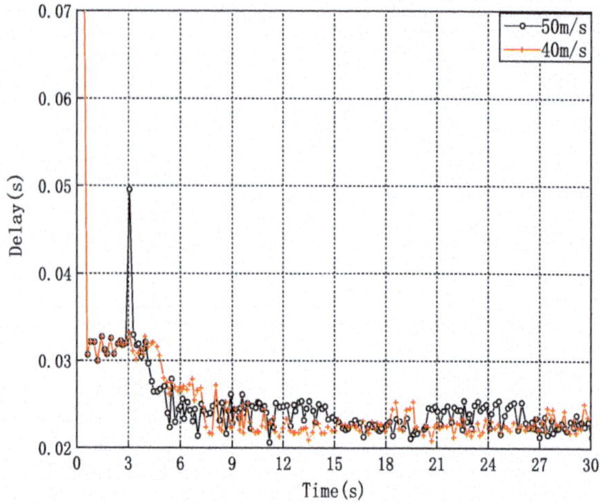

Fig. 9.38 End-to-end delay at sink node speeds of 60 and 50 m/s

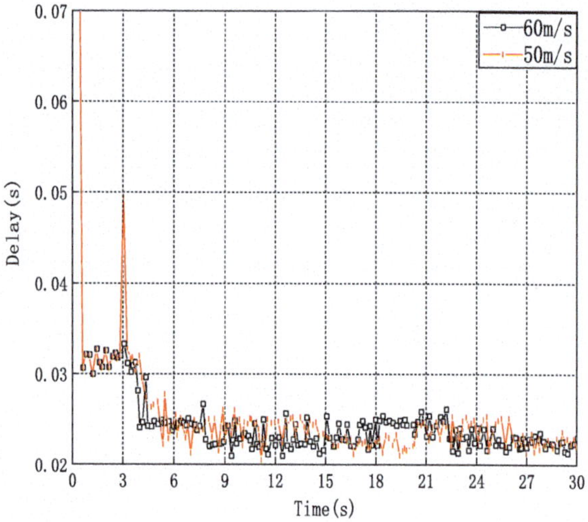

References

1. Linxing C (2009) Wireless sensor network technology and application. Electronic Industry Press, Beijing
2. Tilak S, Abu-Ghazaleh NB, Heinzeiman W (2002) Infrastructure tradeoff for sensor networks. In: Proceedings of the 1st ACM international workshop on wireless sensor networks and applications, Atlanta, pp 1547–1552
3. Pottie GJ, Kaiser WJ (2000) Embedding the internet: wireless integrated network sensors. In: Communications of the ACM, New York, pp 51–58
4. Lin T H, Kaiser W J, Pottie G J. Integrated low-power communication system design for wireless sensor networks.IEEE Communications Magazine,2004,42(12):142–150
5. Elson J, Estrin D (2001) Time synchronization for wireless sensor network. In: Proceedings of the 15th parallel and distributed processing symposium, San Francisco, pp 1965–1970

6. Deng J, Han R, Mishra S (2003) INSENS: intrusion-tolerant routing in wireless sensor network. In: The 23rdIEEE international conference on distributed computing system, Colorado, pp 1547–1552
7. Tanachaiwiwatl S, Dave P, Bhindwale R et al (2003) Secure location: routing on trust and isolating compromised sensors in location-aware sensor network. In: Proceedings of ACM, Los Angeles, pp 324–325
8. Schenenauer LE, Gligor VD (2002) A key-management scheme for distributed sensor network. In: The 9th ACM conference on computer and communication security, Seattle, p 1007–1011
9. Chen H, Perring A, Song D (2008) Random key predistribution scheme for sensor networks using deployment knowledge. In: International conference on information security and assurance, Seoul, Korea, pp 488–49
10. Du W, Deng J, Han YS et al (2003) A pairwise key predistribution scheme for wireless sensor network. In: The 9th ACM conference security, Washington DC, pp 2692–2696
11. Liu D, Ning P (2003) Location based pairwise key establishments for static sensor network. In: The 9th ACM conference on computer and communication security, Washington DC, pp 297–300
12. Heinzelman W, Kuli JK, Balakrishnan H (1999) Adaptive protocols for information dissemination in wireless sensor networks. In: Proceedings of the 5th ACM/IEEE mobicom conference, Seattle, pp 174–85
13. Intanagonwiwat C, Govindan R, Estrin D (2000) Directed diffusion: a scalable and robust communication paradigm for sensor networks. In: Proceedings of the annual international conference on mobile computing and networking, New York, pp 56–67
14. Braginsky D, Estrin D (2002) Rumor routing algorithm for sensor networks. In: Proceedings of the 1st workshop on sensor networks and applications, Atlanta, pp 22–31
15. Fan Y, Chen A, Lu S et al (2001) A scalable solution to minimum cost forwarding in large sensor networks. In: Computer communications and networks, California, pp 304–309
16. Shah RC, Rabaey J (2002) Energy aware routing for low energy ad hoc sensor networks. In: IEEE wireless communications and networking conference, California, pp 350–354
17. Servetto SD, Barrenechea G (2002) Constrained random walk on random graphs: routing algorithms for large-scale wireless sensor networks. In: Proceedings of WSNA 2002, Atlanta, pp 12–21
18. Heinzelman W, Chandrakasan A, Balakrishnan H (2000) Energy-efficient communication protocol for wireless microsensor networks. In: Proceedings of the 33rd Annual Hawaii international conference on system sciences, Maui, pp 3005–3014
19. Lindsey S, Raghavendra C (2002) Pegasis: power-efficient gathering in sensor information systems. In: Proceedings of the IEEE aerospace conference, Los Angeles, pp 1–6
20. Manjeshwar A, Agrawal DP (2001) TEEN: a routing protocol for enhanced efficiency in wireless sensor networks. In: Proceedings of the 15th parallel and distributed processing symposium, San Francisco, pp 2009–2015
21. Subramanian L, Katz RH (2000) An architecture for building self congurable systems. In: Proceedings of IEEE/ACM workshop on mobile ad hoc networking and computing, Boston, pp 321–336
22. Al-Karaki JN, Ul-Mustafa R, Kamal AE (2004) Data aggregation in wireless sensor networks-exact and approximate algorithms. In: Proceedings of the IEEE workshop on high performance switching and routing, New York, pp 241–245
23. Li Q, Aslam J, Rus D (2004) Hierarchical power-aware routing in sensor networks. In: Proceedings of the 29th annual IEEE international conference on local computer networks, Tampa, pp 575–576
24. Ye F, Luo H, Cheng J et al (2002) A two-tier data dissemination model for large-scale wireless sensor networks. In: Proceedings of the 8th annual international conference on mobile computing and networking, Atlanta, pp 148–159
25. Xu Y, Heidemann J, Estrin D (2001) Geography-informed energy conservation for ad hoc routing. In: Proceedings of the 7th annual ACM/IEEE international conference on mobile computing and networking, Rome, pp 70–84

26. Yu Y, Estrin D, Govindan R (2001) Geographical and energy-aware routing: a recursive data dissemination protocol for wireless sensor networks. In: UCLA-CSD TR-01-0023, Los Angeles, pp 1–11
27. Stojmenovic I, Lin X (1999) GEDIR: loop-free location based routing in wireless network. In: International conference on parallel and distributed computing and systems, Boston, pp 3–6
28. Kuhn F, Wattenhofer R, Zollinger A (2003) Worst-case optimal and average-case efficient geometric ad-hoc routing. In: Proceedings of the 4th ACM international symposium on mobile ad hoc networking & computing, New York, pp 267–278
29. Chen B, Jamieson K, Balakrishnan H et al (2002) SPAN: an energy-efficient coordination algorithm for topology maintenance in ad hoc wireless networks. Wirel Netw 8(5):481–494
30. Zhao Y (2007) Research and implementation of wireless sensor network routing protocol. University of Electronic Science and Technology of China Master's Thesis, Chengdu
31. Huang H, Ai Y, Zuo T et al (2010) Design of fine tracking fuzzy control system for free space optical communication. Optoelectron Laser 21(3):366–370
32. Zhao T, Feng Y, Ke X et al (2010) Research on node coverage in "solar-blind" ultraviolet optical communication network. Chin J Opt 8(30):2229–2235
33. Lim H, Kim C (2000) Multicast tree construction and flooding in wireless ad hoc networks. In: Proceedings of the ACM Int'l workshop on modeling, analysis and simulation of wireless and mobile systems, Boston, pp 61–68
34. Reilly DM, Moriarty DT, John A (2004) Unique properties of solar blind ultraviolet communication systems for unattended ground sensor networks. SPIE 5611:244–254
35. Shaw GA, Siegel AM, Model JH (2006) Extending the range and performance of non-line-of-sight ultraviolet communication links. SPIE 6231:244–254
36. Intanagonwiwat C, Govindan R, Estrin D (2000) Direction diffusion: a scalable and robust communication paradigm for sensor networks. In: Proceedings of the 6th annual international conference on mobile computing and networking, Boston, pp 56–57
37. Estrin D (2002) Wireless sensor networks tutorial part IV: sensor network protocols. In: Proceedings of the 8th ACM mobicom, Atlanta, pp 117–123
38. Yu H, Wu C, Guo M (2009) Energy efficiency analysis of mobile sinks wireless sensor network. Comput Eng 5(13):127–129
39. Deng L (2011) Research on routing protocol of ultraviolet light wireless sensor network. Master's Thesis of Xi'an University of Technology, Xi'an
40. Zhao T, Ke X, Deng L et al (2010) Research on flooding routing arithmetic of wireless sensor network based on solar-blind UV light. Optoelectron Lett 6(6):449–453

Appendices

Appendix A

The Main Code of the Monte Carlo Simulation Algorithm of NLOS Ultraviolet Light Transmission

```
clc;
clear all;
count=input('input the count:');
P=0;
  for ii=1:count
%%%%%%%参数初始化(收发仰角，发散角和接收角)%%%%%
fai_T=17*pi/180;
sita_T=20*pi/180;
fai_R=30*pi/180;
sita_R=20*pi/180;
Ksray=0.145;
Ksmie=0.261;
Ka=0.039;
Ks=Ksray+Ksmie;
W0=1;
A=0.01;
g=0.72;
f=0.5;
 Ar=1.77*10^(-10);
 r=0.02;
```


X. Ke, *UV Light Self Organizing Network Theory*, Optical Wireless Communication Theory and Technology, https://doi.org/10.1007/978-981-97-7854-6

```
gama=0.017;
%%%%%%%%%%%%发射端初始位置%%%%%
x=0;
y=0;
z=0;
%%%%%%%%%%%%%%%%%%发射端光束随机产生出射光子%%%%%%%%%%%%%%
posaifai0=unifrnd(0,1);
fai0=2*pi*posaifai0;
posaifaifai=unifrnd(0,1);
cossita0=1-posaifaifai*(1-cos(fai_T/2));
sita0=acos(1-posaifaifai*(1-cos(fai_T/2)));
sinsita0=sqrt(1-cossita0*cossita0);
 ux=cos(fai0)*sinsita0;                    %%%%%%%%%%%%%初始化方向余弦
uy=sin(fai0)*sinsita0;
uz=cossita0;
%%%%%%随机步长及新的坐标。ux，uy, uz 为发射光子的方向余弦值%%%%%
 posais=unifrnd(0,1);
 S=-log(posais)/(Ks+Ka);
 x1=x+S*ux;
 y1=y+S*uy;
 z1=z+S*uz;
 absorb=W0*(1-Ks/(Ks+Ka));
 W=(W0-absorb);
 posaig=unifrnd(0,1);
          temp=(1-g*g)/(1-g+2*g*posaig);
          cossita1=(1+g*g-temp*temp)/(2*g);
 sinsita1=sqrt(1-cossita1*cossita1);
 posaifai1=unifrnd(0,1);
  fai1=2*pi*posaifai1;
     cosfai1=cos(fai1);
  if(fai1<pi)
   sinfai1=sqrt(1-cosfai1*cosfai1);
  else
      sinfai1=-sqrt(1-cosfai1*cosfai1);
  end
%%%%%%%%%%%%%%%%%%%     产     生     新     的     方     向     余
```

```
弦%%%%%%%%%%%%%%%%%%%%%%%%%%%
            temp3=sqrt(1-uz*uz);
            uxx=sinsita1*(ux*uz*cosfai1-uy*sinfai1)/temp3+ux*cossita1;
            uyy=sinsita1*(uy*uz*cosfai1+ux*sinfai1)/temp3+uy*cossita1;
            uzz=-sinsita1*cosfai1*temp3+uz*cossita1;
 %%%%%%%%%%%%%%%%%%%    更    新    数
据%%%%%%%%%%%%%%%%%%%%%%%%%%%%%
        ux=uxx;
        uy=uyy;
        uz=uzz;
        sitas=unifrnd(0,pi);
        u=cos(sitas);
       PRay=(3*(1+3*gama+(1-gama)*u^2))/(16*pi*(1+2*gama));
       PMie=((1-g^2)/4*pi)*(1/((1+g^2-2*g*u)^(3/2))+f*0.5*(3*u^2-1)/(1+g^2)^(3/2));
       Pu=Ksray*PRay/Ks+Ksmie*PMie/Ks;
       xr=-sin(180*pi/180)*r;
       yr=cos(180*pi/180)*r;
       zr=0;
       %%%%%%%%%%%%%%%%%%%%%%接收端判定%%%%%%%%%%%%%%%%%%%
       sita_z=unifrnd(sita_T-fai_T/2,sita_T+fai_T/2);
        if(sita_R-fai_R/2<=(sitas-sita_z)<=sita_R+fai_R/2)
        BC=sqrt((xr-x1)^2+(yr-y1)^2+(zr-z1)^2);
        AC=sqrt((x1-0)^2+(y1-0)^2+(z1-0)^2);
        AB=sqrt((xr-0)^2+(yr-0)^2+(zr-0)^2);
        p1n=(Pu*Ar)/(4*pi*BC^2);
        p2n=exp(-(Ks+Ka)*BC);
        wn=(1-p1n)*exp(-Ka*S)*W;
        P1=p1n*p2n*wn;
        P=P+P1;
        end
        end
        P=P/count;
        P=(10^9)*P;
      % P=10*log10(1/P);
      fprintf(1,'路径损耗为%f',P);
```

Appendix B.1

Procedure for Writing Optical Orthogonal Codes

```
function A=ooc1
t1=(0:0.01: 99.99)';
a11=zeros(100,1);
a11(1,1)=1;
a11(4,1)=1;
a11(47,1)=1;
a12=[a11;a11;a11;a11;a11;a11;a11;a11;a11;a11];
a13=[a11;a11;a11;a11;a11;a11;a11;a11;a11;a11];
a14=[a11;a11;a11;a11;a11;a11;a11;a11;a11;a11];
a15=[a11;a11;a11;a11;a11;a11;a11;a11;a11;a11];
a16=[a11;a11;a11;a11;a11;a11;a11;a11;a11;a11];
a17=[a11;a11;a11;a11;a11;a11;a11;a11;a11;a11];
a18=[a11;a11;a11;a11;a11;a11;a11;a11;a11;a11];
a19=[a11;a11;a11;a11;a11;a11;a11;a11;a11;a11];
a10=[a11;a11;a11;a11;a11;a11;a11;a11;a11;a11];
a1=[a11;a11;a11;a11;a11;a11;a11;a11;a11;a11;a12;a13;a14;a15;a16;a17;a18;a19;a10];
A=[t1,a1];
function B=ooc2
t2=(0:0.01: 99.99)';
a21=zeros(100,1);
a21(1,1)=1;
a21(6,1)=1;
a21(7,1)=1;
a22=[a21;a21;a21;a21;a21;a21;a21;a21;a21;a21];
a23=[a21;a21;a21;a21;a21;a21;a21;a21;a21;a21];
```

```
a24=[a21;a21;a21;a21;a21;a21;a21;a21;a21;a21];
a25=[a21;a21;a21;a21;a21;a21;a21;a21;a21;a21];
a26=[a21;a21;a21;a21;a21;a21;a21;a21;a21;a21];
a27=[a21;a21;a21;a21;a21;a21;a21;a21;a21;a21];
a28=[a21;a21;a21;a21;a21;a21;a21;a21;a21;a21];
a29=[a21;a21;a21;a21;a21;a21;a21;a21;a21;a21];
a20=[a21;a21;a21;a21;a21;a21;a21;a21;a21;a21];
a2=[a21;a21;a21;a21;a21;a21;a21;a21;a21;a21;a22;a23;a24;a25;a26;a27;a28;a29;a20];
B=[t2,a2];

function C=ooc3
t3=(0:0.01:99.99)';
a31=zeros(100,1);
a31(1,1)=1;
a31(5,1)=1;
a31(71,1)=1;
a32=[a31;a31;a31;a31;a31;a31;a31;a31;a31;a31];
a33=[a31;a31;a31;a31;a31;a31;a31;a31;a31;a31];
a34=[a31;a31;a31;a31;a31;a31;a31;a31;a31;a31];
a35=[a31;a31;a31;a31;a31;a31;a31;a31;a31;a31];
a36=[a31;a31;a31;a31;a31;a31;a31;a31;a31;a31];
a37=[a31;a31;a31;a31;a31;a31;a31;a31;a31;a31];
a38=[a31;a31;a31;a31;a31;a31;a31;a31;a31;a31];
a39=[a31;a31;a31;a31;a31;a31;a31;a31;a31;a31];
a30=[a31;a31;a31;a31;a31;a31;a31;a31;a31;a31];
a3=[a31;a31;a31;a31;a31;a31;a31;a31;a31;a31;a32;a33;a34;a35;a36;a37;a38;a39;a30];
C=[t3,a3];

function D=ooc4
t4=(0:0.01:99.99)';
a41=zeros(100,1);
a41(1,1)=1;
a41(10,1)=1;
a41(38,1)=1;
a42=[a41;a41;a41;a41;a41;a41;a41;a41;a41;a41];
a43=[a41;a41;a41;a41;a41;a41;a41;a41;a41;a41];
a44=[a41;a41;a41;a41;a41;a41;a41;a41;a41;a41];
a45=[a41;a41;a41;a41;a41;a41;a41;a41;a41;a41];
a46=[a41;a41;a41;a41;a41;a41;a41;a41;a41;a41];
a47=[a41;a41;a41;a41;a41;a41;a41;a41;a41;a41];
a48=[a41;a41;a41;a41;a41;a41;a41;a41;a41;a41];
a49=[a41;a41;a41;a41;a41;a41;a41;a41;a41;a41];
a40=[a41;a41;a41;a41;a41;a41;a41;a41;a41;a41];
a4=[a41;a41;a41;a41;a41;a41;a41;a41;a41;a41;a42;a43;a44;a45;a46;a47;a48;a49;a40];
```

Appendix B.2

Generating User Data

```
function [sys,x0,str,ts]=sfun_mseries(t,x,u,flag)

switch flag,
    case 0,
        [sys,x0,str,ts]=mdlInitializeSizes;
    case 3,
        sys=mdlOutputs(t,x,u);
    case {2,4,9}
        sys=[];
    otherwise
        error(['Unhandled flag = ',num2str(flag)]);
end

function[sys,x0,str,ts]=mdlInitializeSizes
sizes = simsizes;
sizes.NumContStates   = 0;
  sizes.NumDiscStates   = 0;
sizes.NumOutputs      = 1;
sizes.NumInputs       = 0;
sizes.DirFeedthrough = 0;
sizes.NumSampleTimes = 1;
sys = simsizes(sizes);
x0   = [];
str = [];
ts = [1 0];

function sys=mdlOutputs(t,x,u)
sys=randint;
```

Appendix B.3

Blind Adaptive RLS Multi-User Detection Algorithm

```
clear all;
clc;
s1=O1;
s2=O2;
s3=O3;
s4=O4;
s=[s1 s2 s3 s4];
A=[A1, A2,A3,A4];
N=1000;
M=1000;
thr=1;
SIR1=zeros(1,N);
SIR2=zeros(1,N);
Y=spectrum(s,b,y);

#######################################################
  SNR=10;
  y=awgn(y,SNR);

  r=0.97;

  C1=RLS(y);
#######################################################
  for j=1:N
      b1(j)=C1(:,j)'*y(:,j);
      if (b1(j)>thr)
            b1(j)=1;
      else b1(j)=0;
      end
```

```
        if b1(j)~=b(1,j)
            Enum=Enum+1;
        end
        BER(i)=Enum/N;
        sir1(j)=(C1(:,j)'*s(:,1))^2;
        sir2(j)=(C1(:,j)'*(y(:,j)-b(1,j)*s(:,1)))^2;
    end
    SIR1=SIR1+sir1;
    SIR2=SIR2+sir2;
end
SIR=10*log10(SIR1./SIR2);
###################################################
plot(BER,'r');
ylabel('误码率');
xlabel('独立试验次数');
figure
plot(SIR);
xlabel('迭代步数');
ylabel('信干比SIR/dB');
```

Appendix C

Main C Program Code of HDFB Algorithm and NDCF Algorithm

Related variables used in the main programs of HDFB algorithm and NDCF algorithm and their meanings.

mac_pib.cw_normal median of CW i-th order in HDFB	mac_pib.cw_low Lower limit of CW i-th order in HDFB
mac_pib.cw_high Upper value of CW i-th order in HDFB	mac_pib.cw_temp1 median of the highest order of CW in HDFB
mac_pib.cw_temp2 The lower limit of the lowest order of CW in HDFB	mac_pib.DATA_on complete data packet send counter
mac_pib.DATA_OK Complete data packet sent successfully count	mac_pib.DATA_Fail Complete data packet sending failure count
mac_pib.switch_on BEB algorithm switch	mac_pib.one_on has a counter for data flow
mac_pib.one_count determines the calculator with exclusive channel	mac_pib.cx_on increment and decrement subroutine call counter
mac_pib.cw contention window value CW	mac_pib.cw_p Judge the contention window level change counter
MAC_CW_MIN Contention window value lower limit	MAC_CW_MAX Contention window value upper limit

```
void IncCW(void)                                        //竞争窗口增加子程序
    {   #ifdef HDFB                                     //预定义的为 HDFB 算法
        if(mac_pib.switch_on == 1)
            {   if(mac_pib.one_on >= 1)
                {   mac_pib.switch_on = 0 ;
                    return ;
                }
                mac_pib.cw = mac_pib.cw * 2;
                if(mac_pib.cw > MAC_CW_MAX)
                mac_pib.cw = MAC_CW_MAX;
                return ;
            }
            if(mac_pib.one_count >= 30)
            {   if(mac_pib.one_on < 1)
                {   mac_pib.switch_on = 1 ;
                }else{
                    mac_pib.switch_on = 0 ;
                }
                mac_pib.one_on = 0 ;
                mac_pib.one_count = 0 ;
            }
            SetCW() ;
            mac_pib.cx_on = mac_pib.cx_on + 1;
            mac_pib.Fail = mac_pib.Fail + 1;
            if(mac_pib.cx_on>=2)
            {   mac_pib.cx_on = 0 ;
                if(mac_pib.Fail>=2)
                {mac_pib.cw = mac_pib.cw * 2 ;}
                mac_pib.Fail = 0 ;
                mac_pib.OK = 0 ;
            }
            if(mac_pib.cw >= mac_pib.cw_high)
                mac_pib.cw = mac_pib.cw_high;
        #else                                           //如果没有预定义就为 NDCF 算法
            mac_pib.cw = mac_pib.cw * 2;
            if(mac_pib.cw > MAC_CW_MAX)
            mac_pib.cw = MAC_CW_MAX;
```

```
    #endif
}
void RstCW(void)                                    //减小竞争窗口子程序
{
    #ifdef HDFB
    if(mac_pib.switch_on == 1)
        {   if(mac_pib.one_on >= 1)
            {   mac_pib.switch_on = 0 ;
                return ;
            }
            mac_pib.cw = MAC_CW_MIN;
            return ;
        }
        if(mac_pib.one_count >= 30)
        {   if(mac_pib.one_on < 1)
            {   mac_pib.switch_on = 1 ;
            }else{
                mac_pib.switch_on = 0 ;
            }
            mac_pib.one_on = 0 ;
            mac_pib.one_count = 0 ;
        }
        SetCW();
        mac_pib.cx_on = mac_pib.cx_on + 1;
        mac_pib.OK = mac_pib.OK + 1;
            if(mac_pib.cx_on>=2)
            {   mac_pib.cx_on = 0 ;
                if(mac_pib.OK>=2)
                {mac_pib.cw = Uint16(mac_pib.cw /sqrt(2.0));      }
                mac_pib.Fail = 0 ;
                mac_pib.OK = 0 ;
            }
        if(mac_pib.cw <=   mac_pib.cw_low )
            mac_pib.cw =   mac_pib.cw_low ;
    #else
        if(mac_pib.cw == MAC_CW_MIN)
        {   if(mac_pib.backoff_time <= (mac_pib.cw /2))
            {mac_pib.cw = mac_pib.cw * 2 ;}
        }else
        {   if(mac_pib.backoff_time >= (mac_pib.cw /2))
            {mac_pib.cw = mac_pib.cw /2 ;}
        }
        if(mac_pib.cw < MAC_CW_MIN)
        mac_pib.cw = MAC_CW_MIN;
```

```
        #endif
}
#ifdef HDFB
void SetCW(void)                          //HDF 竞争窗口等级设置子程序
{
        if(mac_pib.DATA_on>=2)
                {       mac_pib.DATA_on = 0 ;
                        if(mac_pib.DATA_Fail>=2)
                        {       mac_pib.cw_normal = mac_pib.cw_normal / 2 ;
                        }else if(mac_pib.DATA_OK>=2)
                        {       mac_pib.cw_normal = mac_pib.cw_normal * 2 ;
                        }else if(mac_pib.DATA_Fail==1&&mac_pib.DATA_OK==1)
                        {mac_pib.cw_p = mac_pib.cw_p + 1 ;}
                        mac_pib.DATA_Fail = 0 ;
                        mac_pib.DATA_OK= 0 ;
                        if(mac_pib.cw_normal <= mac_pib.cw_temp2)
                                mac_pib.cw_normal = mac_pib.cw_temp2;
                        if(mac_pib.cw_normal >= mac_pib.cw_temp1)
                                mac_pib.cw_normal = mac_pib.cw_temp1;
                        mac_pib.cw_low = mac_pib.cw_normal / 2 ;
                        mac_pib.cw_high = mac_pib.cw_normal * 2 ;
                        mac_pib.cw = mac_pib.cw_low ;
                }
        if(mac_pib.DATA_p>=9)
        {       mac_pib.DATA_p = 0 ;
                if(mac_pib.cw_p>0)
                {       mac_pib.cw_p=0 ;
                        mac_pib.cw_normal = mac_pib.cw_normal / 2 ;
                        if(mac_pib.cw_normal <= mac_pib.cw_temp2)
                                mac_pib.cw_normal = mac_pib.cw_temp2;
                        mac_pib.cw_low = mac_pib.cw_normal / 2 ;
                        mac_pib.cw_high = mac_pib.cw_normal * 2 ;
                        mac_pib.cw = mac_pib.cw_low ;
                }
        }
}
#endif
```

Appendix D.1

Simulation Script for Cross Topology

```
#==========================================
#       Simulation parameters setup
#==========================================
set val(chan)    Channel/Channel_802_11      ;# channel type
set val(prop)    Propagation/TwoRayGround    ;# radio-propagation model
set val(netif)   Phy/WirelessPhy/Wireless_802_11_Phy            ;# network interface type
set val(mac)     Mac/802_11                  ;# MAC type
set val(ifq)     Queue/DropTail/PriQueue     ;# interface queue type
set val(ll)      LL                          ;# link layer type
#set val(ant)    Antenna/OmniAntenna         ;# antenna model
set val(ant)     Antenna/DirAntenna          ;# antenna model
set val(ifqlen)  50                          ;# max packet in ifq
set val(nn)      4                           ;# number of mobilenodes
set val(rp)      WLSTATIC                    ;# routing protocol WLSTATIC
set val(x)       2600                        ;# X dimension of topography
set val(y)       2100                        ;# Y dimension of topography
set val(stop)    100.0                       ;# time of simulation end
set val(ni)      1
#==========================================
#          Initialization
#==========================================
#Create a ns simulator
set ns_          [new Simulator]

#Setup topography object
set topo         [new Topography]
$topo load_flatgrid $val(x) $val(y)
create-god $val(nn)

#Open the NS trace file
set tracefd [open crossc.tr w]
$ns_ trace-all $tracefd

set namtrace [open crossc.nam w]
$ns_ namtrace-all $namtrace
$ns_ namtrace-all-wireless $namtrace $val(x) $val(y)
set chan [new $val(chan)];#Create wireless channel
```

```
#===========================================
#       Mobile node parameter setup
#===========================================
$ns_ node-config -adhocRouting   $val(rp) \
                 -llType          $val(ll) \
                 -macType         $val(mac) \
                 -ifqType         $val(ifq) \
                 -ifqLen          $val(ifqlen) \
                 -antType         $val(ant) \
                 -propType        $val(prop) \
                 -phyType         $val(netif) \
                 -channel         $chan \
                 -topoInstance    $topo \
                 -agentTrace      ON \
                 -routerTrace     ON \
                 -macTrace        ON \
                 -movementTrace OFF \
                 -numif   $val(ni)
#===========================================
#           Nodes Definition
#===========================================
#Create 25 nodes
proc create_node { x y z } {
        global ns_
        Mac/802_11 set dataRate_   11mb
        Mac/802_11 set basicRate_  1mb

        set newnode [$ns_ node]
        $newnode random-motion 0
        $newnode set X_ $x
        $newnode set Y_ $y
        $newnode set Z_ $z
        return $newnode
}

proc create_cbr_connection { from to startTime interval packetSize } {
        global ns_
        set udp0 [new Agent/UDP]
        set src [new Application/Traffic/CBR]
        $udp0 set packetSize_ $packetSize
        $src set packetSize_ $packetSize
        $src set interval_ $interval

        set sink [new Agent/Null]

        $ns_ attach-agent $from $udp0
        $src attach-agent $udp0
        $ns_ attach-agent $to $sink

        $ns_ connect $udp0 $sink
        $ns_ at $startTime "$src start"
        return $udp0
}
```

```
$ns_ node-config   -numif 1
set n0 [create_node 200 200  0]
$ns_ initial_node_pos $n0 20
set a [new Antenna/DirAntenna]
$a setType  0
[$n0 set netif_(0)] dir-antenna $a
[$n0 set netif_(0)] set channel_number_  1
[$n0 set netif_(0)] set Pt_    0.28183815

$ns_ node-config   -numif 1
set n1 [create_node 400 200  0]
$ns_ initial_node_pos $n1 20
set a [new Antenna/DirAntenna]
$a setType 4
[$n1 set netif_(0)] dir-antenna $a
[$n1 set netif_(0)] set channel_number_  1
[$n1 set netif_(0)] set Pt_    0.28183815

$ns_ node-config   -numif 1
set n2 [create_node 300 250  0]
$ns_ initial_node_pos $n2 20
set a [new Antenna/DirAntenna]
$a setType  6
[$n2 set netif_(0)] dir-antenna $a
[$n2 set netif_(0)] set channel_number_  1
[$n2 set netif_(0)] set Pt_    0.28183815

$ns_ node-config   -numif 1
set n3 [create_node 300 50 0]
$ns_ initial_node_pos $n3 20
set a [new Antenna/DirAntenna]
$a setType 2
[$n3 set netif_(0)] dir-antenna $a
[$n3 set netif_(0)] set channel_number_  1
[$n3 set netif_(0)] set Pt_    0.28183815

[$n0 set ragent_] addstaticroute 1 1 1 0
[$n1 set ragent_] addstaticroute 1 0 0 0
[$n2 set ragent_] addstaticroute 1 3 3 0
[$n3 set ragent_] addstaticroute 1 2 2 0

set child_mac [[$n0 set mac_(0)] id];
set parent_mac [[$n1 set mac_(0)] id];
[$n0 set ll_(0)] add-arp-entry 1 $parent_mac
[$n1 set ll_(0)] add-arp-entry 0 $child_mac
set child_mac [[$n2 set mac_(0)] id];
set parent_mac [[$n3 set mac_(0)] id];
[$n2 set ll_(0)] add-arp-entry 3 $parent_mac
[$n3 set ll_(0)] add-arp-entry 2 $child_mac

set cbr0 [create_cbr_connection $n0 $n1 0.0 0.002667 1000]
set cbr1 [create_cbr_connection $n2 $n3 0.0 0.002667 1000]
```

```
#=====================================
#              Termination
#=====================================
#Define a 'finish' procedure
for {set i 0} {$i < $val(nn) } {incr i} {
      $ns_ at 100.0 "\$n$i reset"
}
$ns_ at 100.0 "stop"
$ns_ at 100.01 "puts \"NS EXITING...\" ; $ns_ halt"
proc stop {} {
      global ns_ tracefd namtrace
      $ns_ flush-trace
      close $tracefd
      close $namtrace
      exec nam crossc.nam &
      exit 0
}

puts "Starting Simulation..."
$ns_ run
```

Appendix D.2

Throughput Analysis Procedure for Cross Topology

```
BEGIN {
        i = 0 ;
        k = 0 ;
        end_timeA[0] = 0 ;
        pkt_byte_sumA[0] = 0 ;
        time_length = 10 ;
        start_time = 0;
        stop_time = 100;
    average = 0;
    sum_total = 0;
}
{
        action=$1;
        time=$2;
        node_number = $3 ;
        layer = $4;
        packet_id=$6;
        packet_type=$7;
        packet_size=$8;
    if(action=="r"&&packet_type =="cbr"&&layer == "AGT"&&(node_number == "_1_ " || node_number ==
"_3_" )){
                pkt_byte_sumA[i + 1] = pkt_byte_sumA[i] + 1000;
                end_timeA[i + 1] = time ;
                i = i + 1 ;
        sum_total = i;
        }
}
END {
        printf(" sum_total throughput %f kbps\n" ,   sum_total*8/100 );
```

Appendix E.1

State Machine Functions of the Pvar Routing Protocol

Note: The files modified by the NS2 simulation software in the simulation program are: cmu-trace.cc, cmu-trace.h, Makefile, ns-agent.tcl, ns-lib.tcl, ns-mobilenode.tcl, ns-packet. tcl, packet.h

```
void LAR::recvLAR(Packet *p) {
        struct hdr_lar *ah = HDR_LAR(p);
        assert(HDR_IP (p)->sport() == RT_PORT);
        assert(HDR_IP (p)->dport() == RT_PORT);
        /* Incoming Packets.*/
        switch(ah->ah_type) {
         case LARTYPE_RREQ:
           recvRequest(p);
           break;
        case LARTYPE_RREP:
           recvReply(p);
           break;
        case LARTYPE_RERR:
           recvError(p);
           break;
        case LARTYPE_HELLO:
           recvHello(p);
           break;
        case LARTYPE_POS:
           recvPos(p);
           break;
        default:
           fprintf(stderr, "Invalid LAR type (%x)\n", ah->ah_type);
           exit(1);
        }
```

Appendix E.2

Command() Function of PVAR Routing Protocol, corresponding to the cmd{} of the Otcl object, is the implementation of specific commands

```
int LAR::command(int argc, const char*const* argv) {
    TclObject *obj;
    if(argc == 2) {
        Tcl& tcl = Tcl::instance();
    if(strncasecmp(argv[1], "id", 2) == 0) {
      tcl.resultf("%d", index);
      return TCL_OK;
  }
  if(strncasecmp(argv[1], "start", 2) == 0) {
    btimer.handle((Event*) 0);
    htimer.handle((Event*) 0);
    ntimer.handle((Event*) 0);
    rtimer.handle((Event*) 0);
    ptimer.expire((Event*) 0);
    return TCL_OK;
   }
}
else if(argc == 3) {
  if(strcmp(argv[1], "index") == 0) {
    index = atoi(argv[2]);
    return TCL_OK;
  }
  else if(strcmp(argv[1], "log-target") == 0 || strcmp(argv[1], "tracetarget") == 0) {
    logtarget = (Trace*) TclObject::lookup(argv[2]);
    if(logtarget == 0)
   return TCL_ERROR;
    return TCL_OK;
  }
  else if(strcmp(argv[1], "drop-target") == 0) {
  int stat = rqueue.command(argc,argv);
    if (stat != TCL_OK) return stat;
    return Agent::command(argc, argv);
  }
```

```
            else if(strcmp(argv[1], "if-queue") == 0) {
            ifqueue = (PriQueue*) TclObject::lookup(argv[2]);
              if(ifqueue == 0)
             return TCL_ERROR;
              return TCL_OK;
            }
            else if (strcmp(argv[1], "port-dmux") == 0) {
             dmux_ = (PortClassifier *)TclObject::lookup(argv[2]);
             if (dmux_ == 0) {
                    fprintf (stderr, "%s: %s lookup of %s failed\n", __FILE__,
                    argv[1], argv[2]);
                    return TCL_ERROR;
             }
             return TCL_OK;
             }
     else if (strcmp(argv[1], "set-node") == 0) {
       if( (obj = TclObject::lookup(argv[2])) == 0)    {
             fprintf(stderr, "LARAgent(set-node): %s lookup of %s failed\n", argv[1], argv[2]);
             return (TCL_ERROR);
             }
       node = dynamic_cast< MobileNode * >(obj);
       if (node)            // dynamic cast was successful and didn't return NULL
       {            return (TCL_OK);          }
       else {fprintf(stderr, "Unable to dynamically cast %s to a MobileNode\n",argv[2]);
       return (TCL_ERROR); }
   }
}
   return Agent::command(argc, argv);
}
```

Appendix E.3

Forwarding Process of PVAR Routing Protocol Packets

```
void LAR::forward(lar_rt_entry *rt, Packet *p, double delay) {
struct hdr_cmn *ch = HDR_CMN(p);
struct hdr_ip *ih = HDR_IP(p);
  if(ih->ttl_ == 0) {
#ifdef DEBUG
    fprintf(stderr, "%s: calling drop()\n", __PRETTY_FUNCTION__);
#endif // DEBUG
    drop(p, DROP_RTR_TTL);
    return;
  }
  if (ch->ptype() != PT_LAR && ch->direction() == hdr_cmn::UP &&
        ((u_int32_t)ih->daddr() == IP_BROADCAST)
                || (ih->daddr() == here_.addr_)) {
        dmux_->recv(p,0);
        return;
  }
  if (rt) {
      assert(rt->rt_flags == RTF_UP);
      rt->rt_expire = CURRENT_TIME + ACTIVE_ROUTE_TIMEOUT;
      ch->next_hop_ = rt->rt_nexthop;
      ch->addr_type() = NS_AF_INET;
      ch->direction() = hdr_cmn::DOWN;          //important: change the packet's direction
  }
  else { // if it is a broadcast packet
      // assert(ch->ptype() == PT_LAR); // maybe a diff pkt type like gaf
      assert(ih->daddr() == (nsaddr_t) IP_BROADCAST);
```

```
        ch->addr_type() = NS_AF_NONE;
        ch->direction() = hdr_cmn::DOWN;         //important: change the packet's direction
    }
  if (ih->daddr() == (nsaddr_t) IP_BROADCAST) {
    // If it is a broadcast packet
        assert(rt == 0);
        /*   Jitter the sending of broadcast packets by 10ms   */
        Scheduler::instance().schedule(target_, p,
                                        0.01 * Random::uniform());
    }
    else { // Not a broadcast packet
        if(delay > 0.0) {
          Scheduler::instance().schedule(target_, p, delay);
        }
        else {
        // Not a broadcast packet, no delay, send immediately
          Scheduler::instance().schedule(target_, p, 0.);
        }
    }
  }
```

Appendix E.4

Emulating TCL Scripts

(Figures 7.32, 7.33, 7.34, and 7.35 all use this script).

In the form of parameter passing, all plane routing protocols of ad hoc can be easily imitated directly through the command line.

```
if {$argc !=4} {
        puts "Usage: ns adhoc.tcl   Routing_Protocol Traffic_Pattern Scene_Pattern Num_Of_Node"
        puts "Example:ns adhoc.tcl AODV cbr-50-10-8 scene-50-0-20 50"
        exit
}
set par1 [lindex $argv 0]
set par2 [lindex $argv 1]
set par3 [lindex $argv 2]
set par4 [lindex $argv 3]
# Define options
set val(chan)           Channel/WirelessChannel     ;# channel type
set val(prop)           Propagation/TwoRayGround     ;# radio-propagation model
set val(netif)          Phy/WirelessPhy              ;# network interface type
set val(mac)            Mac/802_11                   ;# MAC type
if { $par1=="DSR"} {
  set val(ifq)            CMUPriQueue
```

```
    } else {
    set val(ifq)            Queue/DropTail/PriQueue     ;# interface queue type
   }
set val(ll)                 LL                          ;# link layer type
set val(ant)                Antenna/OmniAntenna         ;# antenna model
set val(ifqlen)             100                         ;# max packet in ifq
set val(rp)                 $par1                       ;# routing protocol
set val(x)                  1000
set val(y)                  1000
set val(seed)               1.0
set val(tr)                 $par1-$par3.tr
set val(nam)                $par1-$par3.nam
set val(nn)                 $par4                       ;# number of mobilenodes
set val(cp)                 "$par2"
set val(sc)                 "$par3"
set val(stop)               100.0
# Main Program
# Initialize Global Variables
set ns_             [new Simulator]
set tracefd         [open $val(tr) w]
$ns_ trace-all $tracefd
set namtracefd [open $val(nam) w]
$ns_ namtrace-all-wireless $namtracefd $val(x) $val(y)
# set up topography object
set topo            [new Topography]
$topo load_flatgrid $val(x) $val(y)
# Create God
set god_ [create-god $val(nn)]
#   Create the specified number of mobilenodes [$val(nn)] and "attach" them
#   to the channel.
# configure node
# set chan_1_ [new $val(chan)]
        $ns_ node-config -adhocRouting $val(rp) \
                -llType $val(ll) \
                -macType $val(mac) \
                -ifqType $val(ifq) \
                -ifqLen $val(ifqlen) \
                -antType $val(ant) \
                -propType $val(prop) \
                -phyType $val(netif) \
                -channelType $val(chan) \
                -topoInstance $topo \
                -agentTrace ON \
                -routerTrace ON \
```

```
                -macTrace OFF \
                -movementTrace OFF
    for {set i 0} {$i < $val(nn) } {incr i} {
        set node_($i) [$ns_ node]
        $node_($i) random-motion 0       ;# disable random motion
    }
puts "Loading connection pattern..."
source $val(cp)
puts "Loading scenario file..."
source $val(sc)
for {set i 0} {$i < $val(nn) } {incr i} {
    $ns_ initial_node_pos $node_($i) 20
}
# Tell nodes when the simulation ends
for {set i 0} {$i < $val(nn) } {incr i} {
    $ns_ at $val(stop).000000001 "$node_($i) reset";
}
$ns_ at $val(stop).000000001 "finish"
$ns_ at $val(stop).000000001 "puts \"NS EXITING...\"; $ns_ halt"
proc finish {} {
    global ns_ tracefd namtracefd
    $ns_ flush-trace
    close $tracefd
    close $namtracefd
   # exec nam adhoc.nam &
      exit 0
}
puts "Starting Simulation..."
$ns_ run
```

Appendix E.5

Packet Arrival Rate Simulation Script

Figures 7.32 and 7.33.

```
# 使用的是无线 trace 的旧格式
BEGIN {
        sendpacket = 0;
        recvpacket = 0;
}
# 统计发送的数据包个数
$0 ~/^s.* AGT/ {
        sendpacket ++ ;
}
# 统计接收到的数据包个数

$0 ~/^r.* AGT/ {
        recvpacket ++ ;
}
END {
        Printf "cbr send:%d recv:%d, getRatio:%.4f \n", sendpacket, recvpacket,(recvpacket/sendpacket);
}
```

Appendix E.6

Routing Cost Simulation Script

Figures 7.34 and 7.35.

```
# 使用的是无线 trace 的旧格式
BEGIN {
        pkt_route_sum=0;
        pkt_cbr_sum=0;
}
{
        pktsize = $8;
}
$0 ~/^s.* LAR/ {
            pkt_route_sum += pktsize ;
}
# 我只计算了正确接收到的数据包，被丢弃的没有算作有效数据
$0 ~/^r.* AGT.* cbr/ {
            pkt_cbr_sum += pktsize ;
}
END {
        # 单位为 bytes
        printf "cbr_bytes:%d, route_bytes:%d, routecosts:%.4f \n", pkt_cbr_sum, pkt_route_sum,
(pkt_route_sum/pkt_cbr_sum);
}
```

Appendix F.1

QoS Multicast Routing Algorithm Program List Based on Ant Colony Algorithm:

(1)加入节点电量：在更新信息素时考虑节点电量（8.4.3.1节）

```
Delta_Tau=zeros(N,N);%更新量初始化
  for m=1:M   %m 表示蚂蚁，p 表示目的节点，k 表示迭代次数
    %判断第 k 代蚂蚁 m 到达目的节点 p 所用的费用，延时及延时抖动情况
    if COSTS(p,k,m)<inf&DELAYS(p,k,m)<Dmax&DELAYJITTERS(p,k,m)<DJmax
      ROUT=ROUTES{p,k,m};%第 k 代蚂蚁 m 到达目的节点 p 的路径保存在 ROUT 中
      TS=length(ROUT)-1;%跳数
```

```
        %第 k 代蚂蚁 m 到达目的节点 p 的延时与延时抖动的和
        Cpkm=DELAYS(p,k,m)+DELAYJITTERS(p,k,m);
        for s=1:TS
          x=ROUT(s);
          y=ROUT(s+1);
          %更新节点 x 与节点 y 上的信息素值
          Delta_Tau(x,y)=Delta_Tau(x,y)+(Q*El(y))/Cpkm;%EL(y)表示 y 节点的电量
          Delta_Tau(y,x)=Delta_Tau(y,x)+(Q*El(x))/Cpkm;
        end
      end
    end
    Tau=(1-Rho).*Tau+Delta_Tau;%信息素挥发一部分，新增加一部分
```

(2)删除到达各个目的节点路径间的环路:从寻找到达第二个目的节点路径开始判断环路存在是否，然后依据程序删除（8.4.3.2 节）

```
if p>1    %如果不是第一个目的节点
  Path1=ROUTES{(p-1),k,m} %将到达前一目的节点的路径赋给 Path1
end
Path=[Path,to_visit];%路径增加
PD=PD+DD(W,to_visit);%路径延时累计
PC=PC+CC(W,to_visit);%路径费用累计
PDJ=PDJ+DDJ(W,to_visit);
if p>1
  pt=Path1; %pt 表示路径
  c=0;  %c 表示费用
  d=0;  %d 表示延时
  dj=0;  %dj 表示延时抖动
  for i=1:length(Path1)
    if Path1(i)==to_visit  %如果已访问的节点等于到达前一目的节点路径上的节点
      for i1=1:(i-1)  %记录下费用、延时、延时抖动
        c=c+C(pt(i1),pt(i1+1));
        d=d+D(pt(i1),pt(i1+1));
        dj=dj+DJ(pt(i1),pt(i1+1));
      end
      if PC>=c %将现寻找的路径与到达前一目的节点路径上的费用相比较
        to_visit=pt(i);
        qq=S;
        for j=2:length(Path)
          TABU(Path(j))=1;%将已访问节点在禁忌表中标记
        end
        for i2=2:i
          qq=[qq,pt(i2)];
          TABU(pt(i2))=0;
        end
          Path=qq;
```

```
            PC=c;
            PD=d;
            PDJ=dj;
     else
       if pt(i)==pt(end)%判断是否为最后一个节点
         if COSTS((p-1),k,m)~=inf&pt(i)==E(p-1)
           pt=[Path];
           COSTS((p-1),k,m)=PC;
           DELAYS((p-1),k,m)=PD;
           DELAYJITTERS((p-1),k,m)=PDJ;
         end
       else
         COSTS((p-1),k,m)=PC;
         DELAYS((p-1),k,m)=PD;
         DELAYJITTERS((p-1),k,m)=PDJ;
         pt=Path;
         for i3=i:(length(Path1)-1)   %添加访问路径上的费用，延时，延时抖动
           COSTS((p-1),k,m)=COSTS((p-1),k,m)+C(Path1(i3),Path1(i3+1));
           DELAYS((p-1),k,m)=DELAYS((p-1),k,m)+D(Path1(i3),Path1(i3+1));
           DELAYJITTERS((p-1),k,m)=DELAYJITTERS((p-
           1),k,m)+DJ(Path1(i3),Path1(i3+1));
           pt=[pt,Path1(i3+1)];%添加访问路径
         end
       end
     end
   end
 end
 Path1=pt;
end
```

(3)保留信息素给寻找后续目的节点使用:寻找到前一个目的节点后,保留信息素（8.4.3.3 节）

```
%更新信息素完成了一次循环
Delta_Tau=zeros(N,N);%更新量初始化
  for m=1:M
    if COSTS(p,k,m)<inf&DELAYS(p,k,m)<Dmax&DELAYJITTERS(p,k,m)<DJmax
      ROUT=ROUTES{p,k,m};%把每一代每一只蚂蚁的觅食路线给保存在 ROUT 中
      TS=length(ROUT)-1;%跳数
      Cpkm=DELAYS(p,k,m)+DELAYJITTERS(p,k,m);
      for s=1:TS
        x=ROUT(s);
        y=ROUT(s+1);
        Delta_Tau(x,y)=Delta_Tau(x,y)+Q/Cpkm;
        Delta_Tau(y,x)=Delta_Tau(y,x)+Q/Cpkm;
      end
    end
```

```
    end
    Tau=(1-Rho).*Tau+Delta_Tau;%信息素挥发一部分，新增加一部分
end
Tau=10*log(Tau/K);%取到达前一目的节点路径上信息素值
%在寻找下一目的节点前，初始化信息素值
for i=1:N
    for j=1:N
        if Tau(i,j)<1
            Tau(i,j)=1;
        end
    end
end
aa=Tau;
```

Appendix F.2

Program List of Ant Colony Algorithm-Based Unidirectional Link Routing Algorithm

```
Delta_Tau=zeros(N,N);%更新量初始化
Bant_path=E;%将目的节点加到后向蚂蚁路径中
Bant_PC=0;%爬行路线费用初始化
Bant_TABU=ones(1,N);%禁忌表初始化
Bant_TABU(E)=0;%S 已经在初始点了，因此要排除
Bant_CC=C;
for m=1:M
  Cse=0; %费用初始化为 0
  if COSTS(1,k,m)<inf
    ROUT=ROUTES{1,k,m};%把每一代每一只蚂蚁的觅食路线给保存在 ROUT 中
    TS=length(ROUT)-1;%跳数
    w1=E;
    while (w1~=S) %当前节点不是源节点时
      for s=TS:-1:1
        x=ROUT(s);
        y=ROUT(s+1);
        CW2=C(w1,:);
        if CW2(x)~=inf   %单播后向蚂蚁
          Cse=Cse+C(x,y);
          Delta_Tau(x,y)=Delta_Tau(x,y)+Q/Cse;
          w1=x;
        else   %发送广播后向蚂蚁
          BBant_path=w1;%BBant 路径
          BBant_PC=0;%爬行路线费用初始化
          BBant_TABU=ones(1,N);%禁忌表初始化
BBant_TABU(w1)=0;%S 已经在初始点了，因此要排除
BBant_CC=Bant_CC;
for k1=1:N
  if BBant_TABU(k1)==0
    BBant_CC(w1,k1)=inf;
  end
end
BBant_TABU(w1)=0;%已访问过的节点从禁忌表中删除
```

```
CT=BBant_CC(w1,:);
CT1=find(CT<inf);
for j=1:length(CT1)
  if BBant_TABU(CT1(j))==0
    CT(j)=inf;
  end
end
CW5=find(CT<inf);
Len_CW4=length(CW5);
while ((w1~=x)&(Len_CW4>=1))
  PPP=zeros(1,Len_CW4);%赋给 PP 一个 1*Len_LJD 大小的全 0 矩阵数组
  for i=1:Len_CW4
     PPP(i)=(Tau(w1,CW5(i))^Alpha)*(Eta(w1,CW5(i))^Beta); %概率选择公式
  end
  PPP=PPP/(sum(PPP));%建立概率分布
  Pcum=cumsum(PPP);%cumsum 数组元素的逐步累计
  t=rand;
  Select=find(Pcum>=t);%rand 用来生成一个 0 到 1 的随机数
  to_visit1=CW5(Select(1));%下一步将要前往的节点
  BBant_path=[BBant_path,to_visit1];%路径增加
  BBant_PC=BBant_PC+BBant_CC(w1,to_visit1);%路径费用累计
  w1=to_visit1;
  for k1=1:N
    if BBant_TABU(k1)==0
      BBant_CC(w1,k1)=inf;
      BBant_CC(k1,w1)=inf;
    end
  end
  BBant_TABU(w1)=0;%已访问过的节点从禁忌表中删除
  CT=BBant_CC(w1,:);
  CT1=find(CT<inf);
  for j=1:length(CT1)
    if BBant_TABU(CT1(j))==0
      CT(j)=inf;
    end
   end
```

```
            CW5=find(CT<inf);%可选节点集
            Len_CW4=length(CW5);%可选节点的个数
          end
            if w1==x
              if w1==S
                Cse=COSTS(1,k,m);
                Delta_Tau(x,y)=Delta_Tau(x,y)+Q/Cse;
              else
                Cse=Cse+C(x,y);
                Delta_Tau(x,y)=Delta_Tau(x,y)+Q/Cse;
              end
            end
          end
        end
      end
    end
end
Tau=(1-Rho).*Tau+Delta_Tau;
```

Appendix G.1

Simulation Program of UV Light Bottom Transmission Model

```
#include <math.h>

#include <delay.h>

#include <packet.h>

#include <packet-stamp.h>

#include <antenna.h>

#include <mobilenode.h>

#include <propagation.h>

#include <wireless-phy.h>

#include <tworayground.h>

static class TwoRayGroundClass: public TclClass {

public:

        TwoRayGroundClass() : TclClass("Propagation/TwoRayGround") {}

        TclObject* create(int, const char*const*) {

                return (new TwoRayGround);

        }
```

```
} class_tworayground;

TwoRayGround::TwoRayGround()

{

  last_hr = last_ht = 0.0;

 }

double TwoRayGround::TwoRay(double Pt, double Gt, double Gr,    double d)

 {

    double Pr1, Pr2, Pr3, Pr4, Pr5;

    Pr1 = Gt * Gr * Pt * Ar * Ks * Js * Fs * Fs * sin(Fy + Jy);

    Pr2 = 32 * PI * PI * PI * d * 0.001 * sin(Fy) * (1 - cos(PI /12));

    Pr3 = Ke * d * 0.001 * (sin(Fy) + sin(Jy));

    Pr4 = sin(Fy + Jy);

    Pr5 = exp(-(Pr3 / Pr4));

    return (Pr1 * Pr5) / Pr2;

 }
```

```
} class_tworayground;

TwoRayGround::TwoRayGround()
{
  last_hr = last_ht = 0.0;
 }
double TwoRayGround::TwoRay(double Pt, double Gt, double Gr,    double d)
 {
   double Pr1, Pr2, Pr3, Pr4, Pr5;
   Pr1 = Gt * Gr * Pt * Ar * Ks * Js * Fs * Fs * sin(Fy + Jy);
   Pr2 = 32 * PI * PI * PI * d * 0.001 * sin(Fy) * (1 - cos(PI /12));
   Pr3 = Ke * d * 0.001 * (sin(Fy) + sin(Jy));
   Pr4 = sin(Fy + Jy);
   Pr5 = exp(-(Pr3 / Pr4));
   return (Pr1 * Pr5) / Pr2;
 }
 Double TwoRayGround::Pr(PacketStamp *t, PacketStamp *r, WirelessPhy *ifp)
 {
   double rX, rY, rZ;          // location of receiver
   double tX, tY, tZ;          // location of transmitter
   double d;                          // distance
   double hr, ht;       // height of recv and xmit antennas
   double Pr;                   // received signal power
   double bizhi;
   double diffX;
   double diffY;
   r->getNode()->getLoc(&rX, &rY, &rZ);
   t->getNode()->getLoc(&tX, &tY, &tZ);
   rX += r->getAntenna()->getX();
   rY += r->getAntenna()->getY();
   tX += t->getAntenna()->getX();
```

```
    tY += t->getAntenna()->getY();
    d = sqrt((rX - tX) * (rX - tX) + (rY - tY) * (rY - tY) + (rZ - tZ) * (rZ - tZ));
if (rZ != tZ) {
        printf("%s: TwoRayGround propagation model assume flat ground\n",
            __FILE__);
    }
    hr = rZ + r->getAntenna()->getZ();
    ht = tZ + t->getAntenna()->getZ();
if (hr != last_hr || ht != last_ht)
{
        last_hr = hr; last_ht = ht;
#if DEBUG > 3
        printf("TRG: xover %e.10 hr %f ht %f\n",
            crossover_dist, hr, ht);
#endif
    }
double Gt = t->getAntenna()->getTxGain(rX - tX, rY - tY, rZ - tZ, t->getLambda());
double Gr = r->getAntenna()->getRxGain(tX - rX, tY - rY, tZ - rZ, r->getLambda());
#if DEBUG > 3
printf("TRG %.9f %d(%d,%d)@%d(%d,%d) d=%f xo=%f :",
Scheduler::instance().clock(),    t->getNode()->index(), (int)tX, (int)tY,
r->getNode()->index(), (int)rX, (int)rY, d, crossover_dist);
#endif
diffX = rX - tX;
diffY = rY - tY;
bizhi = diffY / diffX;
if(d <= 17.4) {
       if((bizhi >(-0.577) && bizhi < 0) || (bizhi > 0 && bizhi < 0.577)) {
       Pr = TwoRay(t->getTxPr(), Gt, Gr, d);
#if DEBUG > 3
```

```
        printf("Friis %e\n",Pr);
 #endif
          return Pr;
      }
        else {
              Pr = 0;
        return Pr;
    }
 }
  else {
        Pr = 0;
 #if DEBUG > 3
        printf("TwoRay %e\n",Pr);
 #endif
        return Pr;
    }
 }
double TwoRayGround::getDist(double Pr, double Pt, double Gt, double Gr, double hr, double ht, double L, double
lambda)
 {
    return sqrt(sqrt(Pt * Gt * Gr * (hr * hr * ht * ht) / Pr));
 }
```

Appendix G.2

Emulated TCL Script Program

(Figures 9.6, 9.7, 9.9, 9.10, 9.11, 9.12 all use this script):

```
set opt(chan)        Channel/WirelessChannel

set opt(prop)        Propagation/TwoRayGround

set opt(netif)       Phy/WirelessPhy

set opt(mac)         Mac/802_11

set opt(ifq)         Queue/DropTail/PriQueue

set opt(ll)     LL

set opt(ant)              Antenna/OmniAntenna

set opt(god)              off

set opt(x)      100  ;# X dimension of the topography

set opt(y)      100       ;# Y dimension of the topography

set opt(traf)        "../test/sk-30-3-3-1-1-6-64 新.tcl"        ;# traffic file

set opt(topo)        "../test/scen-800x800-30-500-1.0-1 新"          ;# topology file

#set opt(onoff)   "../test/onoff-30-3-3-1-1-500-10.tcl"        ;#node on-off

set opt(ifqlen)      50    ;# max packet in ifq

set opt(nn)          8     ;# number of nodes

set opt(seed)        0.0

set opt(stop)        10    ;# simulation time

set opt(prestop)          19          ;# time to prepare to stop

set opt(tr)     "wireless.tr"     ;# trace file
```

```
set opt(nam)              "wireless.nam"    ;# nam file

set opt(engmodel)         EnergyModel       ;

set opt(txPower)          0.660;

set opt(rxPower)          0.395;

set opt(idlePower)        0.035;

set opt(initeng)          10000.0           ;# Initial energy in Joules

set opt(logeng)           "off"             ;# log energy every 1 seconds

set opt(lm)               "off"             ;# log movement

set opt(adhocRouting)     FLOODING

set opt(duplicate)        "enable-duplicate"

LL set mindelay_          50us

LL set delay_             25us

LL set bandwidth_         0     ;# not used

Queue/DropTail/PriQueue set Prefer_Routing_Protocols    1

# unity gain, omni-directional antennas
# set up the antennas to be centered in the node and 1.5 meters above it

Antenna/OmniAntenna set X_ 0

Antenna/OmniAntenna set Y_ 0

Antenna/OmniAntenna set Z_ 1.5

Antenna/OmniAntenna set Gt_ 1.0

Antenna/OmniAntenna set Gr_ 1.0
```

```
# Initialize the SharedMedia interface with parameters to make
# it work like the 914MHz Lucent WaveLAN DSSS radio interface
Phy/WirelessPhy set CPThresh_ 10.0
Phy/WirelessPhy set CSThresh_ 1.559e-7
Phy/WirelessPhy set RXThresh_ 3.652e-5
Phy/WirelessPhy set Rb_ 2*1e6
Phy/WirelessPhy set Pt_ 0.2818
Phy/WirelessPhy set freq_ 914e+5
Phy/WirelessPhy set L_ 1.0
proc usage { argv0 }   {
    puts "Usage: $argv0"
    puts "\t\t\[-topo topology file\] \[-traf traffic file\]"
    puts "\t\t\[-x max x\] \[-y max y\] \[-seed seed\]"
    puts "\t\t\[-nam nam file\] \[-tr trace file\] \[-logeng on or off\]"
    puts "\t\t\[-stop time to stop\] \[-prestop time to prepare to stop\]"
    puts "\t\t\[-initeng initial energy\] \[-engmodel energy model\]"
    puts "\t\t\[-chan channel model\] \[-prop propagation model\]"
    puts "\t\t\[-netif network interface\] \[-mac mac layer\]"
    puts "\t\t\[-ifq interface queue\] \[-ll link layer\] \[-ant antena\]"
    puts "\t\t\[-ifqlen interface queue length\] \[-nn number of nodes\]"
}
proc getopt {argc argv} {
    global opt
    lappend optlist cp nn seed sc stop tr x y
```

```
        for {set i 0} {$i < $argc} {incr i} {
                set arg [lindex $argv $i]
                if {[string range $arg 0 0] != "-"} continue
                set name [string range $arg 1 end]
                set opt($name) [lindex $argv [expr $i+1]]
        }
}
proc finish {} {
        global ns_ nf opt god_ node_
        $ns_ terminate-all-agents
#        $god_ dump
        $god_ dump_num_send
        $ns_ flush-trace
        if [info exists tracefd] {
         close $tracefd
#       exec rm -f $opt(tr)
        }
if [info exists nf] {
                close $nf
#       exec rm -f $opt(nam)
#                exec nam $opt(nam) &
#       exec gzip $opt(nam)
        }
    exit 0
}
proc cmu-trace { ttype atype node } {
       global ns_ tracefd
   if { $tracefd == "" } {
              return ""
       }
```

```
        set T [new CMUTrace/$ttype $atype]
        $T target [$ns_ set nullAgent_]
        $T attach $tracefd
            $T set src_ [$node id]
            $T node $node
        return $T
}
# Main Program
getopt $argc $argv
# do the get opt again incase the routing protocol file added some more
# options to look for
getopt $argc $argv
if {$opt(seed) > 0} {
        puts "Seeding Random number generator with $opt(seed)\n"
        ns-random $opt(seed)
}
# Initialize Global Variables
set ns_         [new Simulator]
# define color index
$ns_ color 0 red
$ns_ color 1 blue
$ns_ color 2 chocolate
$ns_ color 3 yellow
$ns_ color 4 green
$ns_ color 5 tan
$ns_ color 6 gold
$ns_ color 7 black
#set chan  [new $opt(chan)]
#set prop  [new $opt(prop)]
set topo    [new Topography]
```

```
if { $opt(nam) != "" } {
        set nf [open $opt(nam) w]
        $ns_ namtrace-all-wireless $nf $opt(x) $opt(y)
}
if { $opt(tr) != ""} {
        set tracefd [open $opt(tr) w]
        $ns_ trace-all $tracefd
}
$topo load_flatgrid $opt(x) $opt(y)
#$prop topography $topo
# Create God
set god_ [create-god $opt(nn)]
#$god_ on
$god_ $opt(god)
$god_ allow_to_stop
$god_ num_data_types 1
# log the mobile nodes movements if desired
if { $opt(lm) == "on" } {
        log-movement
}
#global node setting
$ns_ node-config -adhocRouting $opt(adhocRouting) \
                -llType $opt(ll) \
                -macType $opt(mac) \
                -ifqType $opt(ifq) \
                -ifqLen $opt(ifqlen) \
                -antType $opt(ant) \
                -propType $opt(prop) \
                -phyType $opt(netif) \
                -channelType $opt(chan) \
```

```
                -topoInstance $topo \
                -agentTrace ON \
                        -routerTrace OFF \
                        -macTrace ON \
                -energyModel $opt(engmodel) \
                -initialEnergy $opt(initeng) \
                -txPower   $opt(txPower) \
                -rxPower   $opt(rxPower) \
                -idlePower   $opt(idlePower)

# Create the specified number of nodes [$opt(nn)] and "attach" them  to the
# channel.
for {set i 0} {$i < $opt(nn) } {incr i} {
    set node_($i) [$ns_ node $i]
        $node_($i) color black
    $node_($i) random-motion 0        ;# disable random motion
        $god_ new_node $node_($i)
}
if { $opt(topo) == "" } {
    puts "*** NOTE: no topology file specified."
        set opt(topo) "none"
} else {
    puts "Loading topology file..."
    source $opt(topo)
    puts "Load complete..."
}

for {set i 0} {$i < $opt(nn)} {incr i} {
    $node_($i) namattach $nf
# 20 defines the node size in nam, must adjust it according to your scenario
```

```
        $ns_ initial_node_pos $node_($i) 20
 }
# log energy if desired
if { $opt(logeng) == "on" } {
        log-energy
 }
# Source the traffic scripts
if { $opt(traf) == "" } {
        puts "*** NOTE: no traffic file specified."
                set opt(traf) "none"
 } else {
        puts "Loading traffic file..."
        source $opt(traf)
 }
 # Tell all the nodes when the simulation ends
 $ns_ at $opt(prestop) "$ns_ prepare-to-stop"
 $ns_ at $opt(stop).00000001        "finish"
 for {set i 0} {$i < $opt(nn) } {incr i} {
        $ns_ at $opt(stop).00000002 "$node_($i) reset";
 }
 # tell nam the simulation stop time
 $ns_ at    $opt(stop).00000003     "$ns_ nam-end-wireless $opt(stop)"
 $ns_ at $opt(stop).00000004 "puts \"NS EXITING...\" ; $ns_ halt"
 puts $tracefd "M 0.0 nn $opt(nn) x $opt(x) y $opt(y)"
 puts $tracefd "M 0.0 topo $opt(topo) traf $opt(traf) seed $opt(seed)"
 puts $tracefd "M 0.0 prop $opt(prop) ant $opt(ant)"
 puts "Starting Simulation..."
 $ns_ run
```

Appendix G.3

End-To-End Delay Simulation Script (Fig. 9.9).

Figures 9.7 and 9.12.

```
BEGIN {
 highest_packet_id = 0;
}
{
 event = $1;
 time = $2;
 node_nb = $3;
 node_nb=substr(node_nb,2,1);
 trace_type = $4;
 flag = $5;
 packet_id = $6;
 pkt_type = $7;
 pkt_size = $8;
 energy = $14;
 if
( event=="s"&&node_nb==0&&trace_type=="AGT"&&pkt_type=="diffusion"&&packet_id>highest_packet_id )
{
          highest_packet_id = packet_id;
  }
 if
( event=="s"&&node_nb==0&&trace_type=="AGT"&&pkt_type=="diffusion"&&packet_id==highest_packet_id
){
          start_time[packet_id] = time;
          remain_energy[packet_id] = energy;
  }
 }
```

```
END {

  for ( packet_id = 0; packet_id <= highest_packet_id; packet_id++ )

  {

     start[packet_id] = start_time[packet_id];

     remain_energy[packet_id] =   remain_energy[packet_id];

     printf("%f %f\n",   start[packet_id],   remain_energy[packet_id]);

  }

}
```

Appendix G.4

Delay Jitter Simulation Script (Fig. 9.10).

```
BEGIN{

highest_packet_id=0;

}

{

   action=$1;

   time=$2;

   packet_id=$6;

   packet_type=$7;

   packet_size=$8;

  if(packet_id>=highest_packet_id)

       highest_packet_id=packet_id;

         if(action!="d") {

              if(action=="r"&&packet_type=="cbr")

                  end_time[packet_id]=time;

                       if(action=="s"&&packet_type=="cbr")

                           start_time[packet_id]=time;

  } else {

       end_time[packet_id]=-1;

}
```

```
}

END {

for(j=0;j<=highest_packet_id;j++) {

    if(end_time[j]!=-1){

        start=start_time[j];

        end=end_time[j];

        duration=end-start;

    }

  if( start < end )printf("%f    %f\n",start,duration);

   }

}
```

Appendix G.5

Throughput Simulation Script (Fig. 9.11)

```
BEGIN {

    highest_uid = 0;

}

{

        event = $1;

        time = $2;

        node_nb = $3;

        node_nb=substr(node_nb,2,1);

        trace_type = $4;

        flag = $5;

        uid = $6

        pkt_type = $7;

        pkt_size = $8;

    if (event=="s"&&node_nb==6&&pkt_type=="diffusion"&& uid > highest_uid )

    {

            highest_uid = uid;
```

```
        }

        #记录 CBR 封包的发送时间

        if (event=="s"&&node_nb==6&&pkt_type=="diffusion"&& id==highest_uid ) {

                start_time[uid] = time;

         }

         #记录 CBR 封包的接收时间

        if (event=="r"&&node_nb==5&&pkt_type=="diffusion"&&uid==highest_uid ) {

                end_time[uid] = time;

         }

}

 END {

         #初始化抖动延时所需的变量

         last_seqno = 0;

         last_delay = 0;

         seqno_diff = 0;

          #当资料行全部读取后，开始计算有效封包的端到端延迟时间

         for ( uid = 0; uid <= highest_uid; uid++ )

         {

                 start = start_time[uid];

                 end = end_time[uid];

                 packet_duration =end - start;

                #只把接收时间大于发送时间的记录列出来

                if ( start < end )

                {

                        #得到了 delay 值(packet_duration)后计算 jitter

                         seqno_diff = uid - last_seqno;

                         delay_diff = packet_duration - last_delay;

                         if (seqno_diff == 0) {
```

```
                jitter =0;
            }
            else {
                jitter = delay_diff/seqno_diff;
            }
        #将有效封包序号以及延时抖动打印出来
        printf("%d %f\n", uid, jitter);
            last_seqno = uid;
            last_delay = packet_duration;
        }
    }
}
```

Appendix G.6

Average Residual Energy Simulation Script (Figs. 9.7 and 9.12).

```
BEGIN{

    i=0;

    init=0;

}

{

    action = $1;

    time = $2;

    curlth = length($3);

    curlth -= 2;

    cur_node = substr($3,2,curlth);

    trace_type = $4;

    pkt_id = $6;

    pkt_type = $7;
```

```
        pkt_size = $8;
        srclth = index($14,":");
        srclth -= 2;
        src_node = substr($14,2,srclth);
        dstlth = index($15,":");
        dstlth -= 2;
        dst_node = substr($15,2,srclth);
        if(action!="d"&&trace_type=="MAC"&&pkt_type=="diffusion") {
            packetsum[i+1]=packetsum[i]+pkt_size;
            if(init==0) {
                start=time;
                init=1;
            }
            timenow[i]=time;
            i=i+1;
        }
}
END {
printf("%.2f\t%.2f\n",timenow[0],0);
for(j=0;j<i;j++) {
    if(timenow[j]==start) {
        throughput=0;
    } else {
        throughput=packetsum[j]/(timenow[j]-start)*8/64;
        printf("%.2f\t%.2f\n",timenow[j-1],throughput);
    }
}
printf("%.2f\t%.2f\n",timenow[j-1],0);
}
```

Note The program correspondingProcedure for Writing Optical Orthogonal Codes to the simulation results in Sect. 9.3.2.3 can refer to the above program. The script program in Appendix II needs to be modified according to the UV light directional diffusion routing protocol. The end-to-end delay, delay jitter, throughput and average residual energy can be Refer to appendαix three.

The manufacturer's authorised representative in the EU is Springer Nature Customer Service Centre GmbH, Europaplatz 3, 69115 Heidelberg, Germany. If you have any concerns regarding our products, please contact ProductSafety@springernature.com

Printed and bound by CPI Group (UK) Ltd, Croydon, CR0 4YY
30/09/2026
02225302-0015